The Design of *Iowa*-Class Battleships Vol. 1

A Visual Tour of Battleship USS *New Jersey*

John M. Miano

A Visual Tour of Battleship USS *New Jersey*

Colosseum Builders, Inc.
ISBN:

Cover Photograph: Nat'l Archives

Table of Contents

Preface

This is the first in a planned series of books that will address the structure of *Iowa*-class battleships. While the history of these ships has been extensively documented, details of how the ships were built have largely been neglected. One of the clearest indications of how poorly their structure has been documented is that no model kit has ever come close to getting the hull form of *Iowa*-class battleships correct.

This book presents a walk-through of *Iowa*-class battleships that focuses on the structural features of the ship. The tour is guided using ship plans that are illustrated with photographs of key features. For visitors or potential visitors, this book explains what you see on the ship. This tour also includes areas that are both accessible and inaccessible to the public. There are many areas of the ships that are dangerous or extremely hard to get to and, therefore, are not open to visitors. This book takes the mystery out of those areas without the reader having to climb sixty-foot ladders, squeeze through manholes, or crawl through oil soaked passages.

The four *Iowa*-class battleships were all different when built and they changed over a half-century of service. Rather than try to cover all the possible permutations I have focused on one ship—USS *New Jersey*—and one configuration—her final tour—for depiction in the plans. While there are variations among the *Iowa*-class battleships, particularly in compartment usage, the ships are laid out largely in the same way. The photographs come from all four *Iowa*-class battleships and the text describes some of the differences.

Ideally, this book would have used original plans for the guide. Unfortunately, no complete set of plans for one time period has turned up and the plans that are available tend to be difficult to read and would not reproduce well. Consequently, the plans in the book are original. The dimensions are taken from existing plans, plans for ship modifications, and direct measurements of the ship. The actual ship construction was frequently different from the plans. In such cases, the objective was to depict what is actually on the ship. Because the plans in this book depict USS *New Jersey* in its final service configuration, they largely reflect what a visitor to the ship sees today. Nonetheless, changes were made at retirement for ease of maintenance and the museum has made changes to accommodate visitors. These museum changes are not reflected on the plans in the book.

Sadly, it appears all the original drawings for *Iowa*-class battleships were destroyed between 2000 and 2010 and the last complete set of plans is currently inaccessible and is slated to be destroyed as well. The Navy made microfilm copies of the plans but these are incomplete and, in many cases, totally unreadable. While incomplete, many copies of original plans have survived.

Another issue researchers looking at a historical ship face is that they lack the tribal knowledge shared among its designers. Features are often depicted in plans ambiguously such that they can be interpreted in different ways. These probably created no issue with the people building the ship but they create problems for those who study the plans nearly a century later. Fortunately, in the case of *Iowa*-class battleships, one can resolve many such ambiguities by looking at one of the ships.

It is not physically possible to include a picture of every compartment on the ship. I have not included photographs of areas that I felt are repetitive. The multiple photographs of storerooms in this book are intended to illustrate the different types of materials stowed or structural features. The Navy cleaned out many compartments when the ship was retired. The Ground Control Station (GCS) for remotely piloted vehicles or the Tomhawk Computer Room sound interesting but they are just empty rooms. Consequently, there are no photographs of these areas in the book. When there is no photograph of an area of the ship, that means such a photograph would be repetitive or there is nothing really interesting remaining.

The plans are intended to convey the three dimensional nature of the ship. Many of the decks are not flat. Notable examples are the second and third decks that have shelves that do not follow the elevation of the main part of the deck. To reflect these differences in elevation, the deck plans have profiles showing the deck sheer and changes in level of the decks. Armor, level changes, and slopes are also indicated by variations in shading on the plans.

USS *New Jersey* is the most decorated battleship of the United States Navy. She served her country for 48 years from 1943 to 1991. Since 2001, USS *New Jersey* has enjoyed her well-earned retirement as a museum and memorial located in Camden, New Jersey, across the Delaware river from the Philadelphia Navy Yard where she was constructed.

All her completed sisters survived as well and are open to the public. USS *Iowa* is located in San Pedro, California, USS *Missouri* is in Honolulu, Hawai'i and USS *Wisconsin* is in Norfolk, Virginia.

The *Iowa*-class battleships continue to serve several important roles:

- They are memorials to those who have served the country, both in uniform and in supporting functions.
- They are educational institutions that teach students and the general public about history and technology.
- The ships are visitor attractions. One could even say they are places for fun. One can take the family to a museum ship for an experience more memorable and educational than an amusement park—and at a much lower cost. There is already so much open that a serious visitor cannot see everything in a single day.
- The ships are places for research. The ships maintain archives on historical and technical details for those who are serious about learning more about those subjects.

If you are on USS *New Jersey*—or one of the other *Iowa*-class battleships—hopefully this book will enhance your experience. If you have not yet visited, hopefully this book will inspire you to go. Each year, as volunteers work on the ships, the number of areas open to visitors expands.

Most of the areas depicted in this book are not open to visitors. However, some of the *Iowa*-class battleships offer curator's tours that will take visitors to areas not on the regular tour routes. If you want to see some of the more obscure areas of one of these ships, you may wish to inquire about a curator's tour.

For the next book, I plan to delve into more of the construction of *Iowa*-class battleships and will rely on illustrations to amplify what the photographs here show.

A project like this book requires the help of many people. Foremost among these is the curatorial staff of USS *New Jersey*. Curator, Ryan Szimanski, Libby Jones, and Alaina Noland provided support and were always ready to explore any area of the ship. This book would have been impossible without them. In addition, USS *New Jersey*'s YouTube channel was a constant source of ideas for this book.

David Way, curator, and Michael Getscher, chief operating officer of USS *Iowa* were an invaluable source of information for that ship and *Iowa*-class battleships in general. Likewise, Keith Nitka, battleship operations manager for USS *Wisconsin*, provided assistance on that ship.

Crewmember Henry M. Strub III was a constant source of information and did a detailed review of the book. Tracy White provided reference photographs. Larry Trick supplied information on helicopter operations. Ralph and Margaret Miano reviewed all aspects of the book.

Starting the process of sorting the blueprints on USS *New Jersey*.

The State of New Jersey presented the first battleship *New Jersey* with a silver service of 55 pieces made by Tiffany in 1906. That set was transferred to the current USS *New Jersey* when it entered service during World War II. At that time, the state expanded the set to 105 pieces. The silver service was used for formal events hosted by the captain of USS *New Jersey*. The Great Seal of New Jersey decorates this piece.

The U.S. Navy still owns the ship's silver. However, it is now on loan and shared between the USS *New Jersey* Museum and Memorial in Camden for display and Drumthwacket, the official residence of the Governor of New Jersey in Princeton where it is still used. (Ryan Szimanski)

Introduction

In 1921, representatives of the victorious powers (U.K., U.S., France, Italy, and Japan) gathered in Washington, D.C. to negotiate a treaty to limit their own navies. The resulting Washington Treaty was signed in early 1922. It imposed limits on the size of each participants' navy, limited battleship gun size to sixteen-inches, limited battleships to 35,000 tons displacement, banned new battleship construction for ten years, and existing battleships could not be replaced until they were 20 years old. Because the U.K. did not have any battleships with sixteen-inch guns, the treaty included an exception permitting that country to build two new battleships: *Nelson* and *Rodney* (1922–1927). Ships designed to fit within the treaty limits were known as *treaty battleships*.

Two more naval treaties followed. In 1930, the same powers gathered in London, producing the London Naval Treaty. This treaty imposed restrictions on smaller warships (submarines, cruisers, destroyers). In 1935, there was another meeting in London that resulted in the Second London Naval Treaty. This treaty imposed a limit of fourteen-inch guns on battleships. However, the treaty contained an escalator clause that increased the gun size limit to sixteen inches and battleship size to 45,000 tons if any of the signatories of the Washington Naval treaty did not accept the Second London Treaty by 1937. That escalator clause was triggered when both Italy and Japan refused to sign the Second London Naval treaty.

USS *Iowa* and USS *New Jersey* were ordered in 1939 and were designed to conform to the 16-inch gun and 45,000 ton limits of the London Naval Treaty. The major design difference between them was the conning tower at the flag (03) level was armored on USS *Iowa*. *Iowa*-class battleships were unusual among American battleships in that the design tradeoffs among armor, armament, and speed resulted in ships with high speed. Historically, the U.S. Navy had sacrificed speed for armor and armament. At 33 knots, *Iowa*-class battleships were the fastest battleships ever built. High speed allowed *Iowa*-class battleships to operate with aircraft carriers and resulted in a long operational life. However, high speed was the result of happenstance and operating with aircraft carriers was not a design consideration. The planned successor to the *Iowa*-class—the *Montana*-class—were to be five knots slower. War was raging by the time construction had started, rendering the naval treaties meaningless but the ships were not redesigned to avoid delay. USS *Iowa* was laid down at the New York Navy Yard in June 1940 and was commissioned in February 1943. USS *New Jersey* was laid down at the Philadelphia Naval Yard in September 1940 and commissioned in May 1943.

The second pair of *Iowa*-class battleships, USS *Missouri* and *U.S.S Wisconsin*, were ordered in June 1940. Freed of the treaty limitations, the navy was able to increase some armor and plate thicknesses for these ships. While there were some small differences, the *Missouri* and *Wisconsin* were the most identical of the *Iowa*-class as constructed and they tend to be difficult to tell apart in pictures from that era. Both were commissioned in 1944.

Iowa-class battleships were designed during a transitional period in shipbuilding when the industry was shifting from riveted to welded construction. The design suggests there was concern that welding would make the hull too stiff. The hull is largely of rivet construction, except in areas that would have been difficult to rivet or would not make the hull more rigid. The superstructure is largely welded.

Under pre-war planning, there were to have been four *Iowa*-class battleships followed by five of the larger *Montana*-class. Due to the wartime need for ships and the resources that would have been demanded by the *Montana*-class, the navy decided to built another pair of *Iowa*-class battleships instead: USS *Illinois* and USS *Kentucky*. These differed from their predecessors by having improved torpedo protection and greater use of welding in place of riveting in construction. Both ships were started in 1942 but construction progress suffered from interruptions resulting from war priorities. *Illinois* was canceled in 1945. Much greater progress was made on *Kentucky* and work continued on and off until 1950. In 1956, USS *Wisconsin* destroyed her bow when it collided with the destroyer USS *Eaton* in a fog. This damage was repaired by cutting off the bow of the *Kentucky* and grafting to *Wisconsin*. USS *Kentucky*'s propulsion machinery was scavenged and installed in USS *Sacramento* and USS *Camden*. USS *Kentucky* was scrapped in 1959.

Iowa-class battleships were the product of craftsmen. In many ways the construction of *Iowa*-class battleships was closer to that of wooden ships than to that of shipbuilding today. As an example, joints were machined into steel in the same manner that woodworkers use. The handcrafted construction resulted in ship dimensions that were slightly off from the designed specifications.

Frames

Locations along the lengthwise axis of a ship are identified using a measurement called a *frame*. The location where the waterline and the bow intersect is called the *forward perpendicular* (FP). The intersection at the stern is called the *aft perpendicular* (AP). On *Iowa*-class battleships the distance between the FP and AP is divided into 215 four-foot frames. Frame 2 is the location eight feet aft of the FP and Frame 215 corresponds to the AP.

Frame as a unit of measurement is a distinct, but related, concept to the structural transverse frames of the ship. Most frame locations have a structural frame. However, this is not always the case as there are frame locations that lack structural frames and many structural frames are located at fractional frame locations.

	Actual	Designed
Length Overall (1940's)	887'-6 $\frac{3}{8}$"	887'-3"
Length Overall (1980's)	887'-8 $\frac{3}{8}$"	
Waterline Length	859'-10 $\frac{1}{4}$"	860'
Beam	108'-1 $\frac{3}{8}$"	108'-3"

USS *New Jersey* Dimensions

Decks

USS *New Jersey* has nineteen levels that are colloquially called *decks*. Technically however, a deck is a level that extends the full length of the hull. A partial level within a hull is called a *platform*. In the superstructure, an elevation is called a *level*. With the hull, the levels are:

- Main Deck (1)
- Second Deck (2)
- Third Deck (3)
- First Platform (4)
- Second Platform (5)
- Third Platform (6)
- Hold (7)

These decks are referenced by name or by number. *Hold* and *seventh deck* are different names for the same thing. There are twelve levels in the superstructure numbered sequentially from bottom to top as 01 (oh one) through 012 (oh twelve). The zero prefix distinguishes the superstructure levels from decks within the hull.

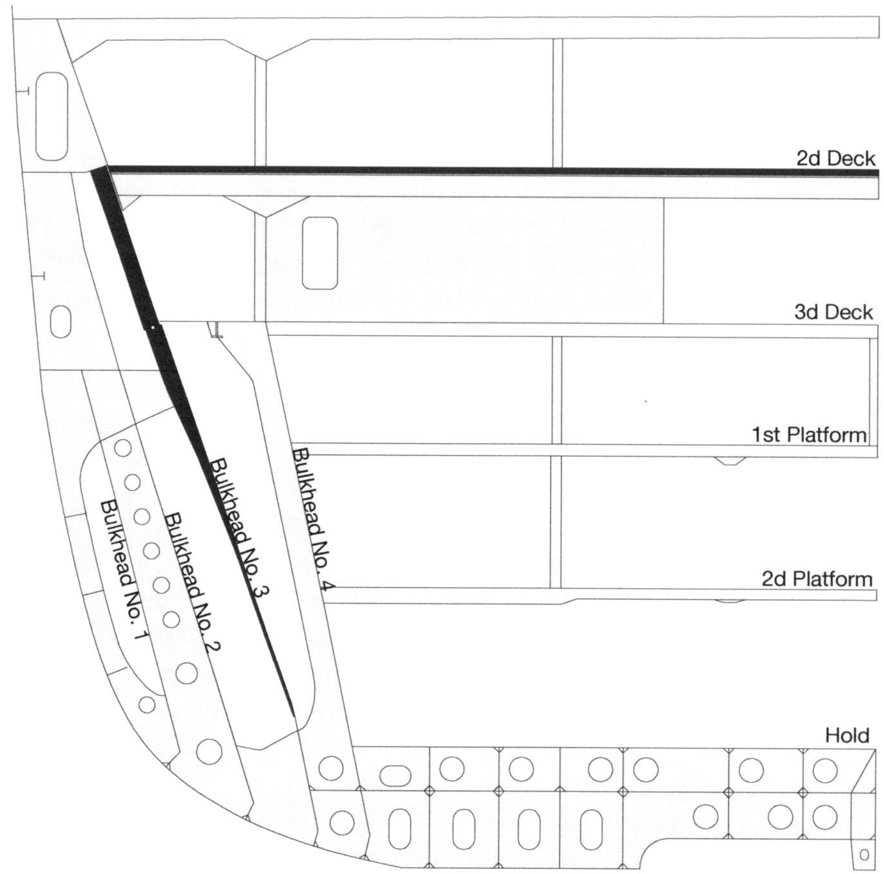

There is no such thing as a typical cross section of *Iowa*-class battleships because the hull shape changes continuously throughout the length of the hull. This is a cross section at Frame 154 $\frac{3}{4}$ looking aft. This is forward of turret no. 3 where the physical frames start to shift to half frame locations to align with the turret.

The protective structure of *Iowa*-class battleships consumes much of the ships' internal volume. Within the hull shell there are four torpedo bulkheads to protect against underwater explosions. The armor cross section forms an inverted U with the deck armor 4.75" and the side armor 12.1" at the top, tapering to 1" at the bottom.

There is an additional level that is deceptively called the *splinter deck* located below second deck. The splinter deck extends between turret no. 2 and turret no. 3 over the engineering spaces. The splinter deck is part of the armor protection of the ship and it amounts to merely a crawl space.

In many cases decks are not flat. Main deck, 01 and 02 levels have camber, meaning their cross section is curved. On the *Iowa*-class, their curve is a parabola where the deck is lower at the edges than the centerline to drain water. Main deck to second platform have sheer, meaning they are curved along the centerline of the ship. On *Iowa*-class battleships these decks curve upwards at the bow and stern. In addition, there are discontinuities where a deck has different elevations or openings at the same frame location. The deck plans include profile views and the plans use shading to indicate changes in elevation.

Bull's Eyes

Every compartment in USS *New Jersey* has a yellow sign called a *bull's eye* to identify it for damage control purposes. If a sailor detected a problem, he would use the bull's eye to report the location. Originally, bull's eyes were red and white circles. The Navy changed the format to use yellow boxes but the name stayed.

A bull's eye on USS *New Jersey* has four lines of information (although other ships may only have three). The first line is the post-1949 compartment designation. The compartment designation has four fields separated by hyphens: the deck, the first frame of the compartment; position relative to the centerline (0 = centerline, 1 = first compartment to starboard, 2 = first compartment to port, 3 = second compartment to starboard, *etc.*), and the compartment type. Bull's eyes specify decks by number, rather than name. The second line gives the range of frames for the compartment. The third line identifies the division responsible for the compartment. The last line is the compartment number in the naming scheme used prior to 1949. It was maintained for compatibility with documents that used the old naming scheme. The old naming scheme divided the ship into three divisions that are the first character in the designation: "A" (stem to Frame 87), "B" (Frame 87 to Frame 151), and "C" (Frame 151 to stern). A watertight compartment number follows with the first digit being the deck number. Finally, there is a letter indicating the type of compartment. Unlike under the new scheme, Non-tight compartments within the same tight compartment shared the same numeric identifier.

An example of a Bull's Eye. The compartment is on third deck, extends from frame 79 to 84, is the third compartment to starboard, and is a control center. This is then the Forward Internal Communications Room.

Letter	Type of Compartment
A	Stowage spaces
AA	Cargo holds
C	Control centers for ship and fire-control operations (normally manned)
E	Engineering control centers (normally manned)
F	Oil stowage compartments (ship use)
FF	Oil stowage compartments (cargo)
G	Gasoline stowage compartments (ship use)
GG	Gasoline stowage compartments (cargo)
J	JP-5 fuel (ship use)
JJ	JP-5 fuel (cargo)
K	Chemicals and dangerous materials (other than oil and gasoline)
L	Living spaces
M	Ammunition spaces
Q	Miscellaneous spaces not covered by other letters
T	Vertical access trunks
V	Voids
W	Water stowage spaces

Compartment Types Used in Bull's Eyes

Compartment Checkoff Lists

As one moves throughout the ship one finds that each compartment has a checkoff list displayed in a prominent location. The compartment checkoff list identifies the compartment, its use, all the fixtures it contains that are related to damage control, and the conditions when they should be closed. These are the possible conditions:

- Condition X-Ray: Lowest protection used in port during peacetime.
- Condition Yoke: Normal at sea operation
- Condition Zebra: Highest level of protection
- William Fittings: Can be open during all conditions but closed in the event of nuclear or biologic contamination.

The compartment checkoff lists remain in place from the ship's retirement in 1991. One can usually determine the final use of a compartment by examining the checkoff list. Unfortunately, changes in compartment use that did not affect damage control were not always updated on the checkoff list.

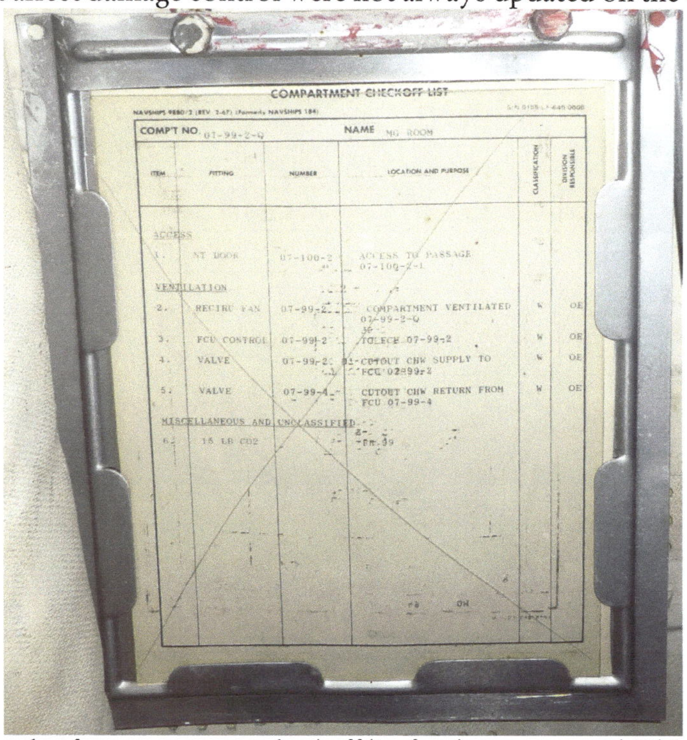

An example of a compartment checkoff list for damage control. This particular list is for the Motor Generator room at the 07 level.

Pipes and Valves

Iowa-class battleships have a complex network of piping that conveys various fluids and gasses around the ship. Valves and control levers are color coded to indicate what a particular pipe contains. Where pipes are not covered with insulation, they are often color coded as well.

Fluid	Color
Steam	White
Potable Water	Dark Blue
Feed Water	Light Blue
Salt Water	Green
Fire Plug	Red
Aqueous Film Forming Foam	Red–Green Stripes
JP-5 (Jet Fuel)	Purple
Fuel Oil	Yellow
Lubrication Oil	Yellow–Black Stripes
Sewage	Gold
Hydraulic	Orange
High Pressure Air	Dark Gray
Medium Pressure Air	Dark Gray–Tan Stripes
Low Pressure Air	Tan
Oxygen	Light Green
Nitrogen	Light Gray
Helium	Buff

Pipe and Valve Colors

Armor

When *Iowa*-class battleships were designed, the Navy only considered steel to be armor when it was over four inches thick. *Iowa*-class battleships were protected on the all-or-nothing principle under which the vital areas of the ship were contained within an armored citadel. The citadel consists of deck armor at the top, belt armor at the sides, and transverse armor at the ends. Combined, these form an inverted box. Between Frame 50 and Frame 166 the citadel protects the main armament, magazines, and machinery spaces. In this area the top of the citadel is at second deck. Aft of that the citadel drops down to third deck where the citadel extends to Frame 203 to protect the rudder machinery. *Iowa*-class battleships are somewhat unusual in that nearly all of their armor is contained entirely within the hull shell. The only citadel armor that is visible from the outside is the turrets, barbettes under the turrets, and the conning tower.

The Navy classified plate armor as either *Class A* or *Class B*. Class A armor was treated to harden the outer face while Class B armor was homogenous in hardness. The names are misleading in that they suggest one type was better than the other. In reality, the relative effectiveness of Class A and Class B armored depended upon the type of incoming projectile and the angle it struck. Class A armor generally was used where it was expected projectiles would strike directly and Class B was used where it was expected projectiles would strike obliquely. The belt armor on USS *New Jersey* and USS *Iowa* was made of Special Treatment Steel (STS). STS is homogenous and, therefore, is a type of Class B armor, but it was a slightly different formulation than the Class B armor used in the rest of the ship and on USS *Missouri* and USS *Wisconsin* for the belts.

Class A armor was difficult to incorporate into a ship's structure because heat from a welding torch could crack the hardened face. In contrast, Class B armor could easily be used structurally. That said, within *Iowa*-class battleships both Class A and Class B armor were used structurally in some places and as nonstructural dead weight in others.

Area	Type	Thickness
Second Deck	Class B	4.75"
Third Deck	Class B	5.6"
Third Deck over Steering Gear	Class B	6.2"
Upper Side Belt	Class A	12.1"
Lower Side Belt	STS or Class B	12.1" tapering at knuckles to 1"
Lower Belt FR166–FR172	Class B	13" variable taper to about 4.5"
Lower Belt FR 172–FR 189	Class B	13.5 variable taper to about 5"
Lower Belt at Steering Gear	Class A	13.5"
Forward Transverse	Class A	11.3" at the top with a taper at a knuckle to 8.5" (14.5" tapering to 11.7" USS *Missouri* and USS *Wisconsin*)
Aft Transverse	Class A	11.3"
Conning Tower Sides	Class B	17.3"
Conning Tower Roof	Class B	7.25"
Conning Tower Floor	Class B	4"
Conning Tower Tube	Class B	16"
Barbettes	Class A	17.3" at sides tapering to 11.6" at front and back
Turret Front	Class B	17"
Turret Sides	Class A	9.5"
Turret Rear	Class A	12"
Turret Roof	Class B	7.25"

Armor on *Iowa*-class Battleships. These thicknesses do not include backing plates and reflect installed thicknesses rather than mill thicknesses.

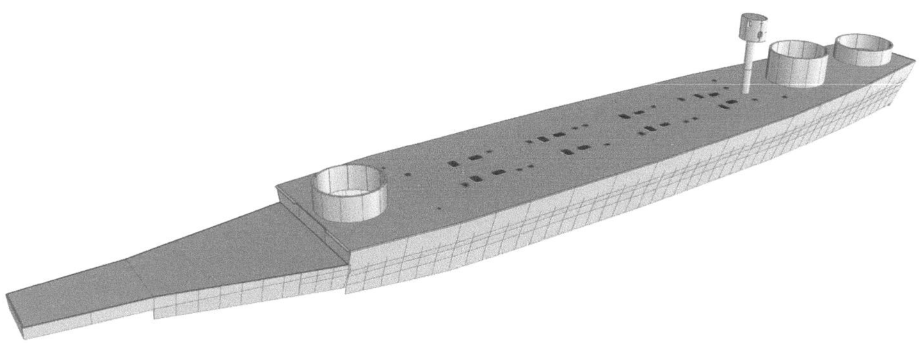

An illustration of the citadel armor. Aft is to the left. Armored grates at second deck protect the air intakes and exhaust uptakes for the boilers. The aft extension of the citadel protects the steering gear.

Electric-Hydraulic Systems

Several sections in this book refer to *A-ends* and *B-ends*. These warrant a brief explanation. Mechanical functions in *Iowa*-class battleships that required fine control are powered by electric-hydraulic systems. These are used to drive all the mechanical systems of the 16-inch turrets that do not operate at a constant speed, the 5-inch mounts, the rudder, anchor wildcats, and the former 40mm gun mounts. In an electric-hydraulic system there is an electric motor, often with reduction gears, that runs at a constant speed. The drive shaft of the motor powers what is called a *Waterbury-type pump*. A similar pump is frequently used in automobile air conditioning compressors.

The output of the pump is called the *A-end* of the system. Two pipes emerge from the A-end that are filled with hydraulic fluid. Normally the pump idles so the pressure within the two pipes is the same at the A-end. The pump can easily be engaged to the motor to create a continuous range of pressure differentials at the A-end. Positive or negative pressure can be directed towards either output pipe, allowing the system to operate in forward or reverse.

The other ends of the output pipes are at the *B-end* of the system. The B-end uses the pressure differential to either drive a piston or a hydraulic motor. When driving a motor, the hydraulic fluid flows within a closed loop.

The hydraulic electric systems have a number of advantages over direct electric drive. The biggest advantage is instantaneous response. They do not have to overcome inertia within an electric motor to start up. Because hydraulic fluid is incompressible, fine changes in pressure at the A-end are immediately transmitted to the B-end. They also allow the system to be spread out, which is a major advantage in a cramped space like the turret. The major drawback of hydraulic systems is that they always leak.

Weapons

16-Inch Main Battery

The main armament of the *Iowa*-class is nine Mk.7 16-inch/50-caliber guns in three 3-gun turrets. This never changed. The U.S. Navy used the simple system of numbering weapon mounts from bow to stern, right to left. The main battery then consisted of turret no. 1, turret no. 2, and turret no. 3. *Iowa*-class battleships were originally planned to use the 16 inch/50 caliber gun Mk.2 that had been designed and built for the *South Dakota*-class battleships and *Lexington*-class battlecruisers that were canceled or scrapped under the naval treaties. The Navy's Bureau of Ordnance designed the turret and guns and the Bureau of Construction and Repair oversaw the design of the rest of the ship. Late in the design process it was discovered that communications failures between the bureaus had resulted in a turret design that was too large to fit in the ship. The Bureau of Ordnance quickly redesigned the turrets and the lighter Mk.7 gun. The machinery spaces in the turrets are extremely cramped and turret no. 1 barely fits within the hull structure. The Mk.7 gun had a longer barrel than the sixteen-inch guns used in previous American battleships. The bore length with the breech closed is 800 (16×50) inches. The guns can elevate to 45 degrees. Turrets nos. 1 and 3 can depress to -2 degrees and turret no. 2 to 0 degrees. The most commonly used ammunition was 2,700 lb. armor piercing projectiles or 1,900 lb. high capacity projectiles.

The fire control system was highly redundant. A turret could either control itself or be controlled remotely (director control). The main battery was normally controlled remotely from one of two plotting rooms within the citadel. This allowed all the guns to be coordinated from a central location. In the alternative, one of the turrets could take on the role of the plotting room and control the turrets.

Officially, each gun could fire every thirty seconds but such a rate of fire would be hard to sustain—and would often not be desirable. The time of flight was usually over thirty seconds and could be over ninety seconds at extreme ranges. Firing as fast as possible in the shore bombardment role would not allow the troops on the ground to call in aim corrections.

The three sixteen-inch gun turrets are largely the same. The structural differences among them flow from their having different elevations. The deck plans illustrate the specific layout of turret no. 2. However, the illustrative photographs come from all three turrets.

Each turret has a nonrotating section and a rotating section. The nonrotating turret is contained within the hull structure and is hidden from external view. The main component of the nonrotating turret is the turret support that bears the weight of the rotating turret. This is a cylinder from second platform to third deck. From there it tapers to a truncated cone. The rotating turret rests on top of the turret support and rotates on bearings. The armored gunhouses are the visible part of the rotating turret. However, the rotating turrets extend down to the second platform. The armored barbettes visible under the gunhouses are not part of the turret structure. While the gunhouses appear to rest on the barbettes, their only connection is a gas seal.

Turrets nos. 1 and 3 have six internal levels while turret no. 2 has seven levels:

1. Powder flat. This corresponds with the second platform and is the only turret level that is aligned with a deck. The turret can only be entered from within the hull at the powder flat. This level is divided between the rotating and non-rotating turret.
2. Lower projectile storage level (turret no. 2 only). This level is part of the nonrotating turret and forms a balcony over the powder flat.
3. Lower shell flat. Projectile storage. This level is split between the rotating and non-rotating turret.

4. Upper shell flat. Nearly identical to the lower shell flat. This level is split between the rotating and non-rotating turret.

5. Electric deck. Holds most of the turret machinery. This level is entirely within the rotating turret.

6. Pan plate. The foundation of the rotating turret. The pan plate rests on the nonrotating turret and is entirely part of the rotating turret. The rotating turret levels below the pan plate are suspended from the pan plate.

7. Shelf plate. The level of the gunhouse and the only level above the weather deck. This level is entirely within the rotating turret.

The official manning requirements called for 77 men to operate a 16-inch turret. That figure included redundant positions that might be called for in a close-range battleship match to the death. Once that scenario became improbable the size of the turret crew decreased. In the 1980's, there were typically fewer than 50 men manning each turret.

During the 1950's plans were made to equip *Iowa*-class battleships with the W23 nuclear projectile. This projectile was known as *Katie*. Plans show the lower shell platform of turret no. 2 was to have been modified to hold nine disassembled *Katie* projectiles. Some preliminary modifications were made to the ships. Given the obvious problems created by firing nuclear weapons from a gun, the U.S. Military abandoned the concept in the early 1960's.

		Turret					
		1		2		3	
Upper Flat	Rotating	72		72		72	
	Fixed	120		70		100	
	Total		192		142		172
Lower Flat	Rotating	72		72		72	
	Fixed	126		125		126	
	Total		198		197		198
Lower Storage Platform					120		
Hoists			15		15		15
Total in Turret		405		474		385	
Total for Ship		1264					

Designed 16-inch Projectile Stowage Capacity

Secondary Battery

The original secondary battery consisted of twenty 5-inch/38-caliber Mk.12 guns in ten twin turrets with five mounts on each side of the superstructure. This was the same arrangement used in the preceding *South Dakota*- and *North Carolina*-classes. The 5-inch gun was dual purpose, meaning that it could be used against either surface or flying targets. The guns could elevate 85 degrees and depress to -15 degrees. The navy claimed the normal rate of fire was 12–15 rounder per minute. The actual the rate of fire depended upon the skill, strength, and endurance of the loaders. It tended to decline as the loaders tired and ready service ammunition was depleted.

The five-inch guns normally operated under automatic control from one of two plotting rooms in the citadel. When under automatic control, the crew at several positions in the mount were on standby. The mount could also operate under local control where it could take input from a fire control system or operate completely on its own. Four 5-inch mounts were removed in the 1980's to create space for missile platforms. The mounts were numbered fore to aft, starboard to port, with 50 added to the value (51, 52, ... 60) to distinguish them from the main battery.

A 5-inch/38-caliber mount on USS *New Jersey* in 1984. The eagle and anchor indicate that the marine detachment manned this mount. Inflatable lifeboats are mounted forward and aft of the mount at the 01 level. The winch for the boat davit is forward on the main deck. Eight tubes for Harpoon missiles and the skip box for loading 20mm CIWS ammunition are directly above the mount. The fueling at sea (FAS) control platform is at the far left at the 01 level. (Nat'l Archives)

Light Anti-Aircraft Armament

The U.S. Navy's standard light anti-aircraft weapons for battleships at the start of World War II were the 1.1" quad mount and the 0.50" Browning machine gun in single or twin mounts. The Navy realized before the United States entered World War II that advances in aircraft design had made these weapons inadequate for ship defense. To correct this deficiency, the Navy licensed the 40mm gun from Bofors in Sweden and the 20mm gun from Oerlikon in Switzerland. Both weapons were extensively reworked for U.S. production and service. *Iowa*-class battleships were the first to be designed to mount these weapons.

A 40mm quad mount on USS *New Jersey* in 1944 (Nat'l Archives)

The 40mm Bofors is a recoil-operated machine gun. When fired, the barrel moves backwards and provides the energy for loading the next round. On *Iowa*-class battleships, Bofers were used in water-cooled quad mounts that weighed about 25,000 lbs. They were normally aimed and fired remotely from directors located near the mounts. The 40mm mounts were powered by electric-hydraulic drive. However, they could also be operated manually at the mount in case of a loss of power. The guns were loaded using four-round clips and had a theoretical rate of fire of 120 rounds per minute. The practicable rate of fire was limited by how fast the crew could load clips. USS *New Jersey* had twenty 40mm mounts. The 40mm mounts were removed from USS *New Jersey* in 1968 and from the other *Iowa*-class battleships before their 1980's reactivations. Variants of the Bofors gun remain in service with the U.S. military.

The 20mm Oerlikon is a blowback-operated machine gun weighing about 150 lbs. The firing recoil pushed the bolt backwards to load the next round. The Oerlikon is not powered so it could be placed nearly anywhere on a ship. The Oerlikon is fed by a 60 round magazine. It has a theoretical rate of fire of about 450 rounds per minute but the practicable rate of fire is limited by the speed in which the crew can replace magazines. USS *New Jersey* had forty-nine 20mm mounts for most of World War II. Early in the, the Oerlikon was the most effective anti-aircraft gun in the U.S. Navy, accounting for nearly half of enemy aircraft shot down. Its effectiveness declined late in the war as its projectile was not large enough to destroy kamikaze aircraft. Twin mounts were introduced late in the war to address the decline in effectiveness. USS *New Jersey* replaced eight single mounts

This is a pair of 20mm guns at the 01 level. Ammunition boxes are located behind the guns. Containers for spare barrels are attached to the bulwark. During the 1980's this platform mounted a saluting gun. (USS *New Jersey* Word War II Cruise Book)

with twin 20mm mounts during its spring 1945 refit. In 1950 USS *New Jersey* carried sixteen twin 20mm mounts and all of the single mounts were removed. All the 20mm mounts were removed from USS *New Jersey* in 1952. The Oerlikon is still in service around the world.

As part of their 1980's reactivation, all *Iowa*-class battleships received four 20mm Phalanx Close in Weapons System (CIWS) mounts for missile defense. These mounts incorporate a self-contained radar system that operated a six-barreled Gatling gun capable of firing 3,000 rounds per minute.

Infantry Weapons

Except for USS *New Jersey* during the Vietnam War, *Iowa*-class battleships carried a detachment of marines. The marines were equipped with standard infantry weapons of the day, such as the 0.50" Browning machine gun. In addition, the ship had armories of light weapons for security or use by a landing force. These are not included in the ships' armament. During the 1980's *Iowa*-class battleships had mounting points for Browning machine guns along the deck. USS *New Jersey*'s sailors took on the security and roles normally handled by marines during the Vietnam War.

Missiles

When *Iowa*-class battleships were reactivated in the 1980's, two types of missiles were added to their armament. First, Harpoon is a turbojet powered, anti-ship missile with a range in excess of 67 miles. The warhead is about 500 lbs. Four launchers carrying four Harpoon missiles each were installed giving a total of sixteen missiles. The variant used in *Iowa*-class battleships has a rocket booster that fires the missile from its launch tube and accelerates it to a speed where the turbojet can take over. Land attack capability has been added to Harpoon missiles but this was not available in time for *Iowa*-class battleships.

Second, the Tomahawk is a turbojet-powered land attack missile with a range of over 1,000 miles. The Tomahawk carries a 1,000 lb. warhead and is capable of carrying a nuclear warhead. Plans show *Iowa*-class battleships were capable of carrying nuclear-armed Tomahawk missiles but it is unlikely there were ever equipped this way. Tomahawk missiles were stored in eight armored box launchers with four missiles each for a total of thirty-two missiles.

Intended Weapons

40mm projectiles from the Bofors guns were not large enough to hold VT (proximity) fuzes. During the 1960's, it was planned to replace the 40mm quad mounts with 3-inch twin mounts on USS *New Jersey*. The Gun Fire Control System (GFCS) Mk. 56 for the 3-inch guns was installed but the guns never made it to the ship. During the 1950's the GFCS Mk. 56 could have been used to control the 5-inch guns for anti-aircraft defense but that need never arose. The GFCS Mk.56 was removed in the 1980's.

In the 1990's all four *Iowa*-class battleships were prepared for mounting two Mk.38 25mm chain guns for close defense. However, there was a shortage of these weapons so they were only installed on USS *Missouri* and USS *Wisconsin*. Nonetheless, two mounts were installed in their intended locations on USS *New Jersey* after the ship became a memorial.

Engineering Spaces

Iowa-class battleships have four alternating fire and engine rooms. Each engine room drives a propeller shaft. The fire rooms extend from the hold to splinter deck while the engine rooms only extend up to third deck. There are no decks within the machinery spaces. Instead, there are platforms at different levels to give crew access to the machinery. Because two-dimensional plans do not readily convey the layout of the machinery spaces, this book maps locations within them using simplified thee-dimensional models depicting fire room and engine room no. 2.

Vertical Measurement

The *baseline* is a theoretical horizontal reference line along the length of a ship's hull from which vertical measurements are taken. *Iowa*-class battleships follow the normal practice of defining the baseline as running along the top of the flat keel.

Plate Thickness

The thickness of an armor plate was specified in decimal inches. This book follows that convention. Other plate thicknesses were specified as pounds per square foot where there is about forty pounds per inch of thickness. A plate specified as 60# mean that it was $1\frac{1}{2}$" thick. In general, this book translates pounds per square foot to fractional inches..

At Commissioning
May 1943

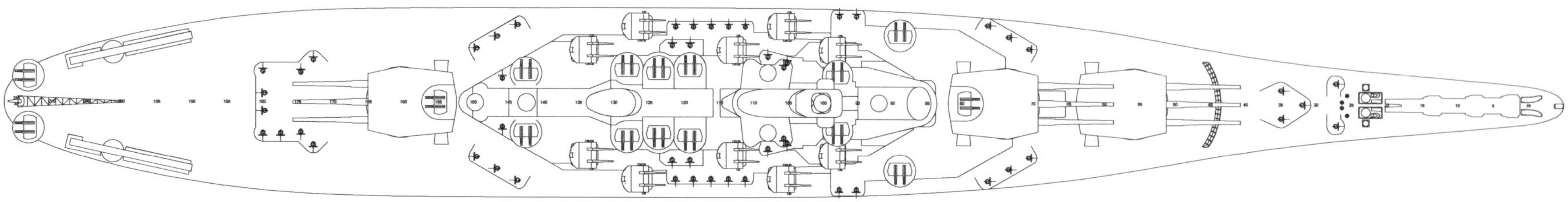

July 1943

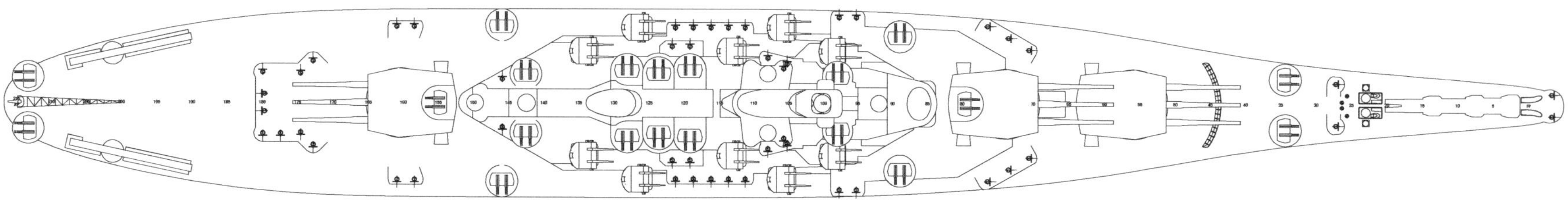

June 1945

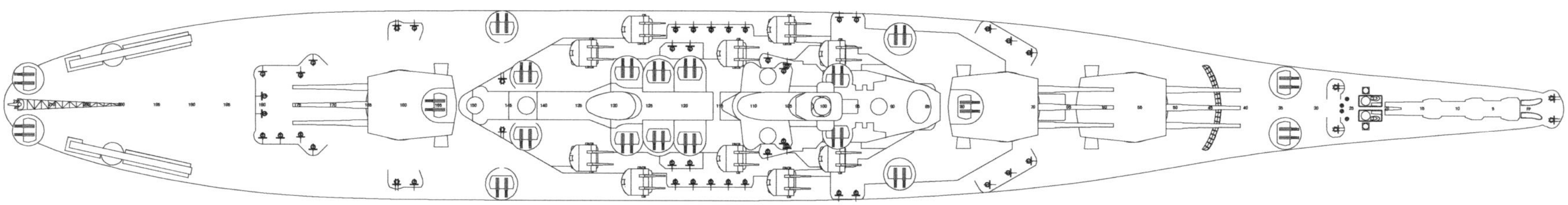

1951

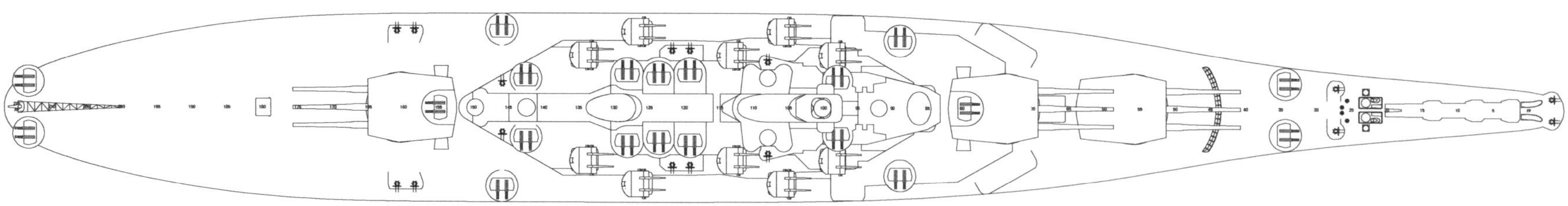

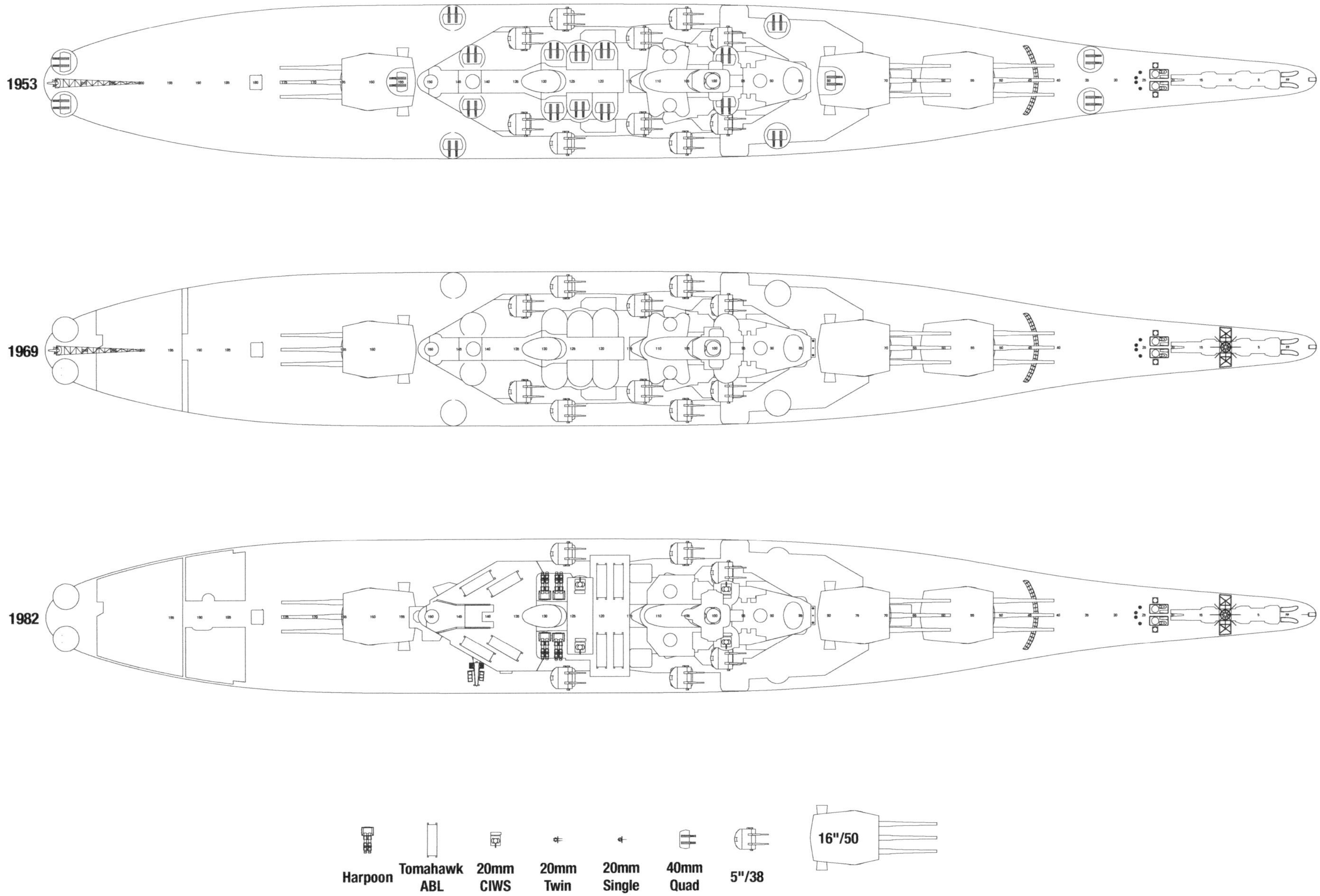

Major Changes to USS *New Jersey*'s Armament Over Time
1"=100'

This is USS *New Jersey* shortly after she was commissioned in 1943. At this point she has an open bridge around the conning tower. Instead of a bow bulwark, there is canvas attached to the railings. There is also a cluster of three 20mm Oerlikon guns forward of the 16-inch turrets that were soon replaced by two 40mm mounts. The square structure on the roof of turret no. 2 is one of the gun directors for 40mm guns. Several types of directors were used during World War II. The director could control multiple guns allowing concentrated fire on a single target, had a more sophisticated sight than those on the mounts, and kept the person aiming away from the vibration of the guns.

The chain running from the bow chock into the water was used for towing paravanes. The chain entered a tube at the lower bow where it returned to the main deck to form a loop. Paravanes were used to cut mine cables. Nearly all World War II photographs of the bow of *Iowa*-class battleships show such a chain. By the Korean War the concept of using battleships as minesweepers had fallen out of favor and the paravane chains were removed. (Naval Heritage and History Command)

USS *New Jersey* was in this configuration for most of her World War II service. The group of 20mm guns forward of turret no. 1 were removed. A flared bow bulwark was added to house two 20mm mounts. More 20mm mounts were added at the sides of turret no. 3. Four 40mm mounts were added at the main deck. The bridge was also enclosed. (Naval Heritage and History Command)

This is the curved bridge that was a distinctive feature of USS *New Jersey* until the Spring of 1945 when she received the current flat front bridge. (Naval Heritage and History Command)

U.S.S New Jersey in 1950 preparing for the Korean War. Note the flat-faced bridge and the flared bow bulwark. The rangefinder hood on turret no. 1 is still in place but it was removed before recommissioning. The aircraft catapults were removed at this time to make room for helicopters to land at the stern. (Naval Heritage and History Command)

After her first tour in Korea USS *New Jersey* returned to the shipyard in the United States. This is USS *New Jersey* during her second Korean tour in 1953. All the 20mm guns, the rangefinder on turret no. 1 and the flared bulwark are gone. (Naval Heritage and History Command)

USS *New Jersey* in her 1969 Vietnam configuration. External Changes include:

A. A discone antenna has been added at the bow for ship-to-ship communication.

B. Forward 40mm tubs removed and all 40mm mounts removed.

C. 40mm tubs converted into swimming pools.

D. 35-foot twin whip antennas added to the front of the bridge.

E. Addition of ears to the forward fire control tower to hold electronic countermeasures (ECM) equipment.

F. Helipad added with four-foot ramp for moving boats to the stern crane. Note the line up line on the helipad is oriented athwartship so aircraft can avoid the crane. In this configuration USS *New Jersey* could only handle one helicopter at a time.

(Naval Heritage and History Command)

This is USS *New Jersey* in 1983 after her third and final recommissioning. External changes include:

A. Radars added to the top of each turret to measure projectile velocity.
B. Swimming pools removed.
C. Four 20mm CIWS mounts installed.
D. Ears on forward fire control tower cut down to 011 level. Two SQL-32 electric countermeasures and AS-301BA/WSC-1(V) Satellite Communications (SATCOM) antennas added to the 011 level. Ship's whistle moved to the 011 level.
E. Platform created for Super Rapid Bloom Offboard Countermeasures (SR-BOC) mortars.
F. New tripod foremast bristling with sensors and antennae. The large revolving antennae are SPS-10 (surface search) forward and SPS-49 (air search) at the platform center. The saucer shaped antenna at the top of the mast at the aft end of the platform is for tactical air navigation (TACAN).
G. New boat davits and bulwarks port and starboard.
H. Four 5-inch mounts were removed and replaced with platforms and blast shields for eight armored box launchers that each contained four Tomahawk cruise missiles.
I. Two Harpoon anti-ship missile launchers holding four missiles each were added to each side for a total of sixteen missiles.
J. Mainmast replaced by smaller structure at the fore end of the aft smoke pipe and SATCOM Antenna added to the aft end.
K. Fueling at sea (FAS) kingpost was added to refuel escorts more efficiently.
L. Boat winch platform added at the 02 level.
M. Helicopter control station and glide slope indicator platform added at the 02 level.
N. Planking covered with nonskid material.
O. 42-foot ramp added so that helicopters could be moved off the helipad and stored. Each ramp has a fire fighting station.
P. Helipad markings and lights added for diagonal approaches from the port stern.
Q. Crane removed.

Many internal changes were made at this time as well. These changes included converting the boilers to use distilled fuel; adding collection and holding tanks (CHT) for sewage; upgrading the crew accommodations; and replacing the admiral's suite with the combat engagement center (CEC). (Nat'l Archives)

This is USS *New Jersey* in her final service configuration. The most visible changes are to support Pioneer remotely piloted vehicles (RPV). The port stern tub has a bladder tank for 100LL (100 octane, low-lead) aviation fuel for RPVs. The tank is on a track so that it can be thrown overboard by pulling a lever in an emergency. There is a large dome antenna on the mainmast and a cylindrical antenna on the helicopter service center for controlling RPVs. A hangar for eight Pioneer RPVs is below the turret no. 3 guns. The diagonal stanchions indicate where the aft missile platform was extended to accommodate tracks for a gantry crane to reload Tomahawk Armored Box Launchers. Also visible in this view are the scupper pipes long the hull and the bolster at the stern around the openings for towing Nixie decoys. These were among the original 1980's modifications. Internally, a new admiral's suite was constructed. The Navy removed the RPV antennae and hangar when the ship retired. (Nat'l Archives)

This forward director tower as a museum and memorial in 2020. The forward leg of the forward tripod mast only extends down to 010 level on USS *New Jersey*. It goes to 08 level on her sisters. Air intakes in the stack cowling have been plated over for preservation. The rangefinders on the Directors Mk.37 have also been enclosed for preservation. The circular air intake at 06 level was an intake for cooling the radar equipment at 07 level. The ship's whistle is visible forward of the circular SATCOM antenna.

On the foremast platform the forward antenna is an SPS-10 surface search radar. At the center of the platform is an SPS-49 air search radar. The stub mast at the aft end of the platform has a TACAN (tactical air navigation) antenna at the top.

An aft view of the forward stack in 2019. The binnacle compass is visible above the canvas at 04 level. All the metal within twelve feet of the compass is made of brass. That includes the platforms and an elliptical area of plating on the uptake housing. Note the pair of fan antennae extending aft from the foremast to the mainmast

What the rings at the bow are for. The main deck substantially overhangs the hull at the bow. The rings allow one to pull close to the hull or to attach painting platforms. The hull shell of *Iowa*-class battleships is largely riveted with some welding. Riveted lap joints are clearly visible in this photograph. At the bow the joints are flush and welded. (Nat'l Archives)

The aft mast on a battleship was called the *mainmast* for historical reasons, even though it was usually smaller than the foremast. Here is the aft stack and mainmast in 2020. The twin fan antennae ends at the mainmast. There is a platform for a SATCOM antenna at the aft end of the update housing. The large air conditioning unit aft of the uptake housing is a museum addition.

The two pipes extending over the top of the uptake housing are exhausts. The larger of the two visible is a boiler atmospheric exhaust. If pressure needed to be quickly released from a boiler, steam could be vented from such an exhaust. The smaller pipe is an auxiliary exhaust.

The forward extension of the mainmast was added in 1991 to house a dome antenna for Pioneer RPVs. The Pioneers were operated by a detail assigned to the ship. The Pioneer equipment left with its detail when USS *New Jersey* was decommissioned in 1991.

A crane for reloading Harpoon missiles is at the side of the uptake housing directly above the missiles. There is an identical crane on the other side. These cranes were added in 1984.

The skip box for hauling CIWS ammunition is parked at the door to the magazine at 04 level. The car is not powered. Instead, it had to be raised and lowered using a portable davit erected at the 05 level. The tracks for the car extend down to 01 level.

USS *New Jersey* with USS *Missouri* behind. The other *Iowa*-class battleships received a wrap around ECM structure at 010 level in the 1980's while USS *New Jersey* retained her ears added in the 1960's. USS *Long Beach* is behind *Missouri*.

The foremast platform looking forward. The base of the SPS-49 air search radar is in the foregound. The museum was in the process of restoring its rotation motors when this photograph was taken. (Ryan Szimanski)

The SPS-10 surface search radar at the forward end of the foremast platform. The museum added the chains to keep the antenna from weathervaning. (Ryan Szimanski)

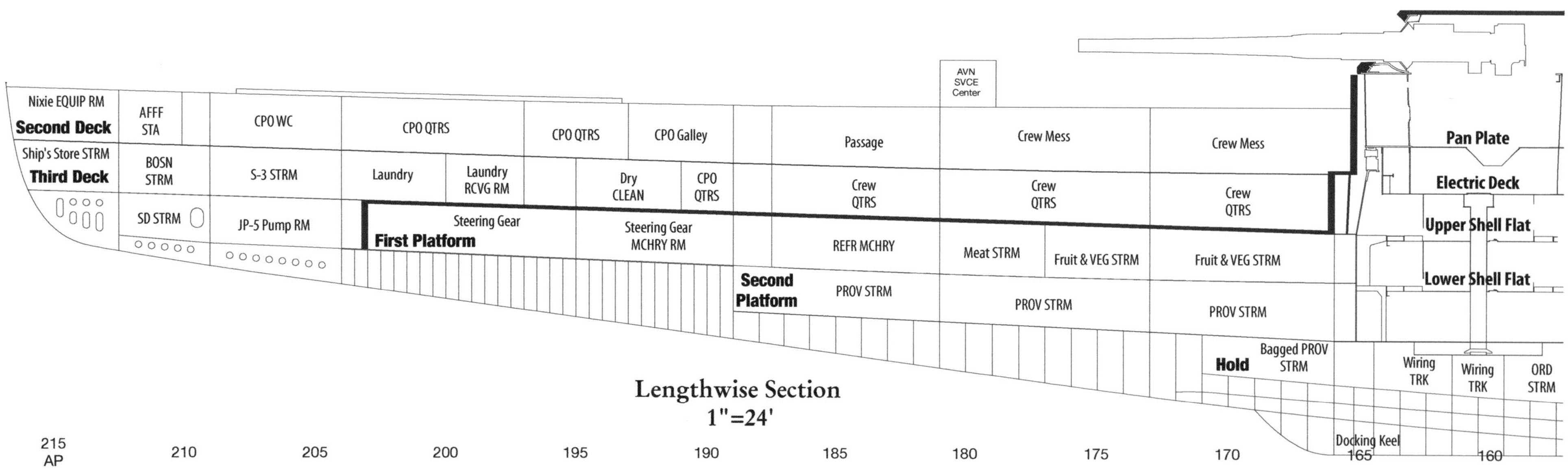

Lengthwise Section
1"=24'

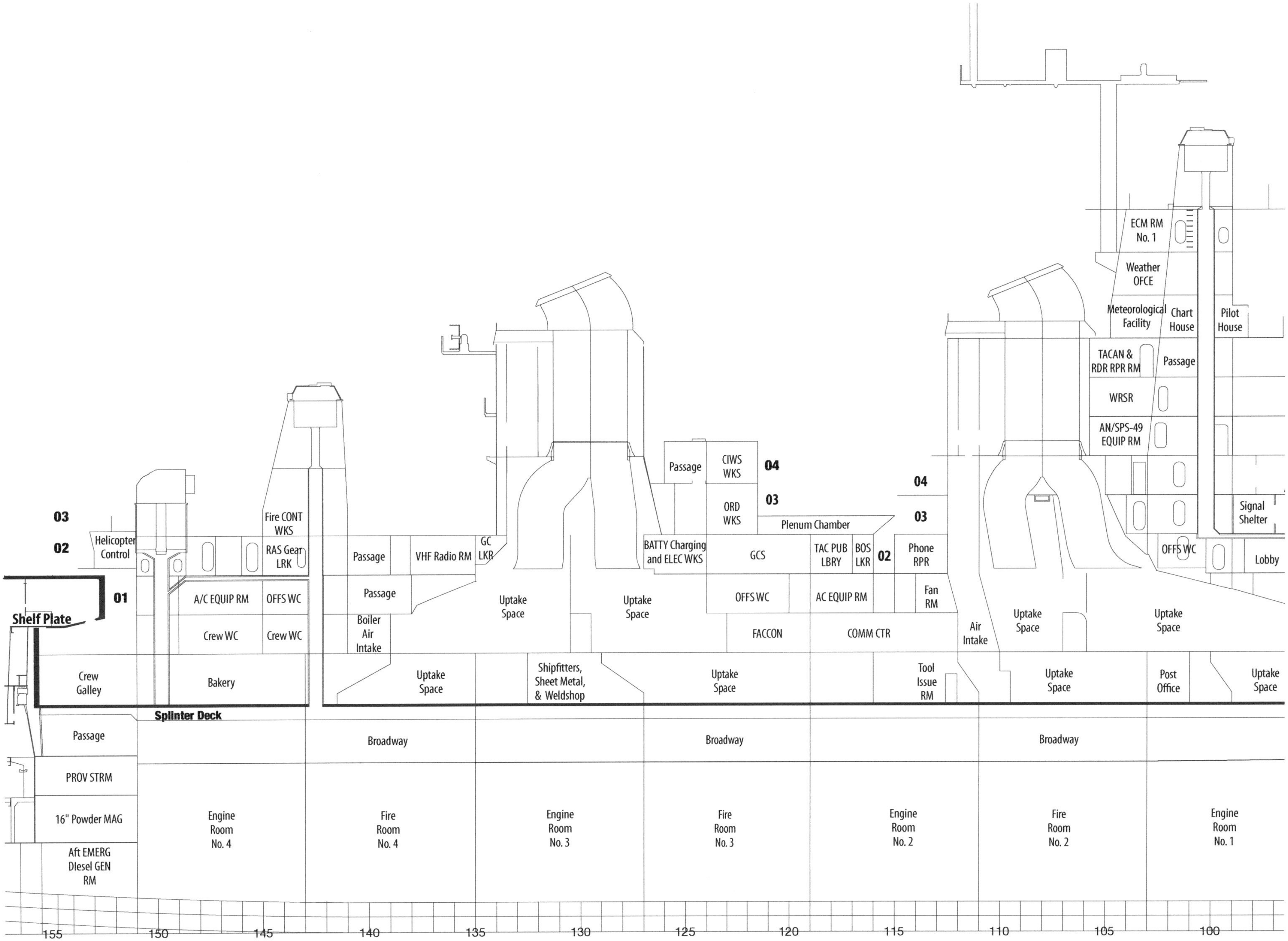
ECM RM No. 1
Weather OFCE
Meteorological Facility
Chart House
Pilot House
TACAN & RDR RPR RM
Passage
WRSR
AN/SPS-49 EQUIP RM
Passage
CIWS WKS
04
ORD WKS
03
Plenum Chamber
04
03
Signal Shelter
03
02
Helicopter Control
Fire CONT WKS
RAS Gear LRK
Passage
VHF Radio RM
GC LKR
BATTY Charging and ELEC WKS
GCS
TAC PUB LBRY
BOS LKR
02
Phone RPR
OFFS WC
Lobby
01
Shelf Plate
A/C EQUIP RM
OFFS WC
Passage
Uptake Space
Uptake Space
OFFS WC
AC EQUIP RM
Fan RM
Air Intake
Uptake Space
Uptake Space
Crew WC
Crew WC
Boiler Air Intake
FACCON
COMM CTR
Crew Galley
Bakery
Uptake Space
Shipfitters, Sheet Metal, & Weldshop
Uptake Space
Tool Issue RM
Uptake Space
Post Office
Uptake Space
Splinter Deck
Passage
Broadway
Broadway
Broadway
PROV STRM
16" Powder MAG
Aft EMERG Diesel GEN RM
Engine Room No. 4
Fire Room No. 4
Engine Room No. 3
Fire Room No. 3
Engine Room No. 2
Fire Room No. 2
Engine Room No. 1
155
150
145
140
135
130
125
120
115
110
105
100

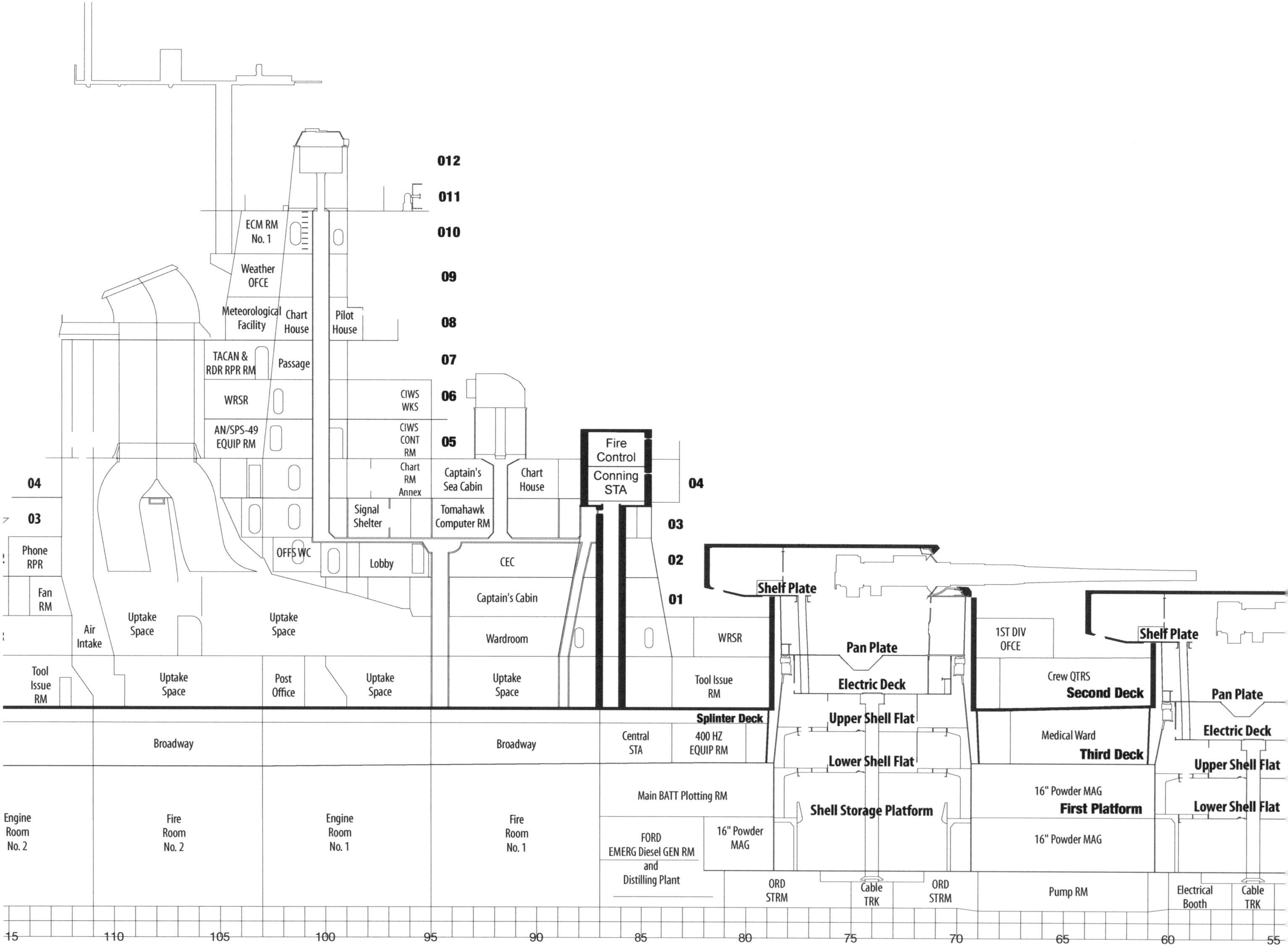
012
011
010
09
08
07
06
05
04
03
02
01
ECM RM No. 1
Weather OFCE
Meteorological Facility
Chart House
Pilot House
TACAN & RDR RPR RM
Passage
WRSR
CIWS WKS
AN/SPS-49 EQUIP RM
CIWS CONT RM
Chart RM Annex
Captain's Sea Cabin
Chart House
Fire Control
Conning STA
Signal Shelter
Tomahawk Computer RM
Phone RPR
OFFS WC
Lobby
CEC
Fan RM
Captain's Cabin
Air Intake
Uptake Space
Uptake Space
Wardroom
WRSR
Shelf Plate
Pan Plate
1ST DIV OFCE
Shelf Plate
Tool Issue RM
Uptake Space
Post Office
Uptake Space
Uptake Space
Tool Issue RM
Electric Deck
Crew QTRS
Second Deck
Pan Plate
Splinter Deck
Upper Shell Flat
Electric Deck
Broadway
Broadway
Central STA
400 HZ EQUIP RM
Lower Shell Flat
Medical Ward
Third Deck
Upper Shell Flat
Main BATT Plotting RM
16" Powder MAG
First Platform
Shell Storage Platform
Lower Shell Flat
Engine Room No. 2
Fire Room No. 2
Engine Room No. 1
Fire Room No. 1
FORD EMERG Diesel GEN RM and Distilling Plant
16" Powder MAG
16" Powder MAG
ORD STRM
Cable TRK
ORD STRM
Pump RM
Electrical Booth
Cable TRK
115
110
105
100
95
90
85
80
75
70
65
60
55

Lengthwise Section
1"=24'

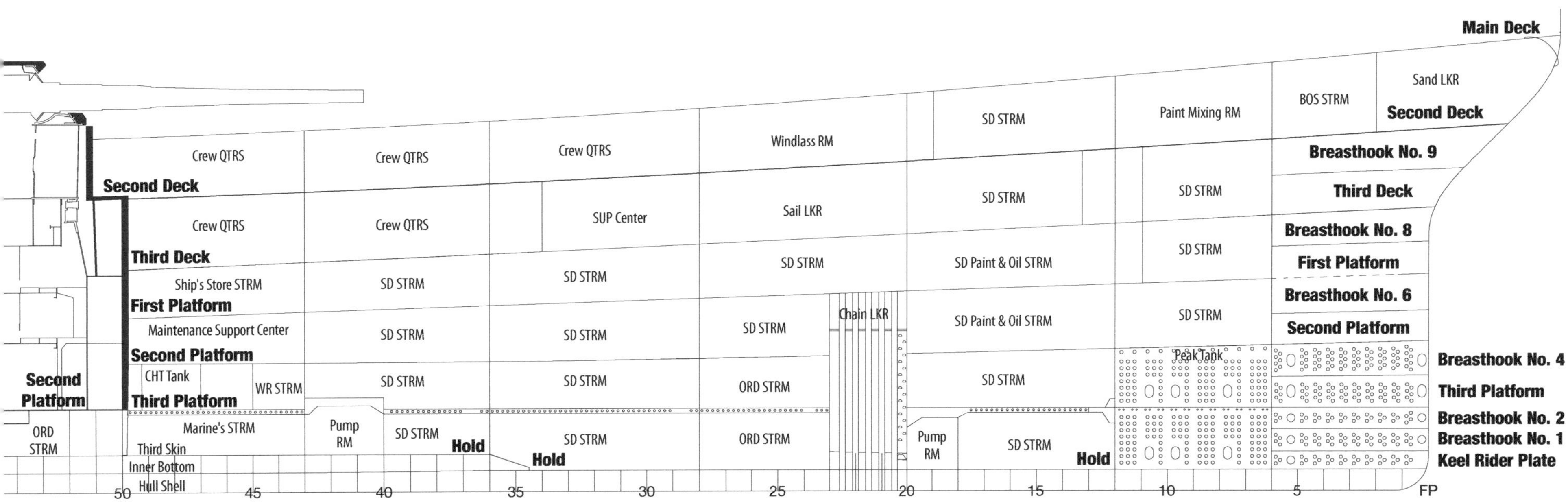

Main Deck

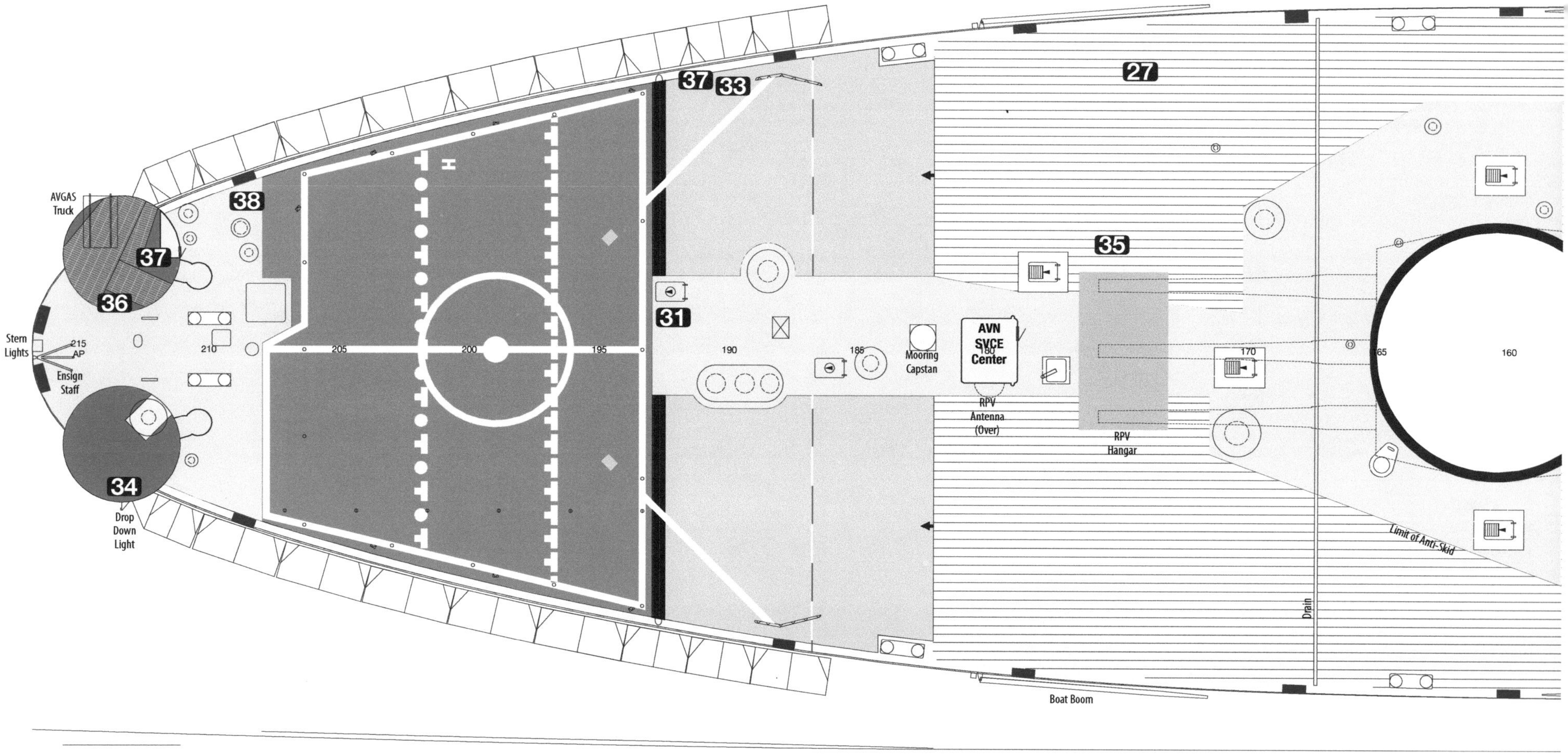

Main Deck
1"=24'

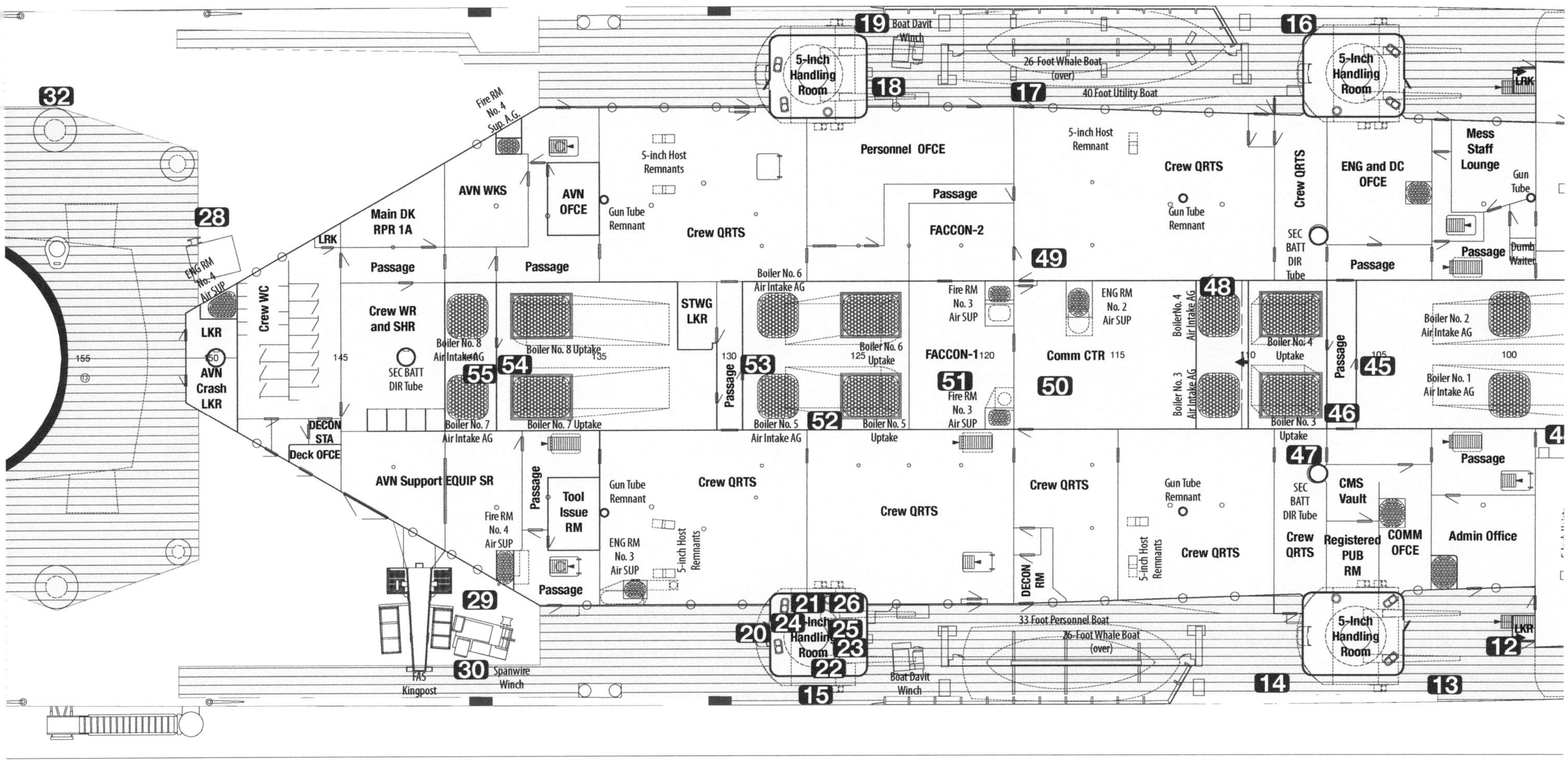
32
28
Fire RM No. 4 Sup. A.G.
ENG RM No. 4 Air SUP
LKR
AVN Crash LKR
Crew WC
LRK
Main DK RPR 1A
Passage
Crew WR and SHR
SEC BATT DIR Tube
DECON STA
Deck OFCE
AVN WKS
AVN OFCE
Gun Tube Remnant
5-inch Host Remnants
Crew QRTS
Boiler No. 8 Air Intake AG
Boiler No. 8 Uptake
Boiler No. 7 Air Intake AG
Boiler No. 7 Uptake
55
54
STWG LKR
AVN Support EQUIP SR
Fire RM No. 4 Air SUP
Tool Issue RM
ENG RM No. 3 Air SUP
29
30
FAS Kingpost
Spanwire Winch
19
Boat Davit Winch
5-Inch Handling Room
18
Personnel OFCE
Passage
FACCON-2
Boiler No. 6 Air Intake AG
Boiler No. 6 Uptake
53
52
Boiler No. 5 Air Intake AG
Boiler No. 5 Uptake
Fire RM No. 3 Air SUP
FACCON-1
51
20
21
24
26
25
23
22
15
17
26-Foot Whale Boat (over)
40 Foot Utility Boat
33 Foot Personnel Boat
49
ENG RM No. 2 Air SUP
Comm CTR
50
DECON RM
Boiler No. 4 Air Intake AG
Boiler No. 3 Air Intake AG
48
Boiler No. 4 Uptake
Boiler No. 3 Uptake
46
47
45
16
SEC BATT DIR Tube
ENG and DC OFCE
Mess Staff Lounge
Gun Tube
Dumb Waiter
Boiler No. 2 Air Intake AG
Boiler No. 1 Air Intake AG
CMS Vault
Registered PUB RM
COMM OFCE
Admin Office
LRK
LKR
12
13
14
4

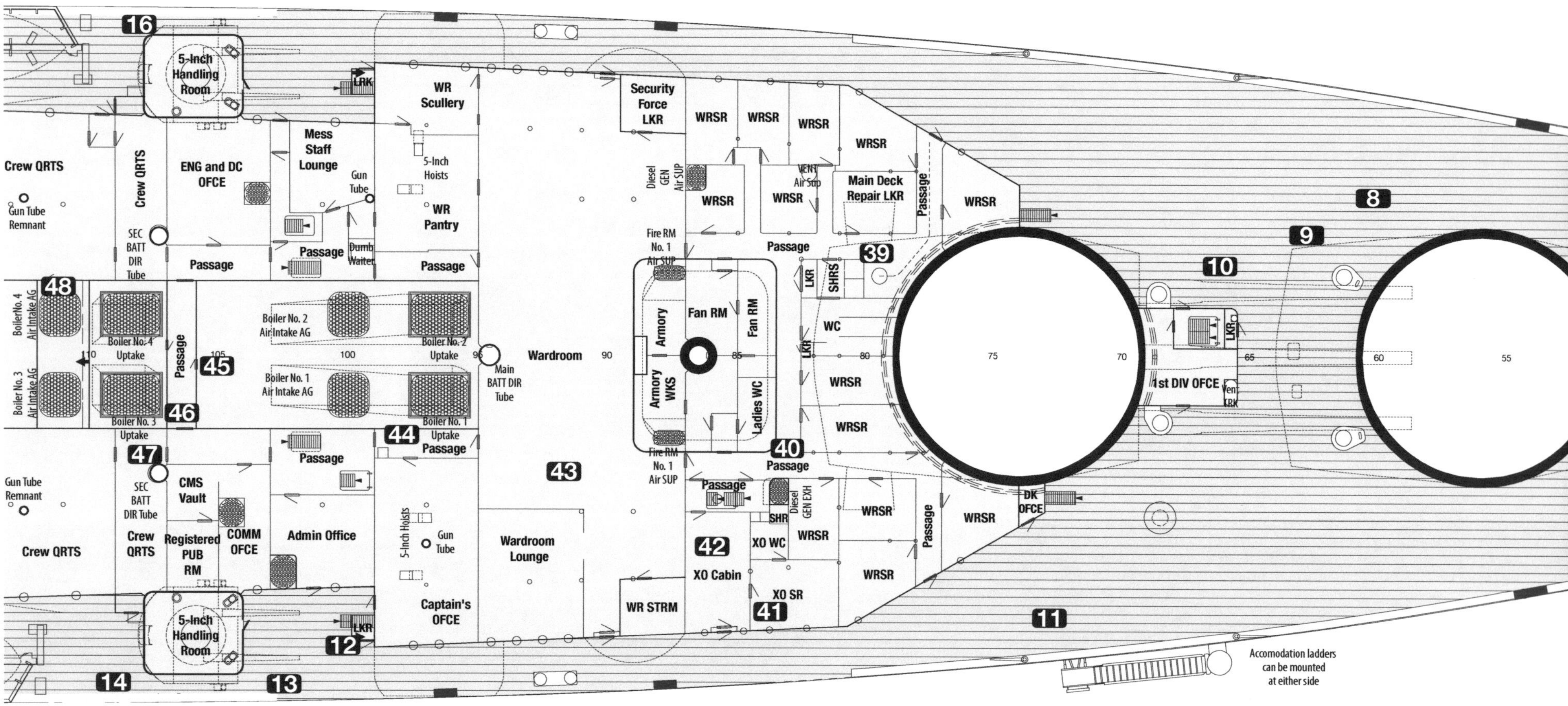

Crew QRTS
Gun Tube Remnant
Crew QRTS
ENG and DC OFCE
Mess Staff Lounge
Gun Tube
WR Scullery
5-Inch Hoists
WR Pantry
Passage
Dumb Waiter
SEC BATT DIR Tube
5-Inch Handling Room
LKR
Security Force LKR
WRSR
Main Deck Repair LKR
Diesel GEN Air SUP
VENT Air Sup
Fire RM No. 1 Air SUP
Armory
Armory WKS
Fan RM
Ladies WC
LKR
SHRS
WC
Boiler No. 4 Air Intake AG
Boiler No. 3 Air Intake AG
Boiler No. 4 Uptake
Boiler No. 3 Uptake
Boiler No. 2 Air Intake AG
Boiler No. 1 Air Intake AG
Boiler No. 2 Uptake
Boiler No. 1 Uptake
Main BATT DIR Tube
Wardroom
1st DIV OFCE
Ven LKR
CMS Vault
Registered PUB RM
COMM OFCE
Admin Office
Wardroom Lounge
Captain's OFCE
WR STRM
XO Cabin
XO WC
XO SR
SHR
Diesel GEN EXH
DK OFCE
Accomodation ladders can be mounted at either side

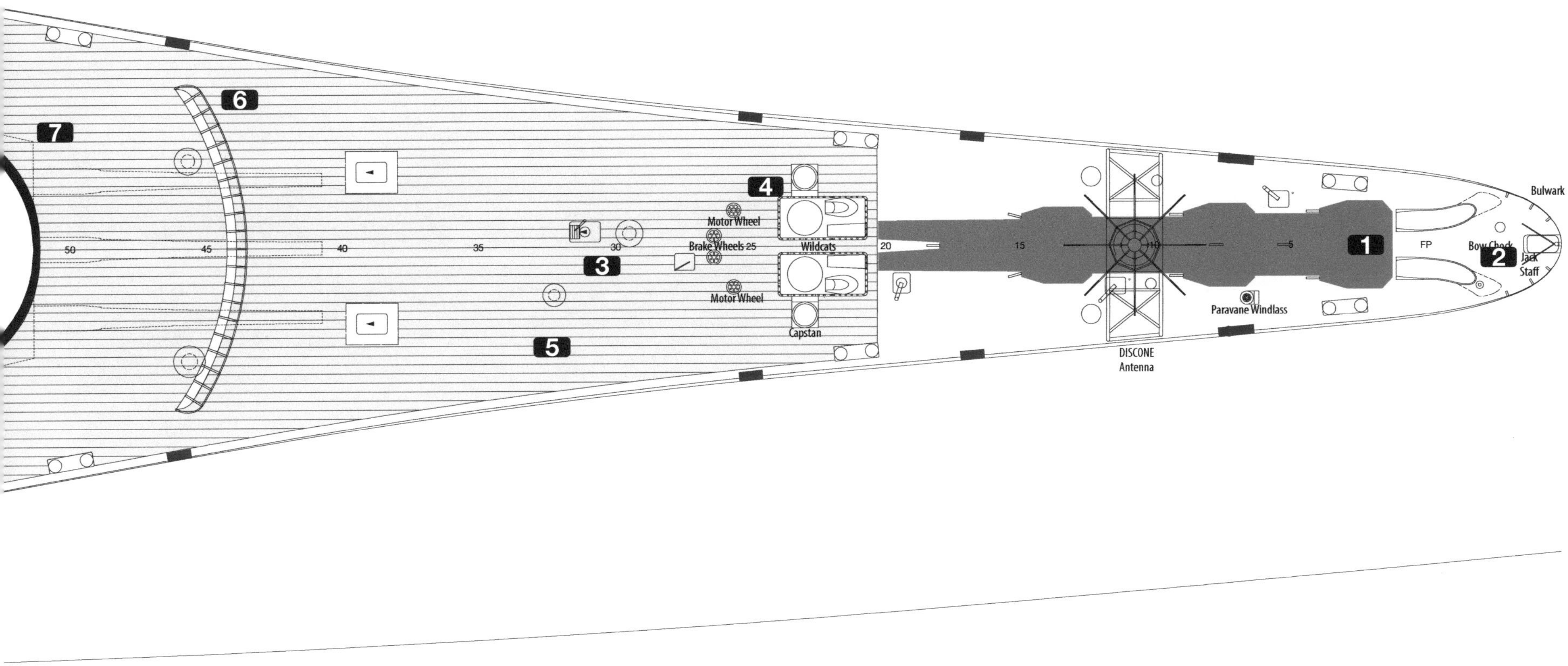

Main Deck
1"=24'

(1) This is the second bow bulwark on USS *New Jersey*. The original was erected in 1943 shortly after commissioning. That bulwark flared outwards at the sides so that it could house two 20mm mounts. After USS *New Jersey* returned from Korea in 1952, the sixteen remaining 20mm twin mounts were removed and this flat-sided bulwark was erected to replace the flared bulwark. The same modification was made to USS *Missouri* when it lost her 20mm mounts.

The jackstaff is attached to the inside face of the bulwark. When the ship was moored or anchored it flew the Navy jack at the bow as shown in the photograph. The forward anchor light is mounted at the top of the jackstaff. There was a steering light mounted just above the support braces but it has been removed. The steering light faced aft and allowed the helmsman to identify the location of the bow at night. The jackstaff can fold down.

Chains run through three openings in the deck. The bow chock is immediately below the jackstaff. This provided a location for running lines to the bow for towing or mooring. During World War II a chain for towing parvanes always ran through the bow chock. The port and starboard hawse pipes for the anchor chains are at the bottom of the photograph.

The security monitoring equipment and the netting at the rails are museum additions.

USS *Iowa* and USS *Wisconsin* retained their flared bow bulwarks even after the 20mm mounts were removed. This is USS *Wisconsin*. The scuttle to the sand locker is aft of the port ladder. (Keith Nitka)

(2) The forecastle viewed from the bow. Shackles for the 30,000 lb. stockless anchors are visible in the hawse pipes. The mesh covers at the forward end are original but the netting is a museum addition. Each chain is 1,122 feet (187 fathoms) long and each link weighs 123 lbs. The bitter ends of the anchor chains are attached to the chain locker at its aft bulkhead.

The anchor chains are connected to chain stoppers attached to the deck. The chain stoppers prevent any force applied to the chain from being transmitted to the machinery and ensure the anchor chains do not inadvertently break free.

There are bitts with roller chocks aft of them at either side of the deck for mooring. Bitts provide a location for securing cables to the ship. The decks are reinforced around bitts and chocks so they can withstand the forces from mooring cables.

Chocks prevent mooring lines from moving across the deck. The openings in the chocks are rounded to minimize chaffing to the cables running through them. Chocks at the starboard side are in use to hold the museum in place. 1.9-inch diameter half round was welded to both sides of the sheer (upper) strake of the hull plating during construction to reduce the likelihood of injury to crew and chaffing to lines by eliminating sharp edges.

The museum has erected a portable J-davit at the bosun's storeroom hatch. Deck hatches that lead to storerooms have sockets for erecting davits that can be used to lower supplies.

The large discone antenna was installed for ship-to-ship communications in the 1960's. The other *Iowa*-class battleships received the same modification in the 1980's. The red band at the base indicates this is a transmitting antenna. Alternately, blue indicates a receiving antenna. A ship's bell was installed at the forward side of the discone antenna.

To the left (starboard side) are fittings that were used for towing paravanes. Paravanes are devices for cutting mine cables. A paravane has a torpedo-shaped body with an attached wing that was ballasted so that it would swim vertically in the water. The ship would tow a paravane on each side of the ship. The wing would force the paravane to swim away from the ship. Naval mines of the World War II era were buoyant and were held below the water by cable linked to an anchor. If a paravane's towing line snagged a mine's anchor cable, the anchor cable and would be pulled towards the paravane and away from the ship. If everything went right, a cutter on the paravane would sever the anchor cable, the mine would float to the surface, and the ship's guns would destroy the mine. Otherwise, the mine might detonate and destroy the paravane after having been pulled away from the ship—an acceptable trade.

The paravane winch is at main deck just inboard of the starboard roller chock. Until the 1980's there was a pipe that ran from the underside of the bow to main deck just forward of the winch. A chain ran in a circular path from that main deck opening, aft around the winch, forward to the roller button next to the hawse pipe, through the bow chock, into the water, into the pipe at the lower bow, and back up to the main deck. The paravane towing cables were attached to this circular chain. By using the winch to pull the chain, the paravanes to be raised and lowered under the water.

The concept of using battleships as minesweepers fell out of favor in the Navy by the Korean War. A ship had to move straight when towing paravanes and had to stop to set them out or retrieve them, creating periods of vulnerability for a warship employing paravanes. In the 1960's the crew refurbished the paravane winch and used it to lower supplies from the main deck to storerooms below. During USS *New Jersey*'s first refit in the 1980's all the paravane piping and underwater fixtures were completely removed because they were leaking.

(3) In this view of the forecastle the entire discone antenna is visible. The term *discone* describes an antenna whose shape is that of a disc at the tip of a cone. In this particular antenna neither the disc nor the cone are solid. The radial bars at the top form the disc and the cone is formed from wires using spreader bars. The ship's forward masthead light is mounted at the top of the antenna.

The wildcats (inboard) and windlasses (outboard) are in the center. The wildcats raise and lower the anchors. The four wheels at the centerline control the wildcats. The wheels at the center set and release the brakes. The outer wheels control speed and direction. The anchors were lowered; not dropped.

The windlasses pulled mooring lines. The windlass rotates at a constant speed. Wrapping a line around a windlass and pulling tight created enough friction to pull the ship along heavy mooring lines.

The cylinder at the corner of the hatch in the foreground is a socket for mounting a J-davit to lower supplies to the storerooms below. The lockers forward and to the starboard side of the bucket vent at the lower left are for storing ready-service ammunition lockers for 0.50" machine guns.

(4) A closer view of a wildcat. The rotating head has teeth that mesh with the anchor chain. The removable hood at the left covers the opening to the chain pipe that leads to the chain locker at the hold.

This is a paravane resting on its side. The weight on the end of the wing causes the wing to be oriented vertically in the water. When moving through the water, the wing could pull the paravane away from the ship. The mine chain cutter is at the top of the paravane. (Naval History and Heritage Command)

(5) A view aft from the forecastle. The circular structure in the deck is the remnant of a 40mm mount. All 40mm mounts were removed in 1968 when the ship was prepared for service in Vietnam. The 40mm tub in this location was removed at that time. Immediately behind that is the coaming for a deck hatch. There is another hatch mirrored on the centerline. The high, 14$\frac{1}{2}$" coaming was used because these hatches were intended to be open as much as possible to provide ventilation to the crew quarters in the days before air conditioning. These hatches were covered when the ship was mothballed.

The breakwater is in front of 16-inch turret no. 1. The right rangefinder hood for turret no. 2 is poking out from an opening in the canvas awning. Turret no. 1 originally had identical hoods. However, they were removed in 1950 along with the rangefinder because they shipped water and the rangefinder had a poor view.

(6) The breakwater was intended to deflect water that reached the deck away from the turret and over the side. *Iowa*-class battleships were extremely wet forward so the breakwater was only marginally effective. The two large bucket vents provide ventilation for the crew quarters directly below. Fire hoses were normally stored on the three platforms on breakwater. There is a fireplug behind the center of the breakwater.

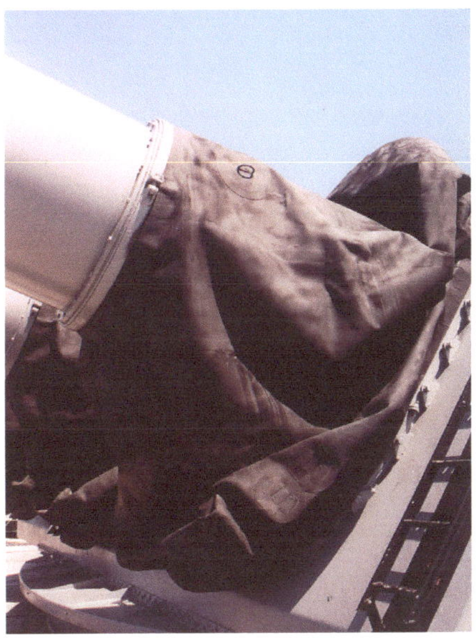

(7) Neoprene bucklers are attached to each 16-inch gun and the turret to keep water and gas from entering the turret through the gunport. There is a metal frame under the buckler that is designed to keep the buckler from tearing by guiding its movement when the gun is fired. The existence of this frame explains why the buckler stands upwards at the top. A clamping ring holds the buckler to the gun. An oval tube presses the outer edge of the buckler to hold it against the turret. The bolts that hold the tube against the buckler are visible near the turret face.

(8) This is the left side of turret no. 1. This part of the turret is called the *gunhouse*. The circular structure below the gunhouse is the armored *barbette*. The barbette extends from the bottom of the turret to the start of the armored citadel at the second deck. There is no structural connection between the rotating turret and the barbette.

The ring at the base of the turret is the gas seal. Hidden from view underneath, there is a circular grease trough attached to the top of the barbette. The rotating turret's gas seal hooks into this trough. The grease prevents water or gases from entering the turret.

The gunhouse rests on the shelf plate. The shelf plate is two inches of special treatment steel (armor) at the front and there are two layers (giving a total of 4 inches) at the back. The division between the two areas is visible at the side of the turret.

The two projections at the side of the turret area the sight hoods used when the turret is in local control. The lower hood is for the pointer who controls elevation and the upper hood is for the trainer who controls traverse. These positions are mirrored on both sides of the turrets. The hoods are 2.5" of cast armor. These hoods are mirrored on the right side of the turret. The various turret hoods are the only places where cast armor was used on *Iowa*-class battleships.

The beams extending past the side of the gunhouse are shell-handling beams. There are two main deck ammunition-loading hatches under the turret bustle. The turret can be rotated so that both shell-handling beams are positioned directly above a hatch. Pulleys were attached to the ends of the beams. These could be used raise or lower ammunition between the magazines and main deck.

The roof is made of five plates 7.25" thick. The plates have scarf joints between them. The outlines of the center roof plate joint are visible on either side of the aft shell-handling beam. The scarf joints are oriented so that this one plate can be removed to allow a crane can move gun components in and out while changing barrels.

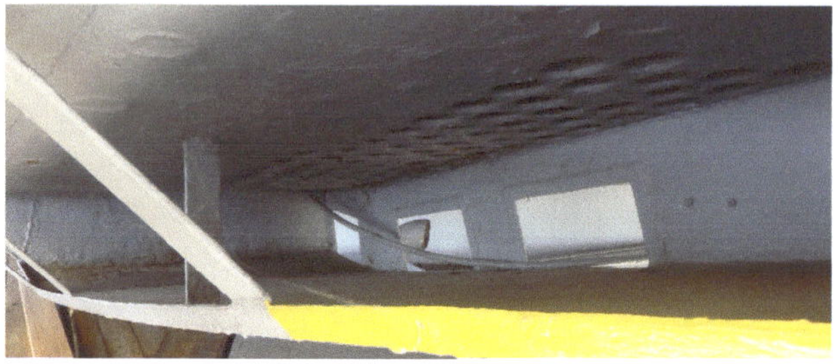

(9) Each turret has four ventilation trunks running along the underside of the shelf plate. This is the left trunk for turret no. 1. The side openings were covered with mesh to keep bugs out when in service. Air flowed from these trunks into the turret through the circular holes cut into the shelf plate. On USS *Missouri* and USS *Wisconsin* the side vent trunks for turret no. 1 were extended up the side of the gunhouse to avoid the problem of water entering.

(10) This structure at the base of turret no. 2 was originally a magazine for 20mm ammunition. After the 20mm guns were removed it became an office for First Division. During the 1980's–1990's First Division was responsible for maintaining the forecastle and operating its mooring machinery. An ammunition-loading hatch is in the deck to the right of the door. There is a similar hatch on the starboard side.

(11) An accommodation ladder in place in the raised position. The accommodation ladder could be lowered to transfer personnel from boats. Normally, the accommodation ladders were stored against the deckhouse. This is USS *Iowa* in 1986. (Nat'l Archives)

(12) Triangular lockers port and starboard conform to ladders to 01 level. These are for awnings and deck gear. There is a similar triangular linen locker conforming to a ladder at 01 level. Could the same designer have planned all three?

(13) This is a refueling port. To refuel at sea, the port would be turned to face inboard (see below). The cover plate was removed and replaced with a flexible extension tube curved to face outboard. A tripod would be set up inboard of the refueling port which would be connected to a spanwire sent over from the tanker. Once the spanwire was in place, a fuel line with a probe would be sent down the spanwire and connected to the extension tube. At this point fuel could flow from the tanker to the battleship.

(14) The boat handling facilities were completely rebuilt during the 1980's upgrade. At main deck, cradles were built for a 33-foot utility boat on the starboard side and a 40-foot utility boat on the port side. A bulwark was erected around the cradles to keep the boats from floating out when the sea came over the deck. USS *Missouri* has a similar bulwark. Simpler and less effective straight-sided bulwarks were built on USS *Iowa* and USS *Wisconsin*.

Large boat davits were installed over the bulwarks. 26-foot motor whaleboats were stored on the davits. The davits have stops mounted on the main deck to keep them from striking the bulwark. The forward starboard stop is lower than the others so the corresponding davit arm has a unique triangular extension.

When the ship was mothballed the davit operating cables were removed and the davit arms were locked in place. In operation, the forward davit arm was connected to a pulley attached to the 02 level kingpost. The aft davit arm cable ran through a pulley attached directly to the deckhouse. A locking bar that was installed during mothballing is below the empty pulley mount. Below that is the spring buffer and the locking arm used to secure the davit when the ship was in service.

This plan shows how a refueling probe from a tanker would be connected to a refueling port.

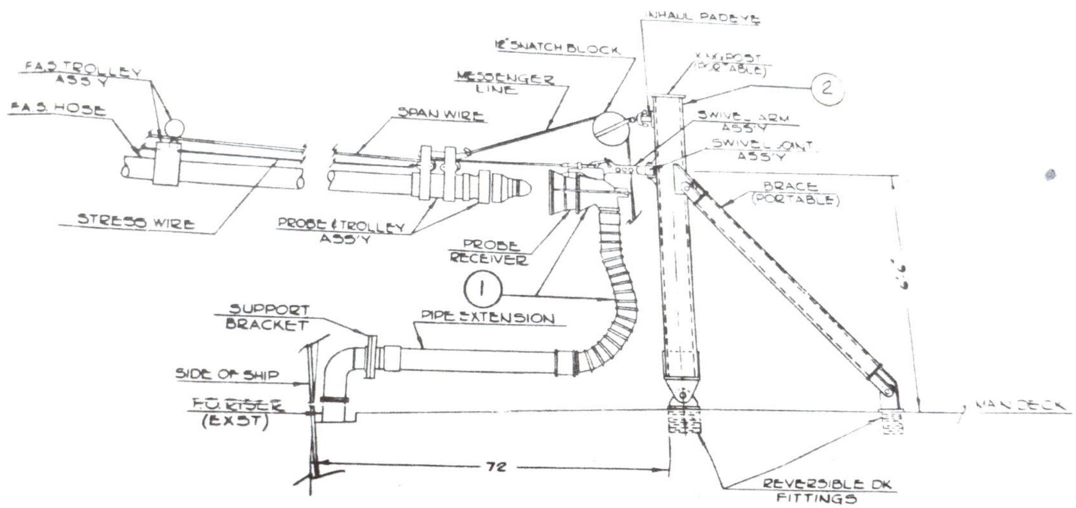

This photograph shows how the 1980's boat davits operated. A motor whaleboat being lowered into the water on USS *Iowa*. The forward davit arm is connected by cabling running through pulley on a kingpost erected on the 02 level. Note the cabling and pulley in place on the kingpost that is missing in photograph (13). (Nat'l Archives)

(15) The starboard boat davit looking forward. The stop for the davit is much taller so there is no extension on the davit arm. The davit winch is to the left. In early testing the forward and aft davit arms could move at different rates causing the strongback between them to separate. Electronic controls were added to the winch to ensure that the two arms moved at the same rate.

(16) The port davit looking aft. Again, note the tall stop for the davit and lack of extension on the davit arm. The davit and bulwark on the port side are longer than those on the starboard side to accommodate a larger boat. The lights are museum additions.

(17) Platforms at the 02 level provide access to the whale boats. The museum cut away the bottom of the access ladder to the platform to protect visitors. This platform is mirrored on the starboard side. The blast deflectors for the Tomahawk missiles are at the top of the photograph.

(18) One of the upgrades made in the 1980's was to add armored ventilation ducts to the 5-inch handling rooms to allow air to circulate. Previously, there had only been air exhausts in the handling rooms. The shape of the duct varies depending upon the location.

(19) Port boat storage looking forward. The davit winch is to the right.

(20) Each 5-inch mount has a handling room directly below it. The handling room is a small magazine that stores ready service ammunition and is protected by $2\frac{1}{2}$ inches of special treatment steel. This is the handling room for the aft 5-inch/38 mount main deck starboard. The handling rooms have common features but they are laid out slightly differently. The large structure at the center is part of the rotating turret. It houses the communications cables and the ammunition hoists.

Ammunition for the 5-inch/38 gun is semi-fixed. That means the powder and primer are contained in a brass cartridge but the projectile is separate. The projectiles weigh about 54 lbs. and the cartridges about 28 lbs. Splitting the ammunition in two parts made it easier for men to move it.

The powder hoists are at the left and the projectile hoists are at the right. The large wheel is for manually driving the powder hoist. The projectile hoist could also be operated manually but its wheel is missing. It would have been over the square projection to the right of the other wheel.

Officially, thirteen crewmen manned each powder handling room. The crew consisted of six powder handlers, six projectile handlers, and a petty officer in charge. However, it could operate with fewer and after World War II the number of gun crewmen tended to be reduced.

(21) Because the guns are in pairs, most of the equipment is in pairs as well. Here are the powder hoists. These are dredger-type hoists that operate like a conveyer belt. The shutters keep the cartridge from falling out as it rises. A wooden dummy is in the right hoist. The ladder at the center gives access to the gunhouse from the handling room. Racks containing canisters for the powder cartridges are in the background. The powder cartridges remained in canisters from they time they were brought on board the ship until they were loaded into the powder hoists.

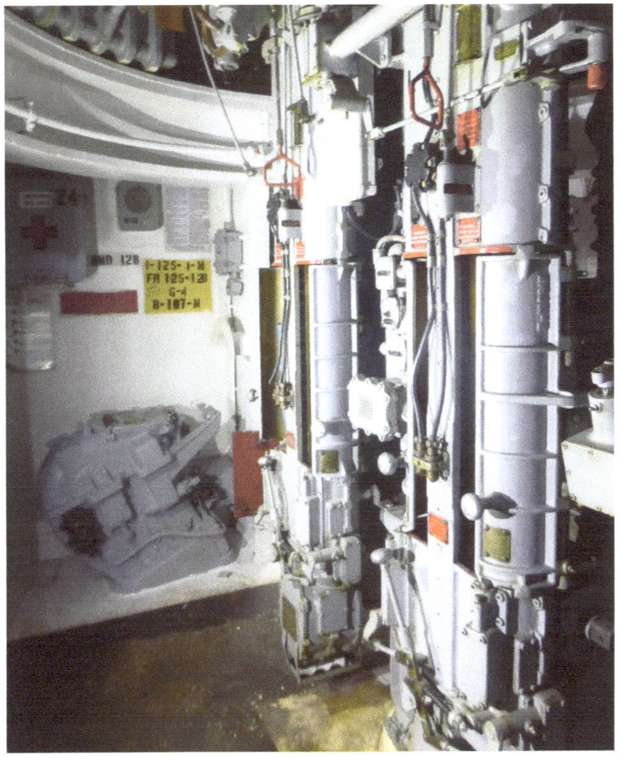

(22) In addition to raising projectiles to the gunhouse, the hoist also sets their fuzes. Anti-aircraft projectiles prior to the introduction of VT (proximity) fuzes had to be timed to explode at the when they would reach their targets. Starshells had to be timed to explode overhead their target. Projectile fuzes had a clockwork timer. Rotating a dial attached to the fuze wound the timer and the amount of rotation determined the time between firing and detonation. The projectiles were oriented nose down in the hoist so their weight would hold them in place as the fuze was wound as ordered by the fire control system.

The projectile hoist unit has four hoists, two for each gun. Dual paths per gun allowed a projectile to be at the gun while another was fuzed, providing a continuous flow of ammunition to the gun.

(23) The handling rooms also contain racks for storing projectiles and powder canisters. The projectiles shown in the handling room are blue (inert) practice rounds.

(24) The 5-inch handing rooms were restocked with ammunition from magazines located at third deck. Each handling room has two dredger-style hoists that connect the handling room to its magazine. Either hoist can transfer projectiles or cartridges within canisters. For more efficient operation, the normal practice was to use one hoist for projectiles and the other cartridge canisters. When loading ammunition on the ship, the hoists could be run in reverse to stock the magazines.

The circular plates in the deck cover decanning scuttles. After a powder cartridge was removed from its canister, the canister was dropped through the scuttle to the deck below.

(25) The mount turns by rotating a pinion gear against a circular rack attached to the ship. Traverse was normally powered by an electric-hydraulic motor. If power failed, the turret could be hand cranked.

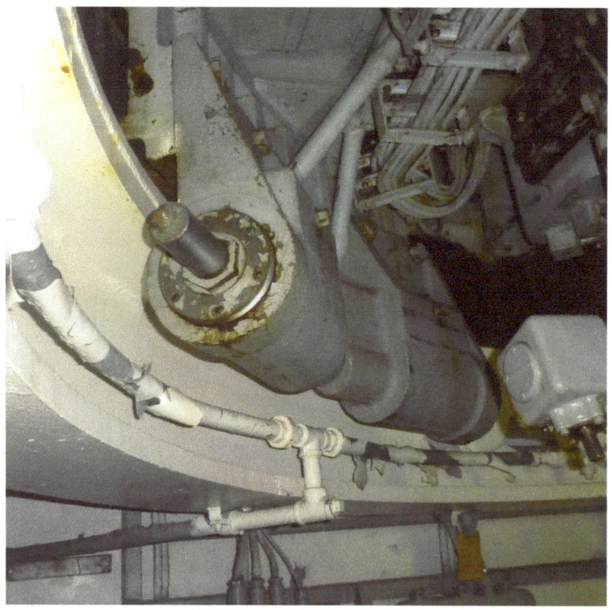

(26) Buffers limited the amount of gun traverse to prevent the ship from firing on itself. The rotation limit is marked by fixed stops. When the mount rotates into a stop, the piston at the end of the buffer depresses causing the traverse motor to be cut off. The buffers are mounted on holding down clips. There is a popular misconception that turrets were just held in place by gravity. This was not the case on *Iowa*-class battleships where all of the turrets are secured from underneath. There are additional holding down clips spaced around the turret ring.

(27) The port side at the stern looking forward. At the time this photograph was taken the teak planking was in the process of being replaced. The museum has since replaced the teak deck in this area. The top of the drain running across the deck was normally flush with the wood deck. The exposed steel shows how the deck plates were riveted together using butt plates and illustrates the sizes of the deck plates. The deck plating at the stern varies from $\frac{5}{16}$" to $1\frac{1}{2}$" in thickness. A J-Davit has been erected in a location used for positioning a boarding ladder at the port side.

Tripods could be erected on roofs of turrets nos. 2 and 3 for replenishment at sea (RAS). Spanwires would run from the supply ship and attach to the tripod. The large half round at the top of the turret was reduce cable chaffing. The large hood extending from the side of the turret houses the rangefinder. A brass door protects the rangfinder from weather. The door slides inward to open.

The aft Mk.37 (lower) secondary battery and Mk.38 (upper) main battery gun directors are visible above the turret. The director Mk.37 has a Mark 25 radar. During the Korean War the Mark 25 radar incrementally replaced the World War II era Mark 12 and Mark 22 radars. The director Mk. 38 has a Mark 13 radar. The Mark 13 radar replaced the original Mark 8 radar during the Korean War. During the mothballing process covers were welded over the rangefinders of the director Mk.37.

The 40mm and 20mm mounts at the left are museum displays. The 40mm gun on display had served on USS *New Jersey* until 1968. The cover over the deck hatch next to the 16-inch turret is also a museum addition.

(28) Additional boats could be stored on dollies at the stern. These were transferred to the water using a large boat boom visible in the background of photograph (27). This winch at the aft end of the deckhouse was used to raise and lower boats from the boat boom.

(30) This is the spanwire winch for the FAS kingpost. A light messenger line was shot to the ship being refueled. Its crew would pull on the message to bring the spanwire across and attach the end to their ship. A fueling hose would run down the span wire to the other ship. The winch was controlled from the FAS control station at the 01 level in the background.

(29) The fueling at sea (FAS) kingpost was added to *Iowa*-class battleships during the 1980's. Large warships often served as tankers for their escorts. The tall kingpost made refueling smaller ships easier by raising the spanwire and the refueling hose well clear of the water.

This is how USS *New Jersey*'s aviation facilities appeared in 1943. The ship had two catapults for launching seaplanes. As shown, these were originally three Vought OS2U Kingfishers. Two were stored on the catapults and one on a dolly. It was also possible to store two aircraft on a catapult but in this configuration it was not possible to use the catapult. The large crane at the stern was used to recover aircraft after they had landed in the water. The catapults were removed in 1950 and the crane was removed prior to the 1982 recommissioning. (Naval History and Heritage Command)

USS *Wisconsin* is unique in having a replenishment at sea (RAS) kingpost aft of the FAS kingpost. (Keith Nitka)

(31) The Aviation Service Center contained power and water supplies for aircraft. This structure was erected in 1950 as a movie projection booth in an arrangement that allowed the stern to be turned into a movie theater for the crew. This was no longer possible when the helipad raised the deck level in 1968. The structure was put to its aviation support use at that time.

The three revolving lights (red, amber, and green) at the top provided a visual indication of the helipad status to aircrews. The fog gong is at the left and the mooring capstan is covered in canvas. The large triple vent exhausts from the aft crew areas. USS *Iowa* is unique in having two large mushroom vents in place of the triple vent. (Nat'l Archives)

The helipad markings had three configurations during the ship's final reactivation. This photograph shows the markings used when USS *New Jersey* was recommissioned in 1982. At that time helicopters made a diagonal approach over the starboard stern. The dropdown light marking the start of the lineup lights is visible at the side of the hull below the end of the diagonal line. The diagonal approach allowed large helicopters to fit on the helipad. The railings in this area were replaced with nets that could be raised or lowered for helicopter operations. Here the nets are in the up position.

The elevated helipad was added in 1968. In the 1980's 42-foot ramps were added to allow helicopters to be moved off the helipad. The large vents and the mooring capstan at the centerline precluded the creation of a full-width ramp. In this photograph an SH-2F Seasprite helicopter has been rolled to a tie down area near turret no. 3. The helipad covered up the original fuel trunk outlet so, in 1968, the port 40mm tub was converted into the refueling station and the fuel lines were rerouted to that position. The starboard tub was converted into a fire fighting station. In the 1980's firefighting stations built at the side of each ramp replaced the one in the stern tub.

The hull cross section at the stern is not a smooth curve. Instead, it has two knuckles. The upper knuckle is visible in the photograph. The other knuckle is below the waterline. Raised boat booms are also visible at each side of the hull. (Nat'l Archives)

The helicopter approach path was changed to over the stern in 1984. This made landing helicopters easier, especially in rough seas. The ensign staff had to be removed prior to helicopter operations and it was stored in brackets at the aft end of the helipad on the starboard side.

There are actually three sets of markings for different helicopter operations. First, the circle and longitudinal line up line are for landing. Second, the two rows of T's are for Vertical Replenishment (VERTREP) operations where the helicopter lowers material to the deck without landing. The Ts indicate the forward limit where the rotor hub can safely be. The forward row is for smaller helicopters and the aft row (alternating with circles) is for larger helicopters. Finally, the large H is for Helicopter Inflight Refueling (HIFR) operation where a helicopter is refueled in the air without landing. The H at the port side marks the spot where the helicopter is to lower a line to haul up a fuel hose. (Nat'l Archives)

(32) Planked areas around turret no. 3 covered with a nonslip epoxy. This mat appears to be bare metal in photographs taken from a distance. This photograph was taken in 1982 just before the ship was recommissioned. An CH-46D helicopter is landing in the background. (Nat'l Archives)

(33) This is the port fire fighting station on the helipad ramp. The hose reel is connected to an aqueous film forming foam (AFFF) generator at second deck stern. The tent over the helipad is a museum addition. The circles in the helipad ramp were welded over tie down locations.

(34) The line up lights for the stern approach are offset to port by twenty-five feet from the centerline. The dropdown bar marking the start of the light path was attached to this frame at the side of the starboard 40mm tube. The light bar was removed when the ship was mothballed and was in storage. The line up lights on the helipad remain, including the ones for the early diagonal approach.

This photograph shows the final markings used on the helipad. While the stern approach made landings easier, large helicopters, such as the CH-53E, did not fit on the helipad when aligned with the centerline. These helicopters had to make a pedal turn right before touchdown so they fit correctly on the helipad. Diagonal lines were added to indicate how much these helicopters had to turn before landing.

Here, a CH-46D helicopter is practicing helicopter inflight refueling. The helicopter has already hauled up the fuel hose that runs to a reel in the port 40mm tub. The top of the ensign staff is visible at its storage location during helicopter operations at the aft end of the helipad. (Nat'l Archives)

(35) During 1983–84 USS *New Jersey* was deployed off Lebanon during its civil war. There, the problem was encountered of how to correct aim against land targets that were not visible from the ship when politics dictated that pilots should not be put in a position where they could be captured. In response the Navy installed RQ-2 Pioneer remotely piloted vehicles (RPVs) and their support equipment on board *Iowa*-class battleships. These were tiny aircraft (452 lbs.) that could transmit live video back to the ship

A ground control station was created at 02 level for flying the Pioneers. Antennae for communications were added to the top of the aviation service center and the mainmast. A hangar that could hold eight disassembled RPVs was installed aft of turret. This photograph shows a Pioneer RPV on its launching stand on board USS *Missouri*. The hangar is immediately behind. (Nat'l Archives)

Pioneers were launched from their stand using rocket assisted takeoff (RATO). The rocket blasted the RPV into the air with no ground run and then fell into the ocean. The RATO rockets were stored in a magazine at the 01 level of the conning tower. This is USS *Iowa*. (Nat'l Archives)

(36) Pioneer RPVs used 100LL aviation gasoline while the ship's aviation fuel system held incompatible JP-5 jet fuel. The system shown here was added to the port 40mm tub to store 100LL fuel. A 500-gallon bladder tank was held in this cart. The Pioneer only held twelve gallons of fuel so this tank did not have to be particularly large. The cart was mounted on inclined rails so that it could be sent overboard in an emergency. The chains shown prevent such an ejection from occurring at the museum. USS *Iowa* is unique in having this system in the starboard tub.

Pioneer RPVs were recovered by flying them into a net erected at the aft end of the helipad. RPV losses were high due to accidents, particularly during recovery. At best, the Pioneer was only moderately successful. However, it did pave the way for later improved RPV's. This is USS *Iowa*. (Nat'l Archives)

(37) Pulling a wire under the fuel cart would allow gravity to pull the cart down the rails and overboard. The wire leads to the lever shown at the right in the port waterway, forward of the fire station.

(38) The main deck at the stern. At the center is the ensign staff that would fly the United States flag when in port. The aft anchor light is at the top of the staff. The ensign staff is removable and was taken down during helicopter operations. Immediately to the right is the housing for the stern lights. These are the standard white light, a blue light for convoy operations, and a light to illuminate the ship's wake. Stern chocks are to either side. The museum's mooring chains run through the chocks and are secured to towing padeyes. The protrusions from the 40mm gun tubs are remnants of ammunition cupolas. Gun directors were mounted on top of these cupolas until the 40mm mounts were removed. After the 1980's, the port tub was used for refueling and the starboard tub was used for storing fenders.

(39) This is the only main deck 16-inch ammunition-loading hatch within the deckhouse. It is positioned near the aft powder magazines for turret no. 2. The overhead trolley moved ammunition from the deckhouse door to this hatch. Once at the hatch, ammunition could be lowered to the magazines below using the winch.

(40) This passage runs along the starboard side of the conning tower support. The view here is aft into the wardroom. The sloped sides of the conning tower support rest on main deck. The bulkhead to the left, added in 1968, is unique to USS *New Jersey*. It encloses ladders leading up to 01 level and down to second deck. The bulkhead improves airflow to the lower decks when the 01 level hatch was open. The brass wheel controls the air exhaust vent at the 01 from the emergency diesel generator room below. Blue tiled areas indicate Officer's Country that was off limits to enlisted sailors except on official business. Much of the ship has been retiled so the museum deck colors do not always match the in service colors.

This is the same area on USS *Wisconsin*. USS *Iowa* and USS *Missouri* are similar. The door to the executive officer's cabin is to the right of the ladders. (Keith Nitka)

(41) The senior officers' cabins were forward of the wardroom. The largest is that of the executive officer (XO). This is the XO's stateroom. It has a bed rather than a bunk. There is a private bathroom to the left.

(42) The Executive Officer also had a cabin that served as his office. The cabin has a door to the main deck that could be opened for ventilation. The door to the left is to the stateroom. The original doorway was expanded by the museum for Americans with Disabilities Act compliance.

(43) The ship's officers (other than the captain) took their meals in the wardroom. Officers paid for their meals through a common pool. The executive officer was the president of the mess who presided over the wardroom. When the XO was absent, he would delegate his duties to the next senior officer. The wardroom was governed by tradition. The uniform of the day was to be worn. During peacetime, seating was arranged by rank. No one sat until the mess president did. Politics, religion, and women were not acceptable topics of conversation and shoptalk was to be avoided until after coffee was served.

The wardroom was used as a battle dressing station during combat. The tables could be used for treating the injured and with lights above illuminating the tables.

Originally, there were 40mm magazines at either side accessed through the wardroom. These were converted in the security force locker (visible here) and a storeroom when the 40mm mounts were removed.

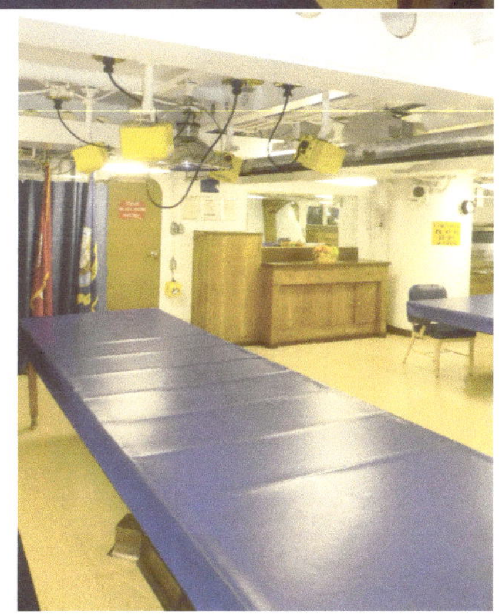

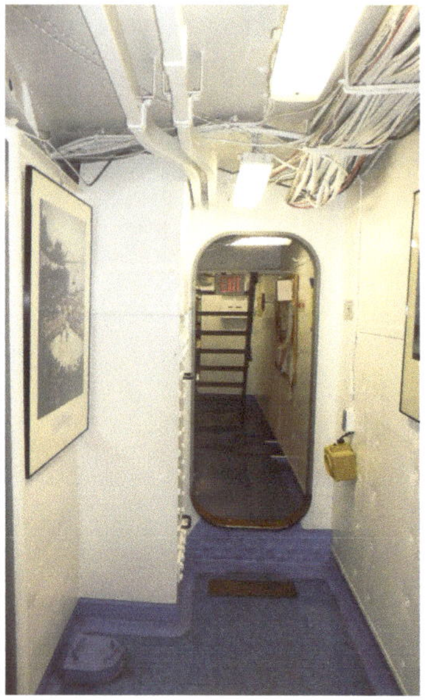

(44) The remnant of an armored trunk that once housed a 40mm ammunition hoist passes through the passage outside the captain's office.

(46) This is the curved passage shown in photograph (45) viewed from inside, looking to port. The entrances to the uptake spaces are in this passage but are obscured by the stowage.

(45) The uptake spaces are open areas that extend to 05 level. This view is up and aft in the uptake space for fire rooms nos. 1 & 2 on USS *Iowa*. The curved bulkhead is part of a passage through the uptake space. The uptakes for fire room no. 1 are forward of this passage and those for fire room no. 2 are aft. The spaces for both uptakes merge above this passage. (Michael Getscher)

(47) This is an armored director tube at the starboard side holding the cabling for the two side Mark 37 directors (Sky 2 and Sky 3). The wiring for each director follows at least two paths to the citadel. A similar tube on port side duplicates the wiring. The tube changes direction below the deck. The toe at the deck is intended to prevent the wiring from chaffing on a corner.

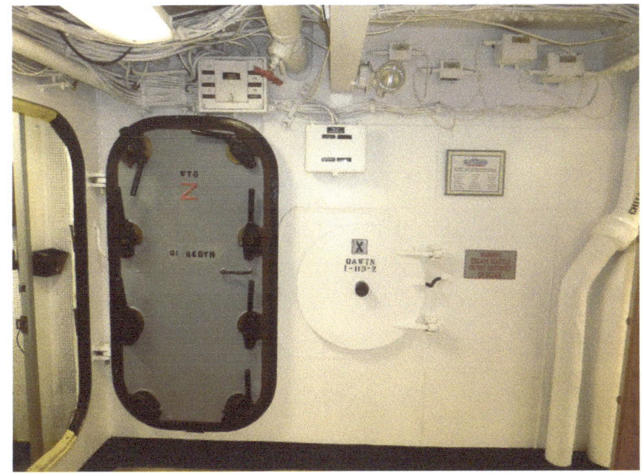

(49) The circular escape hatch in the bulkhead is from facilities control (FACCON). The armored door leads to the communications center.

(48) This armored grate protects the air intake for boiler no. 4. The grate for boiler no. 3 is behind the coaming at the top of the photograph. It is somewhat misleading to describe the main deck intake grates as being for a specific boiler because the intake airflows for the boilers in the same fire room intermingle until passing through second deck.

(50) The communication's center was created in the 1980's. The communications center transmitted messages and distributed received messages. This area was originally crew quarters. One inch of armor was added for protection as part of the conversion.

(51) This is FACCON-1 looking to port at the entrance to FACCON-2. FACCON-1 contained the ships receivers. FACCON-2 contained the decryption equipment. The locked expanded metal door created the need for the escape hatch shown in photograph (49).

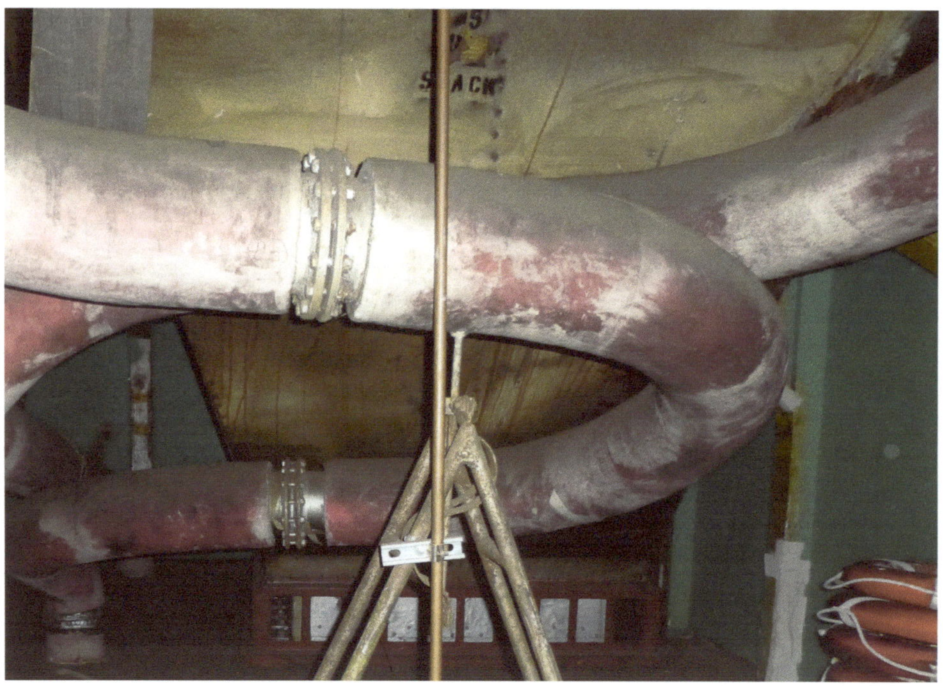

(52) This is the uptake for boiler no. 5. The uptake is doubled-walled to allow the hot flue gases to be cooled by ambient air. The cooling air enters at 05 level and exits through the square openings at the base of the uptake.

(53) The uptakes follow curved paths so they can join together at the smoke pipe. Braces like these support the smoke pipes at curves.

(54) The uptakes are not the only exhausts that run to the smoke pipes. Diesel engine exhaust, trash burner exhaust, and atmospheric exhausts exit the aft end of the smoke pipe. These are atmospheric exhausts heading to the aft smoke pipe on USS *Iowa*. The atmospheric exhausts allow the pressure in the boilers to be vented safely in an emergency. (Michael Getscher)

(55) This is the port side air supply for fire room no. 4 on USS *Iowa* looking upwards from main deck. The intakes for the port and starboard air supplies are at 03 level, forward of the aft fire control tower. The two paths merge at main deck. (Michael Getscher)

Second Deck

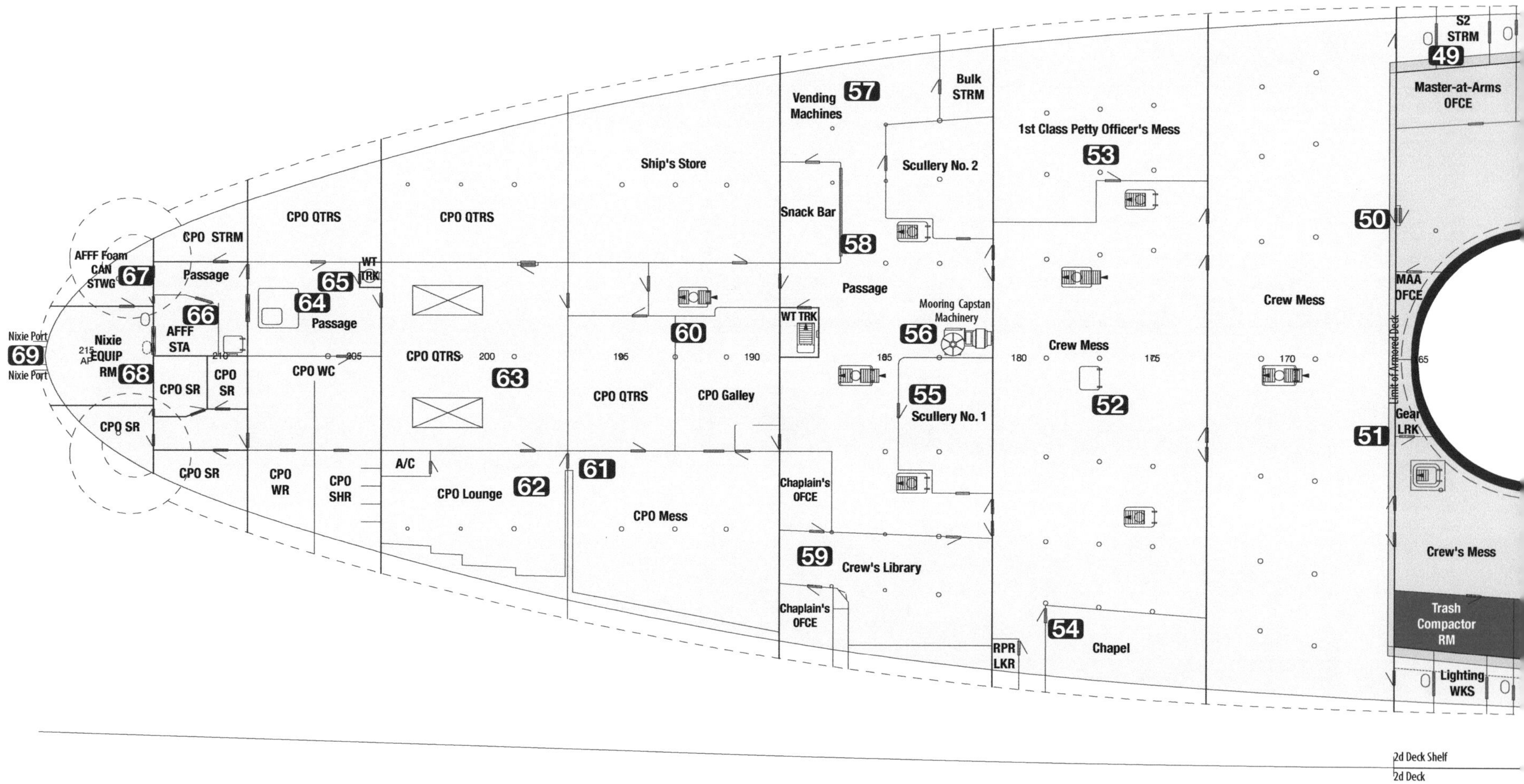

Second Deck
1"=24'

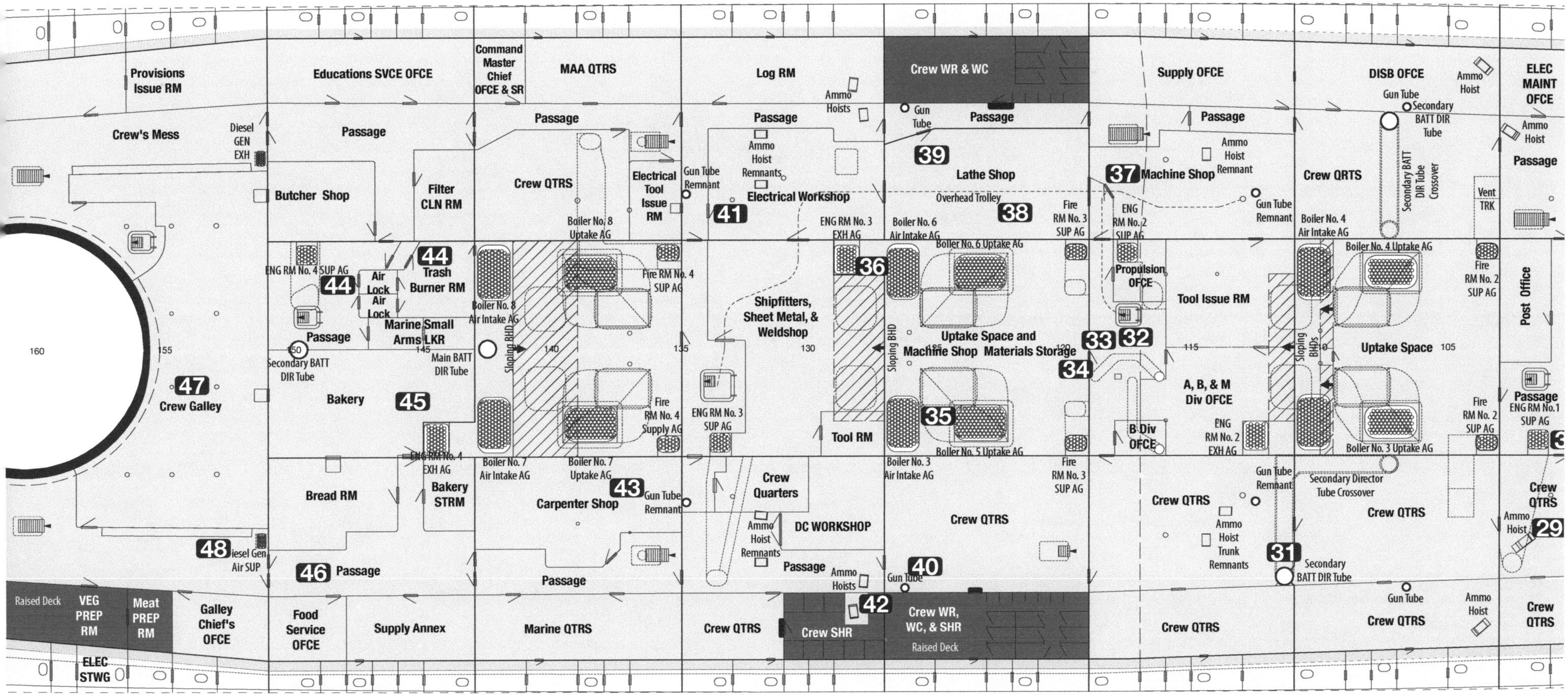

Provisions Issue RM
Educations SVCE OFCE
Command Master Chief OFCE & SR
MAA QTRS
Log RM
Crew WR & WC
Supply OFCE
DISB OFCE
ELEC MAINT OFCE
Crew's Mess
Passage
Butcher Shop
Filter CLN RM
Crew QTRS
Electrical Tool Issue RM
Electrical Workshop
Lathe Shop
Machine Shop
Crew QRTS
Trash Burner RM
Air Lock
Marine Small Arms LKR
Bakery
Shipfitters, Sheet Metal, & Weldshop
Uptake Space and Machine Shop Materials Storage
Propulsion OFCE
Tool Issue RM
Uptake Space
Post Office
A, B, & M Div OFCE
B Div OFCE
Crew Galley
Tool RM
Bread RM
Bakery STRM
Carpenter Shop
Crew Quarters
DC WORKSHOP
Crew QTRS
VEG PREP RM
Meat PREP RM
Galley Chief's OFCE
Food Service OFCE
Supply Annex
Marine QTRS
Crew SHR
Crew WR, WC, & SHR
Raised Deck
ELEC STWG
29
31
32
33
34
35
36
37
38
39
40
41
42
43
44
45
46
47
48

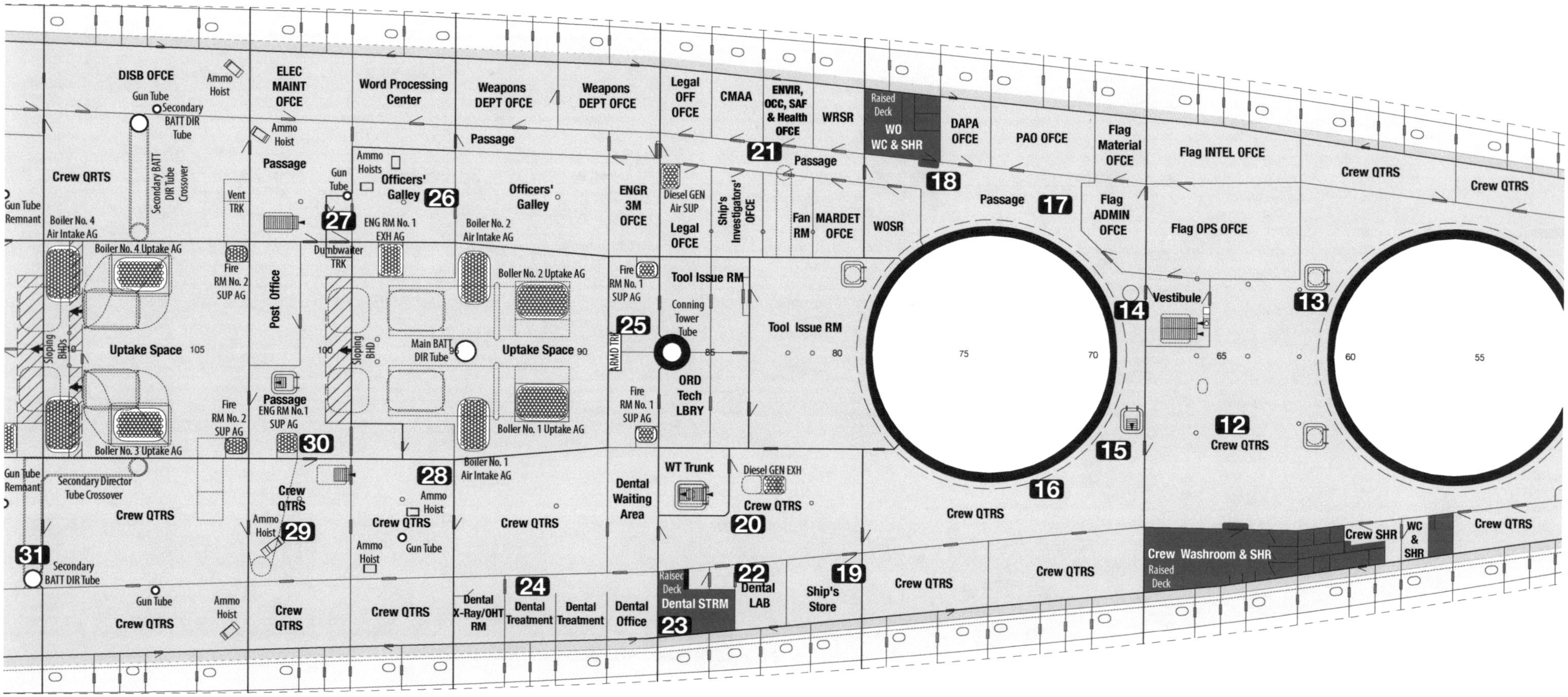
DISB OFCE
Ammo Hoist
Gun Tube
Secondary BATT DIR Tube
Secondary BATT DIR Tube Crossover
ELEC MAINT OFCE
Word Processing Center
Weapons DEPT OFCE
Weapons DEPT OFCE
Legal OFF OFCE
CMAA
ENVIR, OCC, SAF & Health OFCE
WRSR
Raised Deck
WO WC & SHR
DAPA OFCE
PAO OFCE
Flag Material OFCE
Flag INTEL OFCE
Crew QTRS
Crew QTRS
Crew QRTS
Gun Tube Remnant
Boiler No. 4 Air Intake AG
Vent TRK
Passage
Ammo Hoists
Officers' Galley
26
Officers' Galley
Passage
ENGR 3M OFCE
Diesel GEN Air SUP
Ship's Investigators' OFCE
21
Passage
Fan RM
MARDET OFCE
WOSR
18
Passage
17
Flag ADMIN OFCE
Flag OPS OFCE
27
ENG RM No. 1 EXH AG
Boiler No. 2 Air Intake AG
Legal OFCE
Dumbwaiter TRK
Boiler No. 4 Uptake AG
Fire RM No. 2 SUP AG
Post Office
Boiler No. 2 Uptake AG
Fire RM No. 1 SUP AG
Tool Issue RM
Conning Tower Tube
Tool Issue RM
Vestibule
14
13
Sloping BHDs
Uptake Space
105
100
Sloping BHD
Main BATT DIR Tube
Uptake Space
90
ARMD TRK
25
85
80
75
70
65
60
55
ORD Tech LBRY
Passage
ENG RM No. 1 SUP AG
30
Boiler No. 1 Uptake AG
Fire RM No. 2 SUP AG
Fire RM No. 1 SUP AG
15
12
Crew QTRS
Boiler No. 3 Uptake AG
WT Trunk
Diesel GEN EXH
16
Gun Tube Remnant
Secondary Director Tube Crossover
28
Boiler No. 1 Air Intake AG
Crew QTRS
Crew QTRS
Ammo Hoist
Crew QTRS
Crew QTRS
Dental Waiting Area
Crew QTRS
20
Crew QTRS
Ammo Hoist
29
Ammo Hoist
Gun Tube
31
Secondary BATT DIR Tube
Crew SHR
WC & SHR
Crew QTRS
Crew Washroom & SHR
Raised Deck
Crew QTRS
Crew QTRS
19
22
Raised Deck
24
Gun Tube
Crew QTRS
Ammo Hoist
Crew QTRS
Crew QTRS
Dental X-Ray/OHT RM
Dental Treatment
Dental Treatment
Dental Office
Dental STRM
23
Dental LAB
Ship's Store

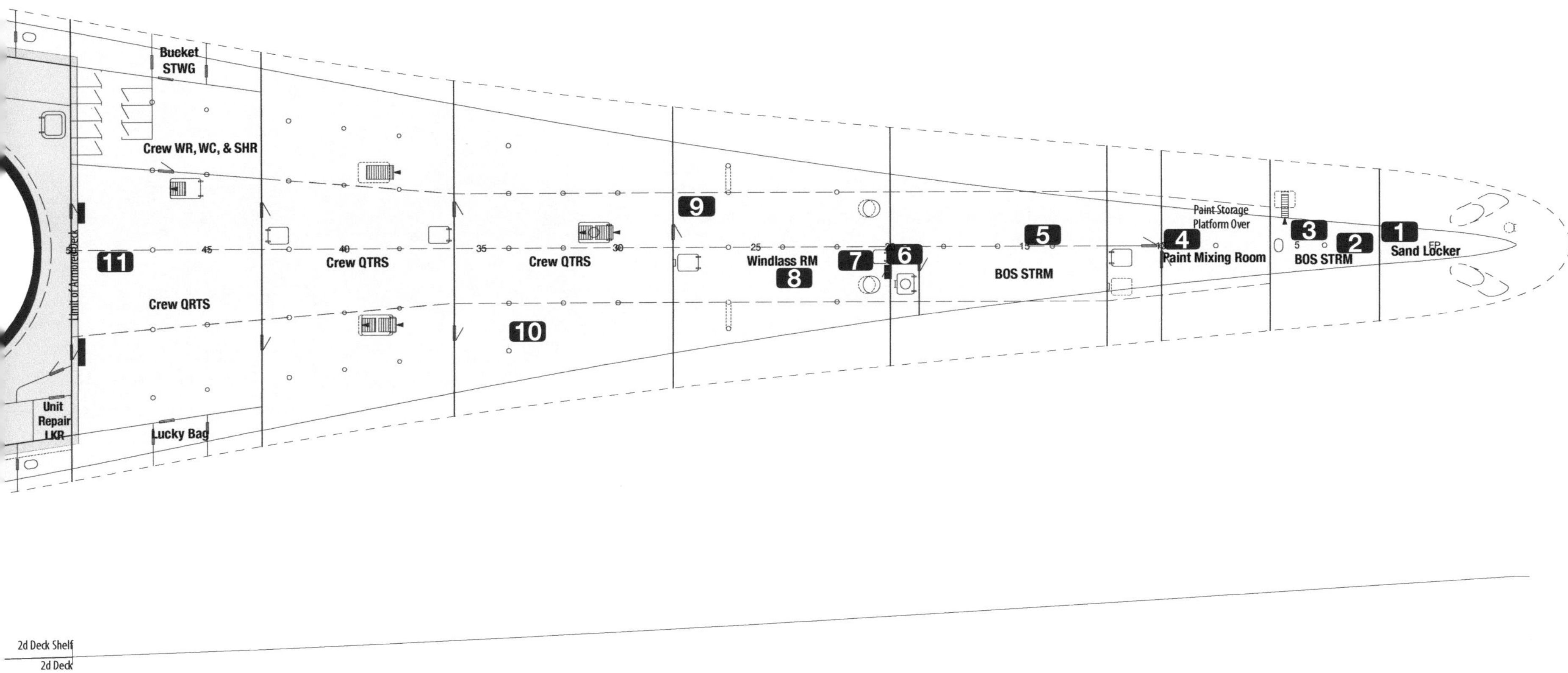

Second Deck
1"=24'

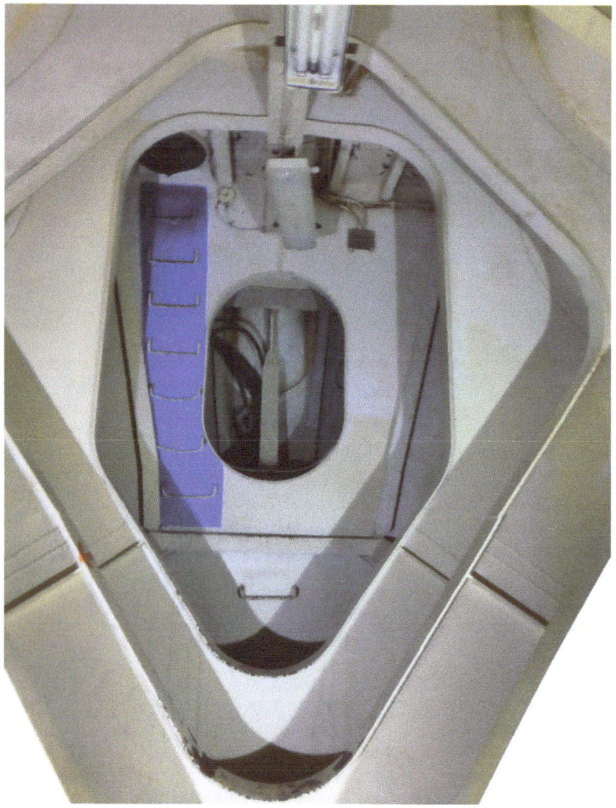

(1) The forwardmost second deck compartment on U.S. Navy battleships was traditionally the sand locker where supplies for holystoning were stored. This photograph is looking forward in the send locker with the bow in the background. The inside face of bow chock is visible at the center. In this area of the ship the frames are spaced every two feet and they are extremely thick to the point where they almost form bulkheads with openings in the center. When USS *New Jersey* was in service the only access to the sand locker was through an eighteen-inch scuttle at the main deck visible above the rung ladder on the bulkhead and there would be about eight feet of sand.

(2) There are two bosun's storerooms on second deck in the bow. The forward storeroom is just aft of the sand locker and was originally used to store paravane equipment. The opening visible in the Frame 2 bulkhead was cut during the mothballing process to allow access for dehumidification equipment and leads to the sand locker. The sheer of the main deck above creates the extremely high overhead.

(3) Like the sand locker, the only access to bosun's storeroom when the ship was in service was from the main deck. This ladder running along the port side of the hull leads to the 30"×30" hatch at the deck.

Historically, wooden decks were cleaned using a sandstone block known as a *holystone*. The holystone was used to wet sand the deck. Water was applied as a lubricant and sand as an abrasive while the holystone was used to rub the two over the deck. The process removed the upper layer of wood to expose fresh, white teak. While the holystoning made the deck look nice, it rapidly eroded the expensive teak. The U.S. Navy officially banned holystoning in 1931. Nonetheless, the practice persisted long after that as this 1943 photograph on USS *New Jersey* demonstrates. (Nat'l Archives)

(4) When USS *New Jersey* was in service, the paint mixing room was the farthest forward one could go along second deck without going up to the Main Deck. This view of the paint mixing room is looking forward. The slats shown above form a partial deck for storage that takes advantage of the high overhead created by the main deck sheer. The narrow pipe running horizontally is for fire suppression and three output nozzles are visible extending downward. In this photograph the museum was using this area to store flags that do not reflect the usage of this area when USS *New Jersey* was in service. The black tubing is for dehumidification and was put in place at retirement. Such tubing can be found throughout the ship. The museum removes such tubing when it refurbishes spaces for display.

(5) This is the second bosun's storeroom on second deck looking forwards. This storeroom contains cables for mooring the ship and held other materials used by the bosun. This view is looking forward.

(6) Ease of movement was a low priority in the design of *Iowa*-class battleships. A watertight door creates a weak spot in a bulkhead and doors in low traffic areas below the waterline have the potential not to be closed at a critical moment. Consequently, getting around an *Iowa*-class battleship often requires more vertical travel than horizontal, even when moving between two adjacent compartments on the same deck. This is an example of one of the trunks that runs from second deck down to the hold. This particular trunk is at Frame 19 and provides the only access to a series of storerooms and a pump room below. The straight path downward allowed supplies to be lowered from the main deck. The main deck hatch has a fixture for a portable J-davit used for that purpose.

(7) The windlass room contains the machinery that drives the wildcats and capstans on the main deck used to raise and lower the anchor and to pull mooring cables respectively. This is the port anchor chain pipe. It runs from the wildcat on the main deck to the chain locker below at the hold. There is another chain pipe mirrored on the starboard side.

(8) The machinery in the windlass room is in duplicate pairs on either side of the ship. This view is looking forward. The large object shown is the starboard reduction gear unit. The vertical shaft in the center turns the wildcat on the main deck to raise and lower the anchor. The shaft at the right turns the capstan. The rectangular unit at the right side is the hydraulic motor that drives the gear unit. Such a motor is called the *B-end* of an electric-hydraulic system. The two pipes coming out of the top of the motor are connected to a hydraulic pump that creates a pressure differential to drive the motor. At the center foreground is a four-way crossover valve. This valve allows the starboard pump to drive the port motor and the port pump to drive the starboard motor. The starboard chain pipe is visible in the background behind the drive shaft. The door in the background leads to the trunk shown in photograph (6). To the right is the hatch to the trunk that gives access to the chain locker.

(9) This view is from the aft end of the windlass room. At the center are the brake wheels for the port and starboard windlass machinery. The rods above the wheels mechanically link these controls to the brake wheels at the main deck. The large unit to the left is the port hydraulic pump that supplies the pressure to drive the hydraulic motors on the gear units. Such a pump is called the *A-end* of an electric-hydraulic system. The wheel on top of the pump controls the direction (raise anchor or lower anchor) and speed of the system. The rod above the wheel mechanically links it with the control wheel on the main deck. The wooden beams in the center are shoring timbers for damage control. Caches of such supplies are located throughout the hull.

(10) Crewmembers slept in bunks stacked three high. The bunks were known as *coffin racks* due to their size and because they were closed except at the front. The curtain created the only privacy ordinary sailors had. Each Crewmember had a locker located near their bunk and there is storage area under the bunk. There is an emergency escape breathing device (EEBD) for each sailor to provide the breathable air to escape in case of fire. The three EEBDs for the bunks are visible next to the upper left locker.

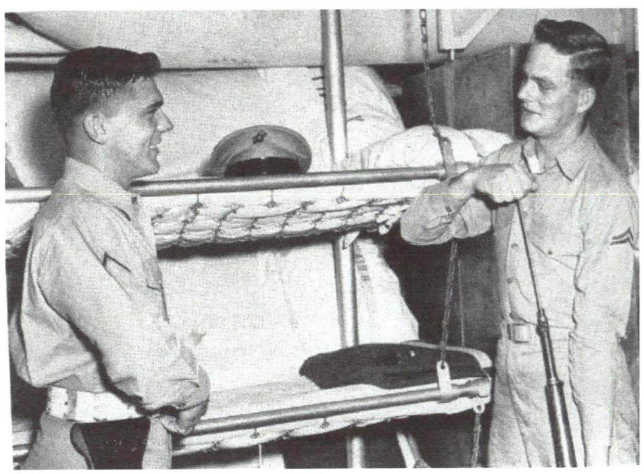

From World War II until the 1980's, sailors slept in open bunks that were stacked three or four high. This is marine's berthing on USS *New Jersey* in 1951. (Korean War Cruise Book 1950–51)

(11) On early battleships the armor protection extended the full length of the ship. The armor was typically thickest around the machinery and magazine spaces amidships and tapered towards the ends. Before World War I the U.S. Navy realized that the destruction of some spaces, such as paint storage, would not affect the fighting capability of a ship. This led the Navy to adopt the all-or-nothing principle of protection in the *Nevada*-class battleships of 1910 whereby the vital areas in the hull are fully protected in an armored box called the *citadel* and the rest of the hull is left unarmored.

On *Iowa*-class battleships, the citadel starts at Frame 50 and the top of the citadel is at second deck. Aft of Frame 50 there is 4.75" of armor pinned to the top of the 1$\frac{1}{4}$" deck creating the horizontal protection for the citadel. The addition of armor on top of the deck creates the step up shown in this photograph. Transverse armor directly below the deck protects the front of the citadel. The transverse armor is 11.3" and tapers from a knuckle to 8.5" on USS *Iowa* and USS *New Jersey*. This was increased to 14.5", tapering to 11.7" on USS *Missouri* and USS *Wisconsin*. A row of 6$\frac{3}{4}$" bolts secures the deck armor to the transverse armor. However, a cover completely conceals the bolts attaching the forward transverse armor.

(12) These are ventilation fans located at Frame 65 that draw in air from the vents located forward of turret no. 2 and circulate it in the crew spaces. Such fans are located throughout the ship. As here, mechanical systems frequently intrude on living spaces.

(13) In order to load and unload ammunition there are deck hatches located around the turrets. The hatches are aligned vertically so, when opened, they create a straight path from the main deck to second platform. Looking up from second deck one can see this hatch in the main deck at on the port side aft of turret no. 1.

(14) This is an example of an ammunition hatch in second deck. This example is located forward of turret no. 2 at the port side.

(15) Entry to the armored citadel is through hatches in the main deck. These hatches provide an illustration of the thickness of the deck armor. This hatch is forward of turret no. 2 on the starboard side and leads to a watertight trunk. The eight rotating dogs at the bottom face secure the hatch. Cables running to the counterbalance springs are visible at either side of the hatch opening. This hatch weight about 1,800 lbs. so could not be opened manually without a counterbalance. One person can open and close the hatch but it still takes considerable effort.

(16) Armored barbettes protect the turret machinery at the sides. The barbettes extend from the underside of the turret gunhouses to second deck citadel armor. Each barbette is made up of between seven and twelve Class A plates arranged around the turret axis. The barbettes are 17.3" thick at the sides and taper to 11.6" at the front and rear. The brittle face-hardened outer surfaces of the barbettes would crack if welded. In order to support the main deck, the plates have slots machined into their edges. Brackets like the one shown were inserted into the slots as the barbettes were assembled. The circular Carlin beam transfers the weight of the deck to the brackets that transfer the force to the barbettes.

(17) This area served a wide range of purposes over the years. Its original function was the warrant officer's mess. The compartment behind the bulkhead at the center was the pantry and the mess area and lounge was forward of that. A door was formerly located at the far left to give direct access from the passage to the mess area. As anti-aircraft guns were removed after World War II the size of the crew complement declined so there was sufficient room for warrant officers to eat in the wardroom. This area was then used for band instrument storage, a chapel, and weight room at various times. Its final use was flag offices.

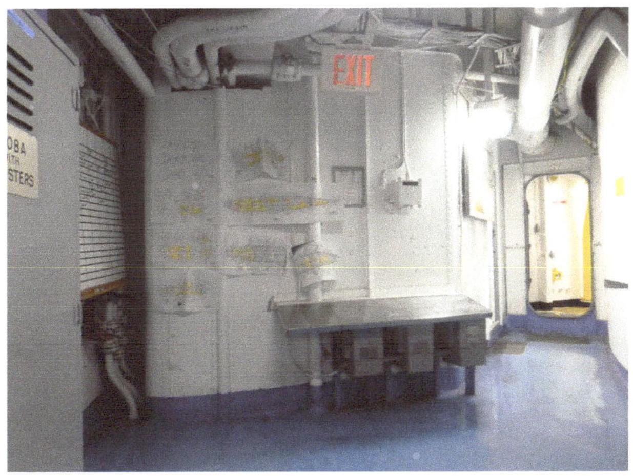

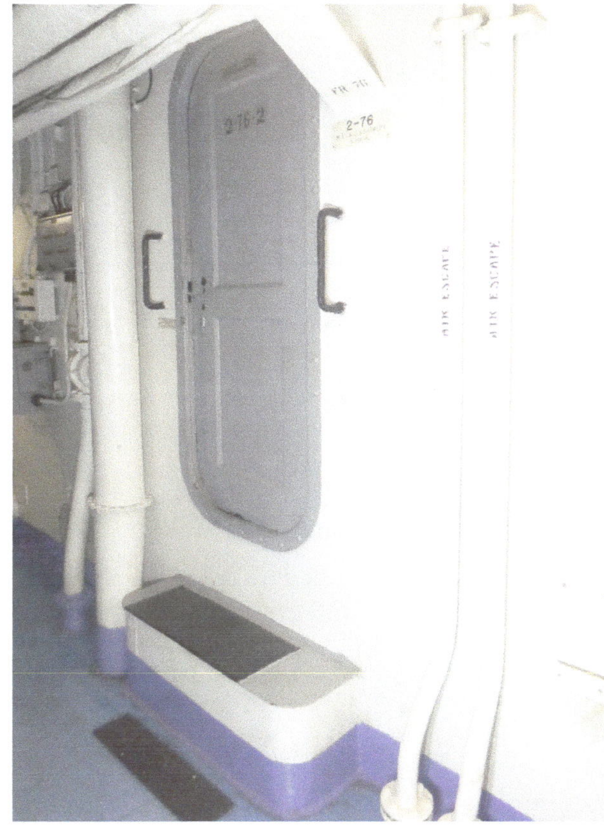

(18) Washrooms and other areas over the citadel that require plumbing are located over raised platforms. Without these platforms, openings would have to be drilled through the armor to run pipes. The washrooms have a step to reach the raised platform. This is the entrance to the warrant officers' washroom. The access panel to the plumbing space is partially visible behind the vertical pipes.

(19) There were two stores for the crew on second deck. These stores were stocked primarily with snacks, grooming supplies, and tobacco products. This is the *Quick Stop*, the smaller of the two stores. The stores sold items at near cost with any modest profits going to the crew activities fund.

(20) The prominent structure at the center of this photograph is the air exhaust trunk from the emergency diesel generator room in the hold. This trunk exits at a mushroom vent on the 01 level. The door in the background leads to a watertight trunk that leads from the 01 level to the forward emergency diesel generator room at the hold. The exhaust from the diesel generators runs through the forward uptake housing.

(21) Various controls for ship functions can be found along passages throughout the ship. The unit at the right in the forward port side passage is for flooding the rocket-assisted takeoff (RATO) magazine on the 01 level. The large distance between this control and the magazine ensured that a catastrophe in the magazine would not prevent access to the flooding controls.

(22) USS *New Jersey* has a full dental treatment facility. The dental capabilities included this laboratory. The bolted manhole at the right leads to a void space that gives access to fuel storage below. This was originally the dental X-ray room.

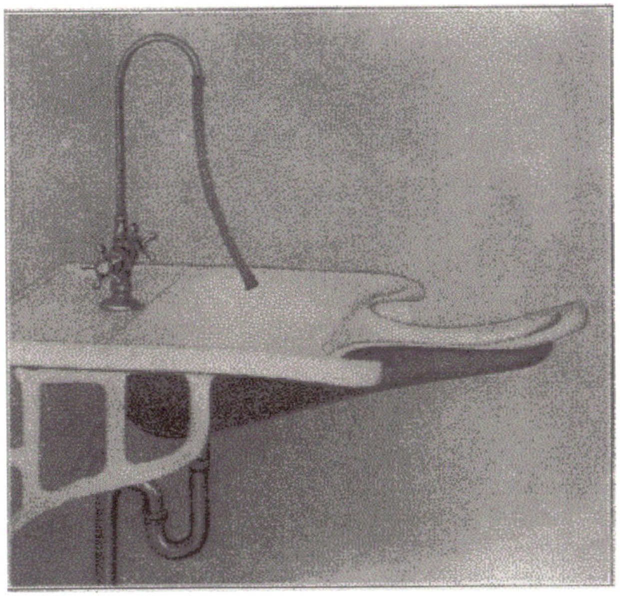

(23) The dental storeroom has an elevated deck because its original function as the treatment room required plumbing. It was originally equipped with two straddle stands like this that the military used to treat social diseases before the introduction of antibiotics. After penicillin became widely available, this mode of treatment faded away. By the Korean War, the plumbing and stands had been removed and this room became an office. It became the dental storeroom in the 1980's. (J. Am. Med. Ass'n, 1918)

(25) There are several armored trunks that run from the superstructure to the armored citadel below second deck. The critical functions in the superstructure, such as steering stations and gun directors, are duplicated and each has redundant wiring routed through multiple trunks. This particular armored trunk was added during the 1980's to protect new wiring added during the modernization. This trunk runs from the citadel to the conning tower and is 1 $\frac{1}{2}$" thick. There are several access points along trunk. One of those openings is shown.

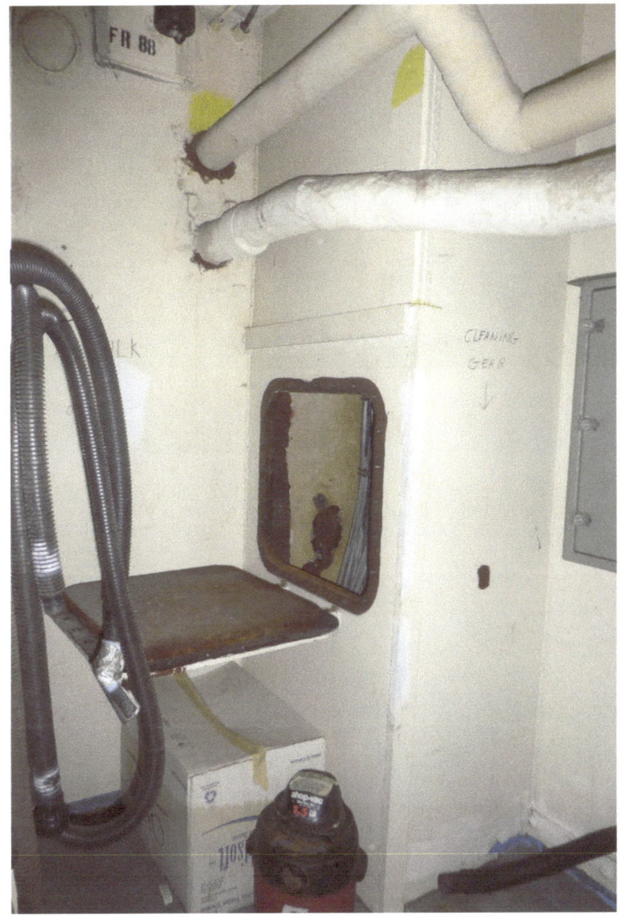

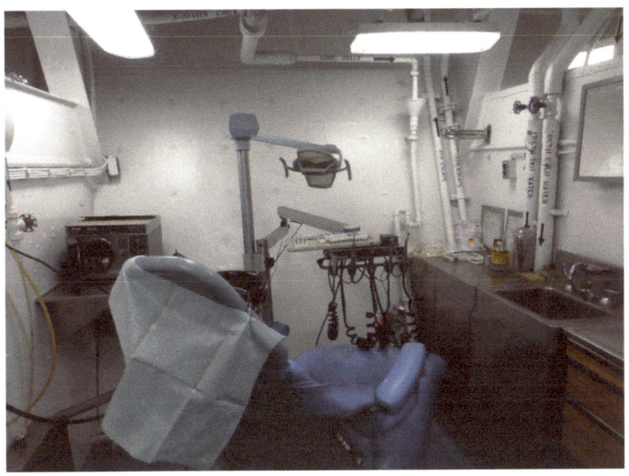

(24) This is one of two examination rooms within the dental office. Other than the exposed pipes, it is very similar to what one would find in a civilian dental office.

(26) Officers had to pay for their own food out of their pay and allowances so they ate separately from the crew, whose meals were paid for directly by the Navy. Officers paid monthly into a common pool that funded their meals. The officers had their own galley on second deck. This photograph is of the officers' galley looking aft. The armored trunks for two five-inch ammunition hoists pass through the galley.

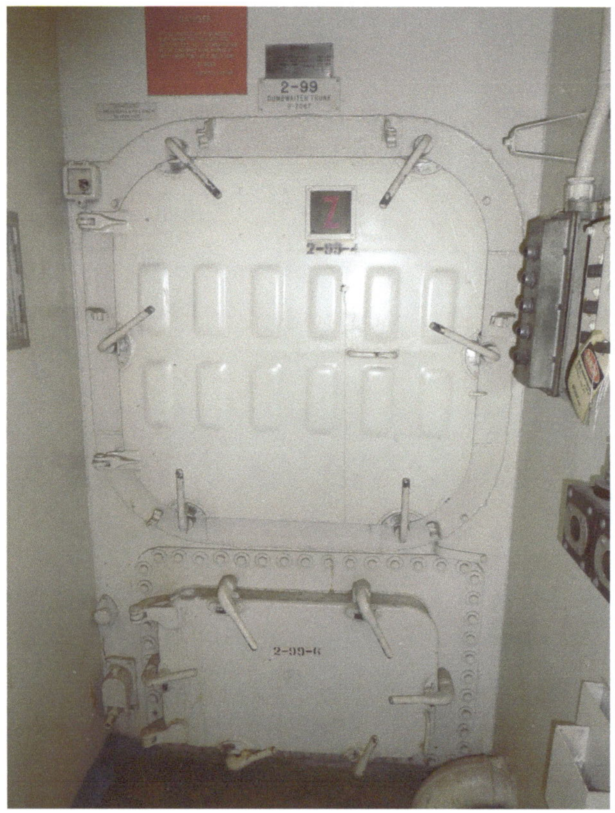

(27) Officers ate in the wardroom on the main deck. This dumbwaiter located at the entrance to the officers' galley, took food from second deck to the pantry outside the wardroom. The dumbwaiter extends to the 02 level so that the officer's galley could be used to prepare food for large functions in the captain's cabin (at the 01 level) or the admiral's cabin (located at the 02 level until the 1980's).

(28) *Iowa*-class battleships had ten five-inch twin mounts until the 1980's. Each mount has two dredger-style ammunition hoists protected by $2\frac{1}{2}$" thick armored trunks that run from the magazines on the third deck to the handling rooms below the guns on the main deck and 01 level. Four of the five-inch mounts were removed in the 1980's to create room for missiles and their projectile hoists were removed. On USS *New Jersey*, all of the armored trunks were left in place except for one that had been located in the machine shop. This photograph shows one of the nineteen remaining armored trunks for the hoists. Twelve of those trunks, including this one, still house a hoist. Some of the former hoist trunks were repurposed as armored cable trunks.

(29) This photograph shows another of the five-inch ammunition hoists. The museum removed the cover on this hoist to show how ammunition passes through it. The dredger-style hoist operates as a continuous conveyer with the prongs at intervals pushing the ammunition. The ammunition for the five-inch guns was semi-fixed, meaning that the projectile and propellant cartridges were loaded separately. Propellant cartridges were stored in protective aluminum canisters until they were ready to be fired. The hoists could move both projectiles and propellant cartridges within their canisters. The hoists could also be run in reverse to load ammunition into the magazines. A blue inert practice projectile is shown in the hoist.

(30) The transverse passage at Frame 100 houses the ship's post office. This view is to port. The intake air for the boilers flows through armored grates at the main deck and second deck. To improve protection, the grates are not aligned vertically. The sloped bulkhead directs the intake air from the main deck to the plenum chambers below second deck. A cache of damage control supplies is at the right next to the sloped bulkhead.

(31) Electrical cables link the gun directors in the superstructure with the plotting rooms at the first platform (forward) and third deck (aft). There are five armored trunks like this one at second deck that protect the director wiring. The director wiring follows redundant paths so, at second deck, each of these trunks contains wiring linked to multiple directors. The director trunks do not follow straight paths. This trunk follows the overhead at the main deck before turning downward here. The director trunks are $2\frac{1}{2}$" thick at second deck.

(32) The overhead trolley system allowed heavy material to be moved throughout the ship. It has track sections on the main, second, and third decks as well as the first and second platform. The tracks were laid out to move equipment among the engineering spaces and workshops as well as to move ammunition among the magazines. The trolley cars could not move vertically so moving material between decks required transferring it from the car at the source deck to a car at the destination deck. The trolley cars have winches to raise and lower heavy items between decks.

This transverse passage at Frame 119 leads to and acts as an extension of the machine shop. This view is to port, towards the machine shop door. The trolley car shown is at the end of a track and directly above a hatch leading to the third deck. A shear cutting machine for metal is in the foreground.

(33) This is the transverse passage at FR119 viewed to starboard. The door shown leads to the uptake space for fire room no. 3. This uptake space is used to store the metal stock for the machine shop. The ladder rungs on either side of the door that appear to lead nowhere provide access to a padeye above the door. A pulley could be attached to the padeye to assist moving materials out of storage so an overhead trolley could move them to a workshop. Similar ladders that appear to lead to nowhere can be found throughout the ship. Look up and there always will be a fixture or critical valve for damage control above such ladders.

(34) This is the uptake space for fire room no. 3 looking aft. This space was used for raw materials stowage for the machine shop. The navy left the stockpile on USS *New Jersey* when she was retired. The uptakes for boiler no. 5 (left) and no. 6 (right) are behind the material storage. The uptake segments at second deck span armored grates at the main deck and armored grates at second deck. The grates are not aligned so that a single projectile would be unlikely to strike both grates. This space also serves as the intake for the boiler room. Intake air enters through armored grates at the main deck. Then the air flows freely through the compartment and it exits through armored grates in second deck (see photograph 36).

(35) This is the connection between the uptake for boiler no. 5 and second deck armored grate. The sloped side below the rounded expansion joint is the only part of the armored grate that is visible. It is surrounded by the bolted plate that secures the grate to the deck. The coaming keeps water from flowing downward through the opening.

(36) This photograph was taken within the uptake space for boiler room no. 3 with the view to starboard. This sloped bulkhead directs air from the main deck armored grates into the uptake space. The main deck armored grates are visible above, to the right of the light. The starboard second deck armored grate is located within the coaming at the far bulkhead. The grating in the foreground is a walkway over the port side armored grate.

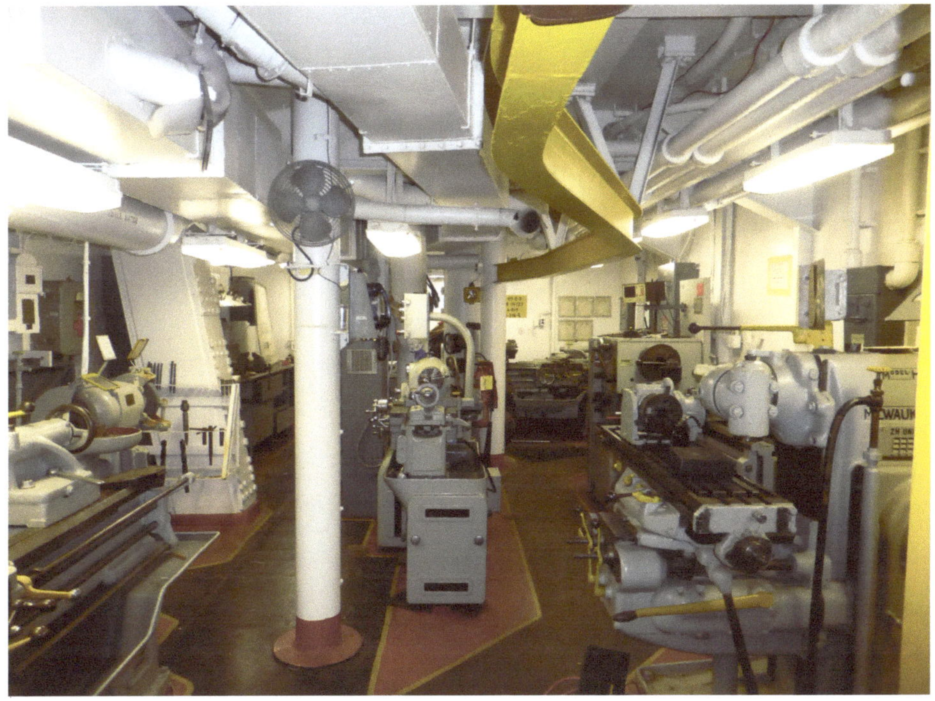

(37) *Iowa*-class battleships were designed to be self-sufficient for maintenance while in operation. They are equipped with shops that could perform any type of repair likely to occur during the normal running of the ship. They could also carry out repairs for other ships. This view of the machine shop looks forward. The compartment is filled with machining tools. All of these are functional and are still used to make repairs on the ship.

At the left is an armored trunk for a five-inch ammunition hoist that was removed in the 1980's. Originally, another hoist ran though the machine ship. However, it was removed in the 1980's to make room for the cooling pipe directly behind the lathe in the center.

(38) Anything that was too big for the machine shop could be fixed in the lathe shop. Most of the equipment here is original to when the ship was first built. At the left is a 25-inch Lodge & Shipley lathe. The view is aft. The doorway leads to the electrical workshop.

(39) This Bullard vertical turret lathe illustrates the scale of the machinery in the lathe shop.

(40) The five-inch guns were normally controlled remotely from plot rooms within the citadel. Wiring runs from the plot rooms to the gun mounts in order to transmit orders to the guns and for the plot room to receive status information from the guns. Between second deck and the gun mounts (at main deck and 01 level), the wiring runs through armored gun tubes. The gun tubes are $2\frac{1}{2}$" thick and have armored access doors. There are ten of these gun tubes that run from second deck to the 5-inch handling rooms at the main deck and 01 level. Four of these tubes became unused when their mounts were removed in the 1980's. The gun tubes are noticeably smaller in diameter than director tubes.

(41) The electrical workshop could repair most electrical parts. A frequent job was to rewind motors. The trunk for a former five-inch ammunition hoist is at the left. It may appear from the photograph that the trolley track prevents the door from closing. However, the trolley tracks have removable segments at door openings to allow the doors to close.

(42) Propellant cartridges for the five-inch guns were stored in protective canisters. Crewmembers in the handling rooms had to remove the cartridges from the canisters before they were placed into the powder hoist that took them to the guns. The empty canisters were ejected from the handling room through decanning scuttles in the deck. The decanning scuttles open into unexpected areas of the main deck and second deck. This decanning scuttle opens into a shower. When the five-inch guns were active, canvas chutes were erected under the decanning scuttles to collect the canisters.

(43) The equipment in the carpenter's shop is more modest in scale than that of the lathe shop. Nonetheless, it is equipped to repair anything constructed out of wood on the ship. The carpenter's shop is not tiled so the armored deck is visible. At some point someone made an unauthorize ship alteration (SHIPALT) by cutting through the watertight bulkhead and installing a window air conditioner.

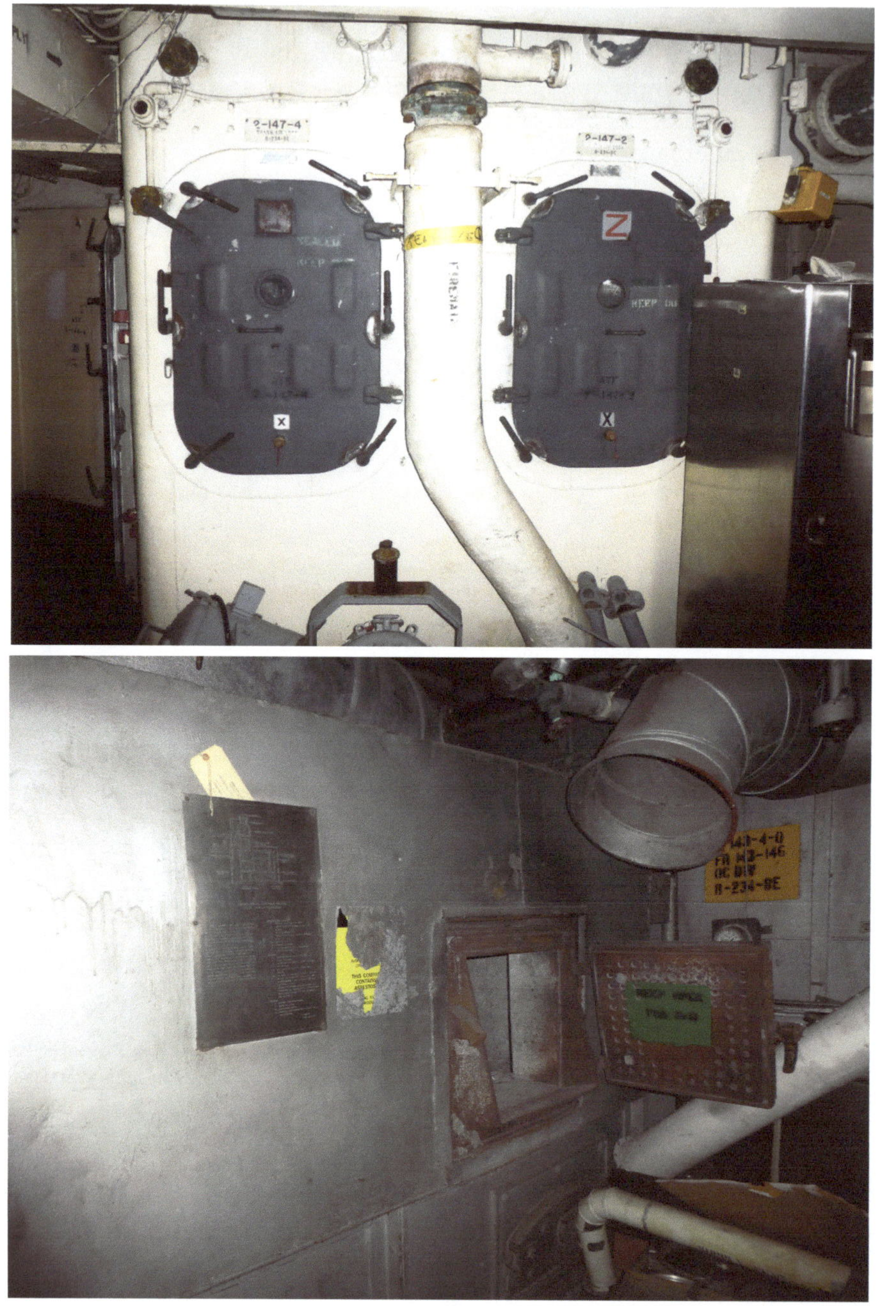

(44) The garbage incinerator was pressurized when in operation. The trash was passed through a pair of airlocks (top) to reach the garbage incinerator (bottom).

(45) This is the bakery on USS *New Jersey*. There is a window with a watertight door in the bulkhead at the center for passing the baked goods into the galley.

(46) Here is the starboard entrance to the crew mess looking aft. The serving line is through the doors to the right. The door at the left is a quick acting watertight door that can be sealed by rotating the wheel in its center. The lever attached to the wheel can assist in sealing. The door at the right is a watertight door. It has eight dogs that have to be rotated individually to seal the door. The doors have hooks to they will not flop around while open in seaway.

There is evidence that, at some time periods, the watertight door was closed and the quick acting watertight door was kept open in condition Yoke. The Z on the door indicates that, in her final deployment, both doors were only closed in condition Zebra.

(47) The crew galley looking to starboard. This is where the food for most of the enlisted sailors was cooked on the ship. The barbette for turret no. 3 is at the right. The starboard serving line is on the other side of the windows visible in the background. There is another serving line on the port side of the galley. The port side mess line served pickup items, such as hot dogs and hamburgers, so it was known as the *fast line*. The starboard line served full meals and was known as the *slow line*. In the 1980's, there was also a salad bar.

(48) The air intake for the aft diesel generator room is located at the start of the starboard serving line. The sloped sides of the armored grate that protects the opening are visible at the deck. The air exhaust for the aft generator room is located at the start of the port side serving line.

(49) The area outside the torpedo bulkhead no. 3 is called *second deck shelf*. The shelf is structurally separate from second deck and has a slightly different profile. The shelf extends over fuel tanks and void spaces arranged to provide underwater protection. This photograph of the shelf was taken in the supply room at the port side, aft end of the shelf. Torpedo bulkhead no. 3 is at the top of the photograph and second deck shelf is at the bottom. The scalloped plate is welded to the top of the 12.1" armor belt and supports the shelf at the inboard side. The slope of the scalloped plate matches the slope at the top of the armor. The straight plate covers the gap between the armor and the torpedo bulkhead. This gap is filled with cement. One gets a glimpse of the belt armor plate in the triangular openings created by the scalloping. This is one of the few places where the belt armor is easily visible.

(50) The 4.75" armor over second deck ends at Frame 166. This creates a step down from the crew mess serving area into the eating area. The armored citadel continues aft from here at the third deck to protect the steering gear. Transverse class A armor extending from the third deck to second deck protects the citadel from projectiles coming from the aft direction. Unlike the forward transverse armor, the aft armor has constant thickness so it does not taper at the bottom. A row of bolts connects the deck armor to the transverse armor below in the manner as at Frame 50. In the area shown here, there is a cover over the bolts. However, that cover does not extend the full length of the joint as it does at Frame 50.

(51) Some of the bolts connecting the deck armor to the transverse armor are exposed along Frame 166. The bolt heads are $6\frac{3}{4}$" in diameter.

(52) The crew's mess between Frame 173 and Frame 181. There is an additional similar mess between Frame 166 and Frame 173. The original mess tables had room for eight with bench seating. Tables seating four with swing out seats replaced these after World War II. During the Vietnam War, stackable chairs replaced the swing out seats. The eating areas were thoroughly updated in the 1980's with four fixed seats at each table.

Originally, there were two more eating areas for the crew. One occupied most of the space between Frame 181 and Frame 189, including the areas housing the crew library, chaplains' offices, and sculleries now. There was a smaller mess area between Frame 189 and Frame 194 where the ship's store is located and remains functioning now. Officers did not eat in the crew's mess except for special celebrations and lowly ensigns who were assigned to make checks on food quality.

(53) The port side of the eating area between Frames 173 and 181 was partitioned to create a lounge and separate dining area for first class petty officers. Their food came from the crew mess but the petty officers could eat and socialize separately.

This photograph shows the original seating arrangement in the crew's mess. It was taken during a dinner opened to guests during 1945 while USS *New Jersey* was in Bremerton. Note the bunks against the bulkhead. (USS *New Jersey*, World War II Cruisebook)

(54) The chapel had been located in various places over the years. This was its final location at the starboard side of the aft mess. The chaplains' offices were nearby, off the crew library.

(55) There are two sculleries in the mess area with industrial dish washing equipment. Along the passageways, the sculleries have slots for depositing flatware and a window for food trays. When USS *New Jersey* was built, the sculleries were located within the current mess areas.

(56) The engine for the aft mooring capstan sits in the passageway aft of the crew mess. The control box is on the bulkhead and the electric motor is below that. The motor drives a worm gear that runs along the side of the unit that turns a worm wheel. The vertical shaft extends from the worm wheel and drives the capstan on the main deck.

(57) The structural elements of the hull are most visible at the aft end of second deck. This photograph shows Frames 183, 184, and 185 at the port side hull shell. Stringer no. 1 runs horizontally to connect the frames.

(58) The Geedunk (snack bar) sold food like snacks, ice cream, hotdogs, and drinks. Like the ship's stores, items at the Geedunk were priced near cost with any profits going to fund crew activities. This area was originally part of the ship's store.

(59) The crew's library looking forward. The chaplains' offices are to either side of the camera location. The crew library originally occupied part of what is now the ship's store, between Frames 194 and 197 and this area was originally part of the crew mess. Since the ship became a museum, the crew library has become a research library containing thousands of books on naval history.

(60) The chief petty officers (CPOs) ate separately from the rest of the crew. They had their own galley and mess. This is the CPO galley looking forward and starboard. The window at the right leads to the CPO mess. This galley is directly below the triple vent exhaust on the main deck. Consequently, there is a mass of exhaust trunking along the overhead. The CPO galley was originally about half its current size, ending around the back edge of the table in the foreground.

(61) This is the CPO mess looking forward. This compartment has a decorative bulkhead inside the hull that conceals the piping running behind it. The furniture shown is not original. This room originally had a vertical ladder at the aft bulkhead accessing a hatch at the main deck. The hatch and ladder were removed when the hatch opening was covered by the helipad in the 1960's. There is no visible sign that the hatch was ever there.

(62) The CPO lounge is aft of the CPO mess. It also has decorative bulkheads that conceal piping. This area was originally used for additional mess seating with a small lounge area at the aft end. As the crew complement was reduced, this entire room became a lounge.

(63) The bunks for CPOs similar to those of the crew. However, the CPOs have more personal storage space. This view is looking aft.

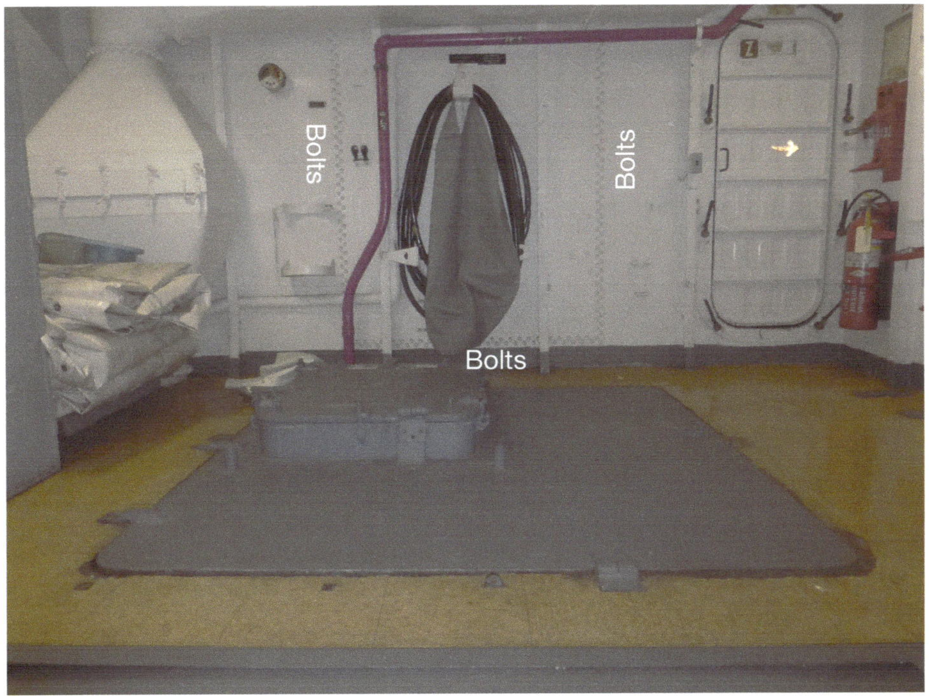

(64) This removable plate is below the large main deck hatch aft of the helipad. It leads to a storeroom on the third deck. The double row of bolts in the bulkhead marks the outline of another removable plate. There are removable deck plates throughout the ship that allow machinery to be replaced. Many of these plates are hidden by tile.

(65) The small, closed door leads to the trunk that is the only access to the aviation fuel tanks. Originally, the trunk ran from the main deck to the storage tanks and the fuel lines ran through it. In the 1960's the trunk's main deck hatch was covered by the helipad and the fuel lines were rerouted to the port 40mm gun tub.

(66) This machinery was used to create aqueous film forming foam (AFFF) for the helipad. Piping runs from here to the firefighting stations on the main deck helipad ramps. The piping on the starboard side runs behind the decorative bulkheads in the CPO lounge and mess. This was originally a 40mm magazine.

(67) The compartment beneath the port 40mm gun tub was used for storing AFFF concentrate canisters. It was originally used to store empty 40mm cases and clips. Chutes from the gun tub above conveyed those into the compartment. The framing in this area follows the curve of the hull. In the rest of the ship the frames are perpendicular to the centerline. One has to squat in this compartment due to the 40mm tub lowering the overhead.

(68) Nixie torpedo decoys were added to *Iowa*-class battleships as part of their modernization in the 1980's. The decoys were towed behind the ship where they emitted sounds to make it appear to a torpedo that the ship was located astern of its actual position. The Nixie room is at the aft end of second deck. A cable could run from either of the winches shown at the center, though the pulley above, and out the stern openings (shown closed at the center). Until the 1980's, *Iowa*-class battleships had a large crane at the stern that was originally used for floatplanes and later for boats. The crane was located directly above this compartment that held the crane machinery. No visible trace of the crane or its machinery remains.

(69) One upgrade made in the 1980's was to add a large bolster at the stern to reinforce the openings cut in the hull for the Nixie decoys. The bolster's surfaces are curved to minimize cable chaffing. (Nat'l Archives)

A Nixie decoy is shaped like a tiny torpedo with a cable running from its nose. This is a Nixie decoy being lowered into the water from USS *John F. Kennedy* (CV-67). (Nat'l Archives)

These plans show how the Nixie towing tables were rigged in the Nixie room. (Plan 802-5773164)

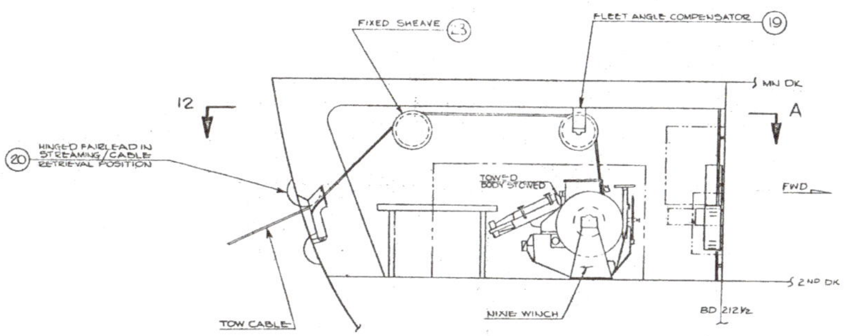

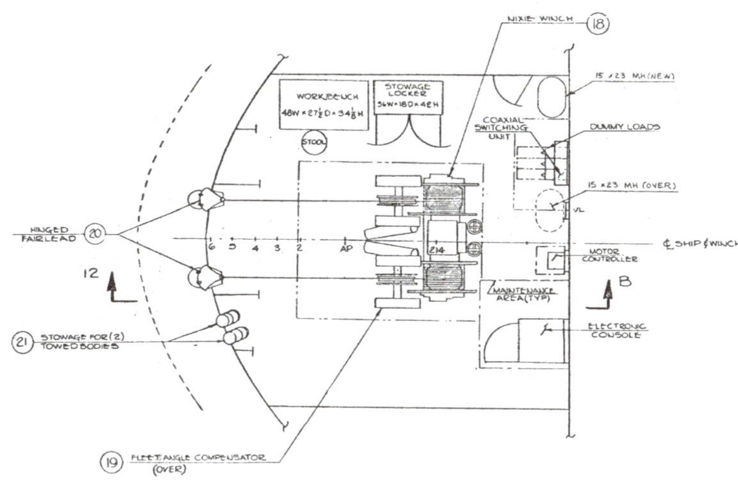

Splinter Deck

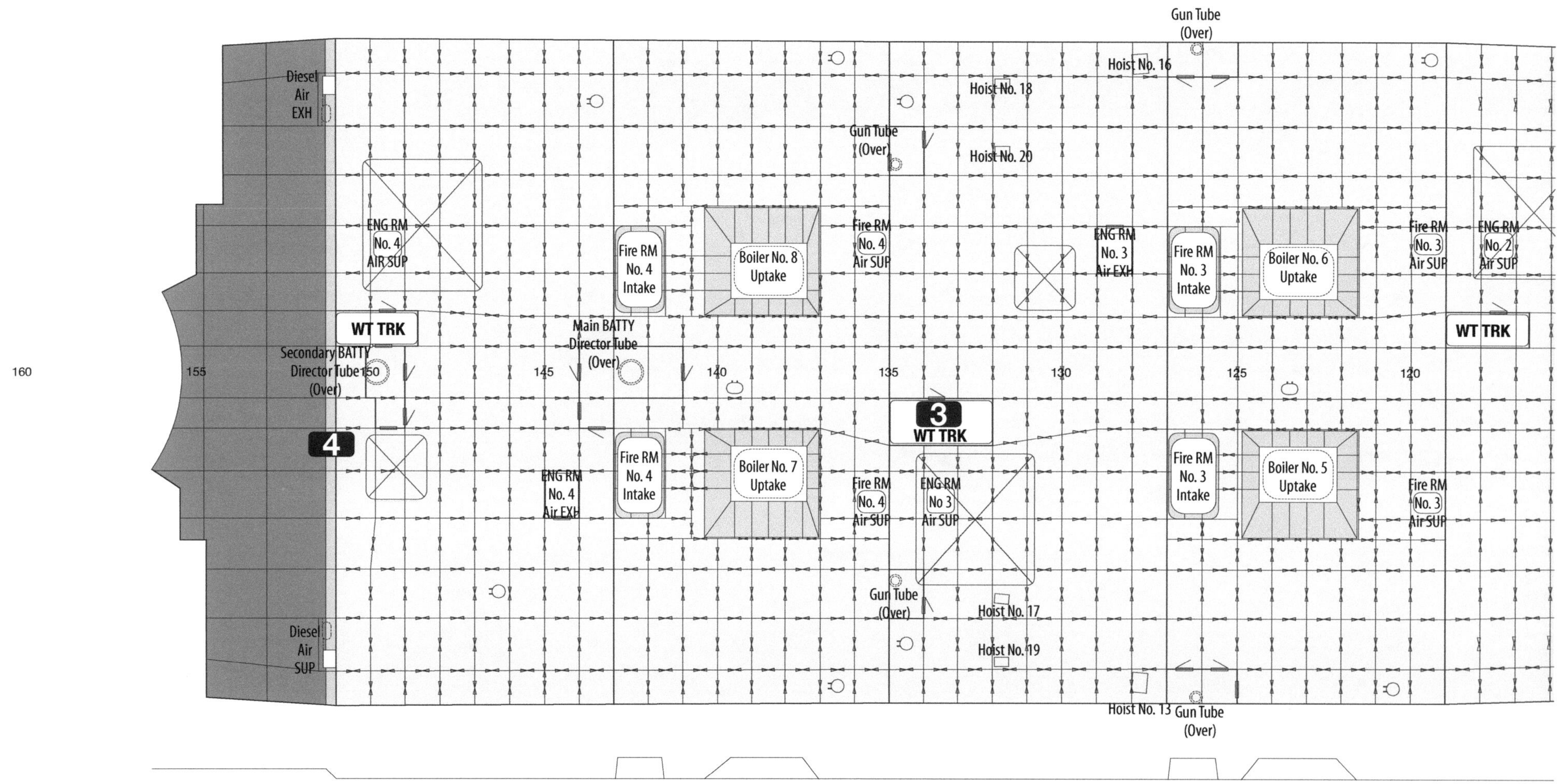

Splinter Deck
1"=18'

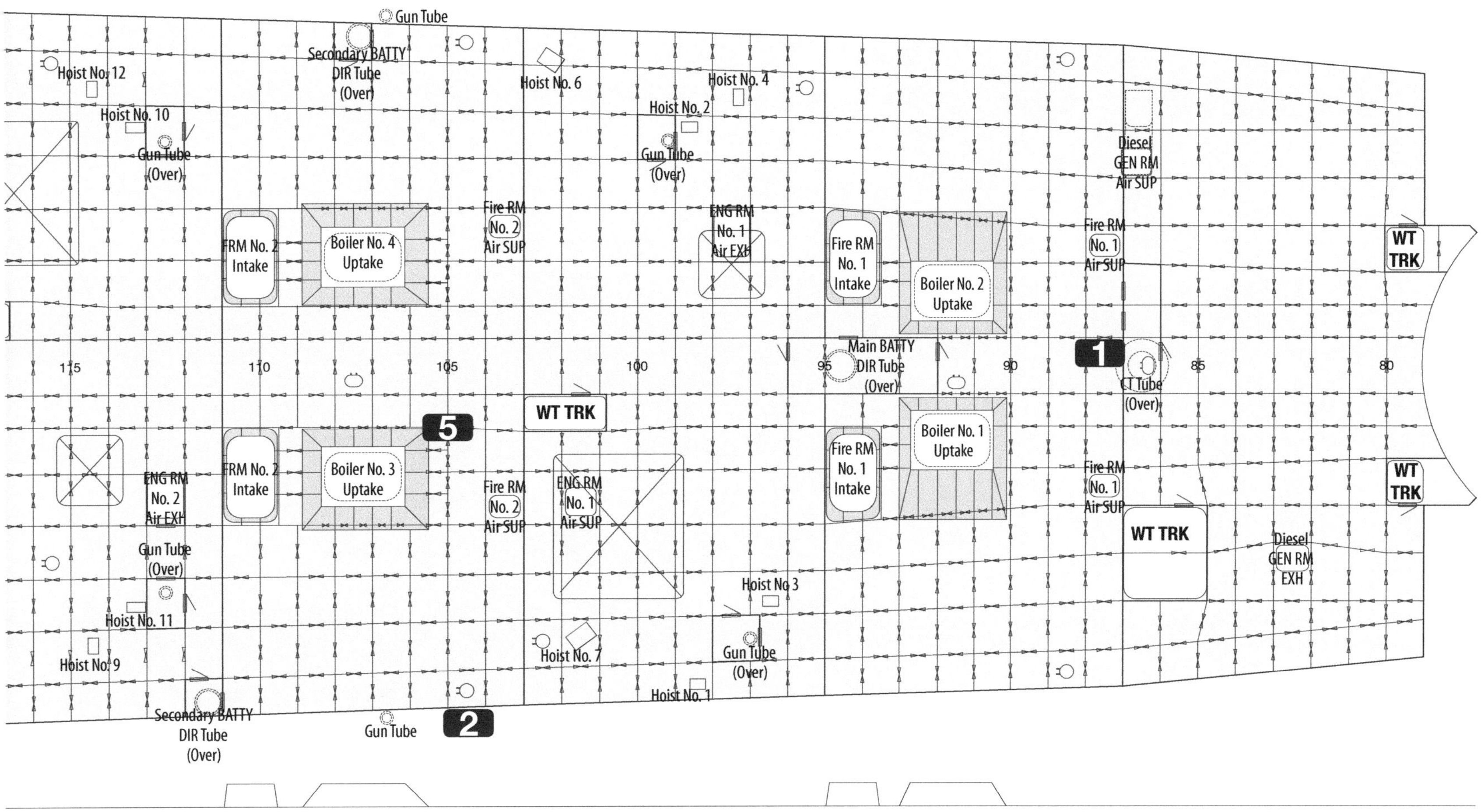
Gun Tube
Secondary BATTY DIR Tube (Over)
Hoist No. 12
Hoist No. 10
Gun Tube (Over)
Hoist No. 6
Hoist No. 4
Hoist No. 2
Gun Tube (Over)
Diesel GEN RM Air SUP
FRM No. 2 Intake
Boiler No. 4 Uptake
Fire RM No. 2 Air SUP
ENG RM No. 1 Air EXH
Fire RM No. 1 Intake
Boiler No. 2 Uptake
Fire RM No. 1 Air SUP
WT TRK
Main BATTY DIR Tube (Over)
1
115
110
105
100
95
90
85
80
75
CT Tube (Over)
WT TRK
5
Boiler No. 1 Uptake
ENG RM No. 2 Air EXH
FRM No. 2 Intake
Boiler No. 3 Uptake
Fire RM No. 2 Air SUP
ENG RM No. 1 Air SUP
Fire RM No. 1 Intake
Fire RM No. 1 Air SUP
WT TRK
WT TRK
Diesel GEN RM EXH
Gun Tube (Over)
Hoist No. 11
Hoist No 3
Hoist No. 9
Hoist No. 7
Gun Tube (Over)
Hoist No. 1
Secondary BATTY DIR Tube (Over)
Gun Tube
2

(1) An armored tube connects the conning tower to the citadel. The tube protects the fire control and steering cabling. This is a view up the conning tower tube from the splinter deck showing the mass of cables running through the tube.

(2) This is a view of splinter deck above the starboard side of fire room no. 2. The term *deck* is misleading because there is only a 30 inches high crawl space. The splinter deck was intended to trap any fragments that a projectile strike might break off the armored second deck. It also serves as a cable trunk. The circular objects on the lower face of second deck are quilting pins that secure armor plates above to the deck. These are similar to rivets except that the top end is welded to the armor. Splinter deck is $\frac{5}{8}$" thick on USS *Iowa* and USS *New Jersey*. The splinter deck was increased to $\frac{3}{4}$" on USS *Missouri* and USS *Wisconsin*.

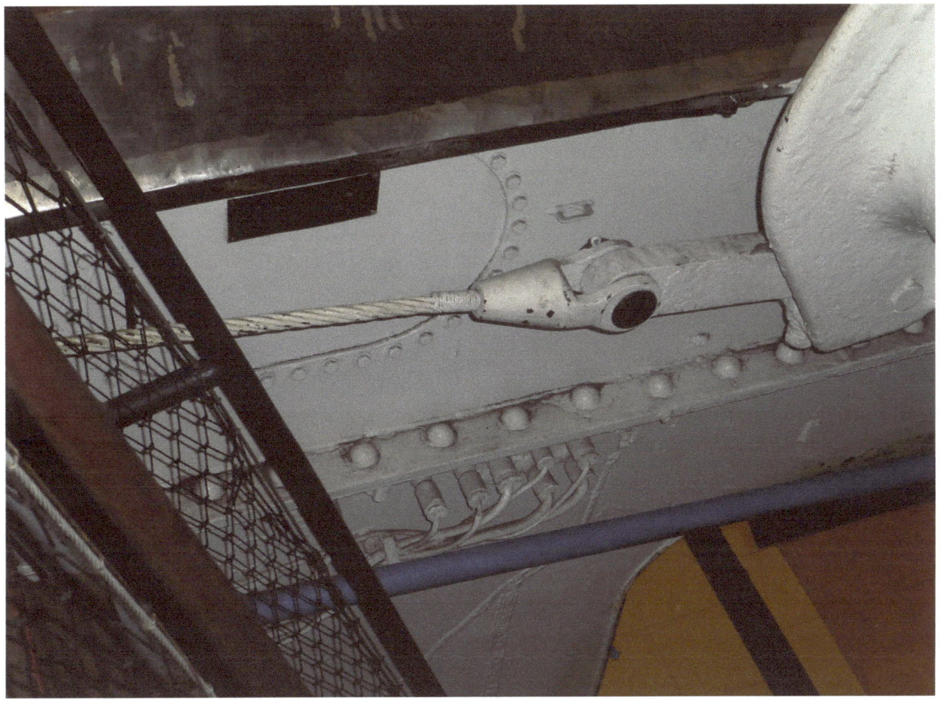

(3) Most of the access points to the splinter deck are in its underside. There are additional access points through bulkheads in watertight trunks that pass through the splinter deck. This bolted plate manhole is located in trunk at Frame 133. The cable passing over the plate is the counterbalance for the armored deck hatch.

(4) Aft of Frame 151 at the end of Broadway, the splinter deck slopes up so that the crawl space shrinks to sixteen inches. This slope is visible in the third deck aft primary plot.

(5) This view upwards shows a recess in the splinter deck in fire room no. 2 that accommodates the uptake.

Third Deck

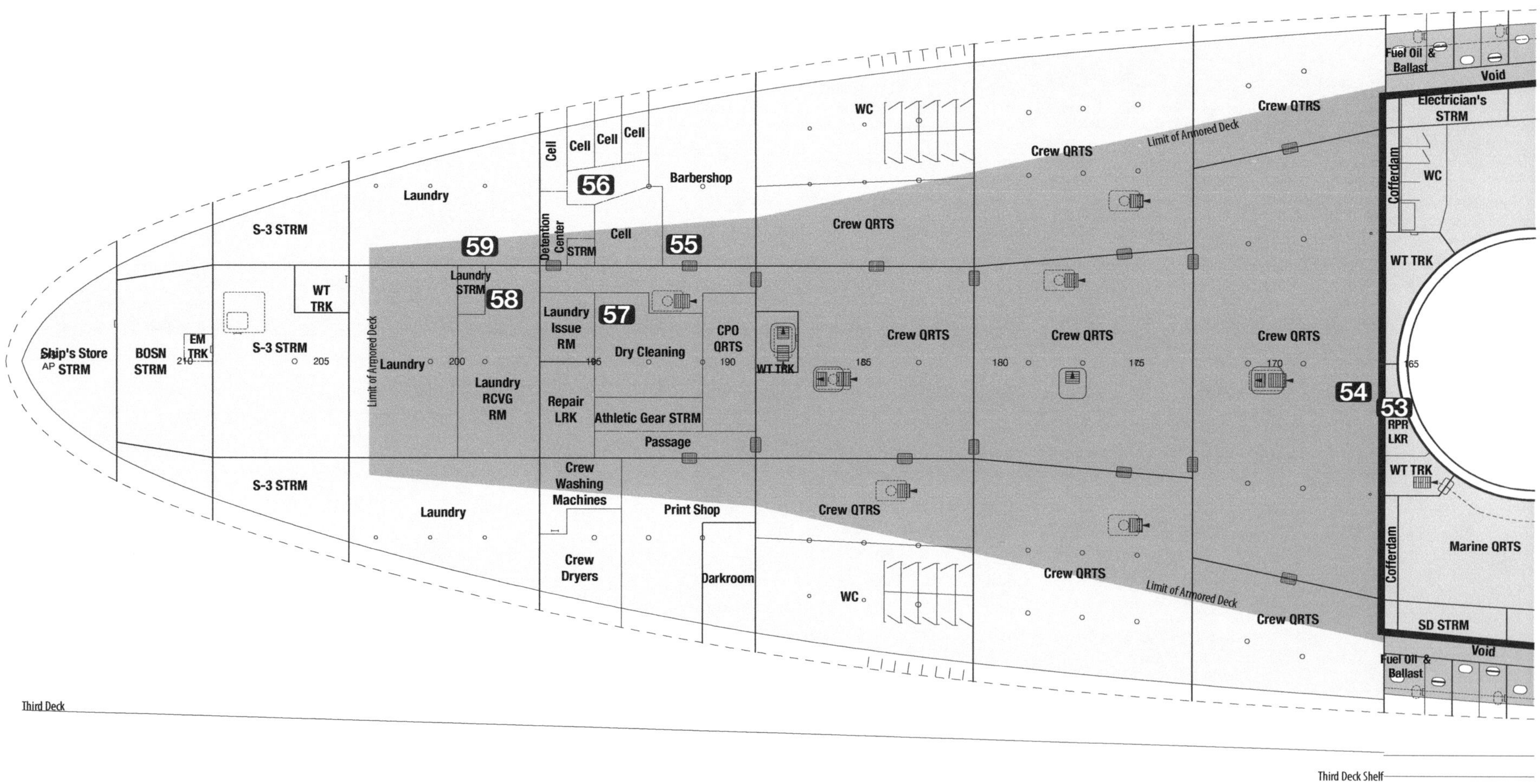

Third Deck
1"=24'

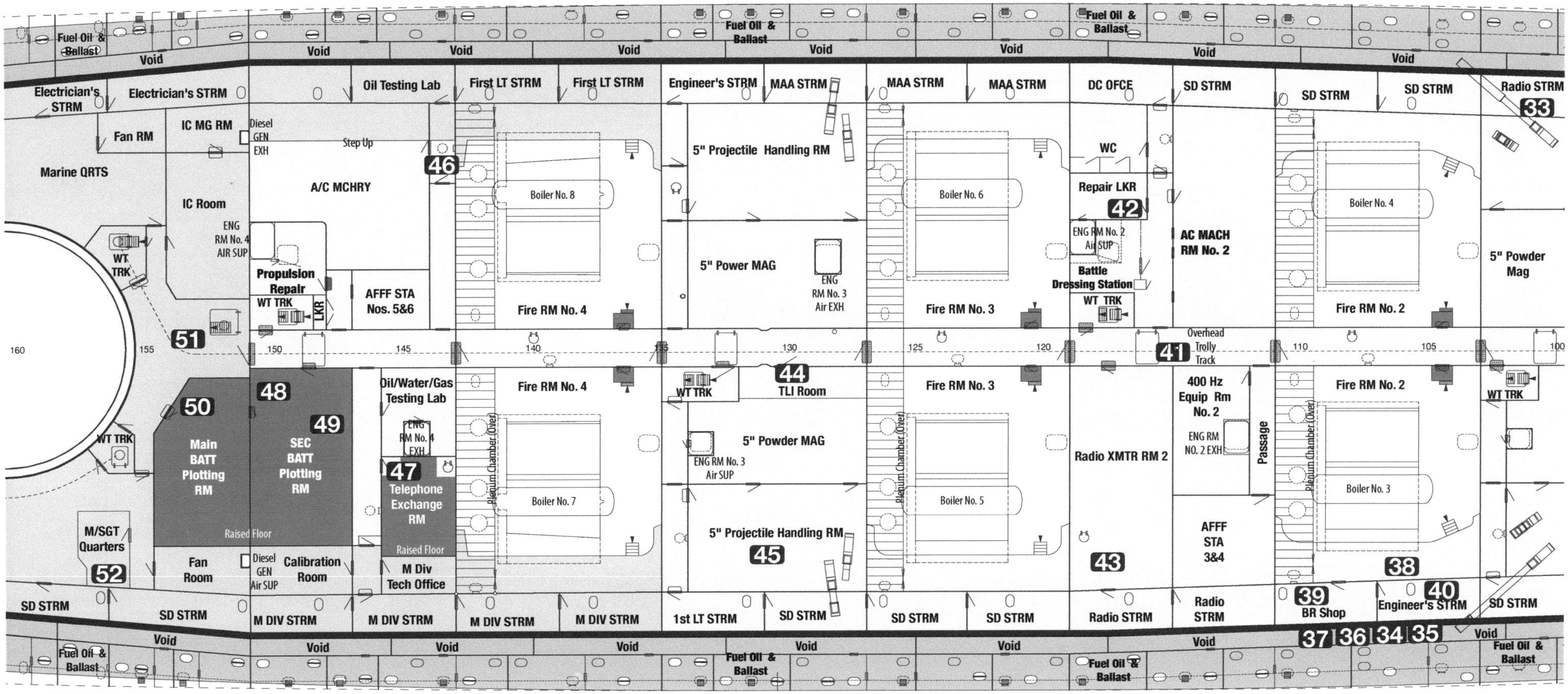

Third Deck at Center

Outer
Inner
Shelf

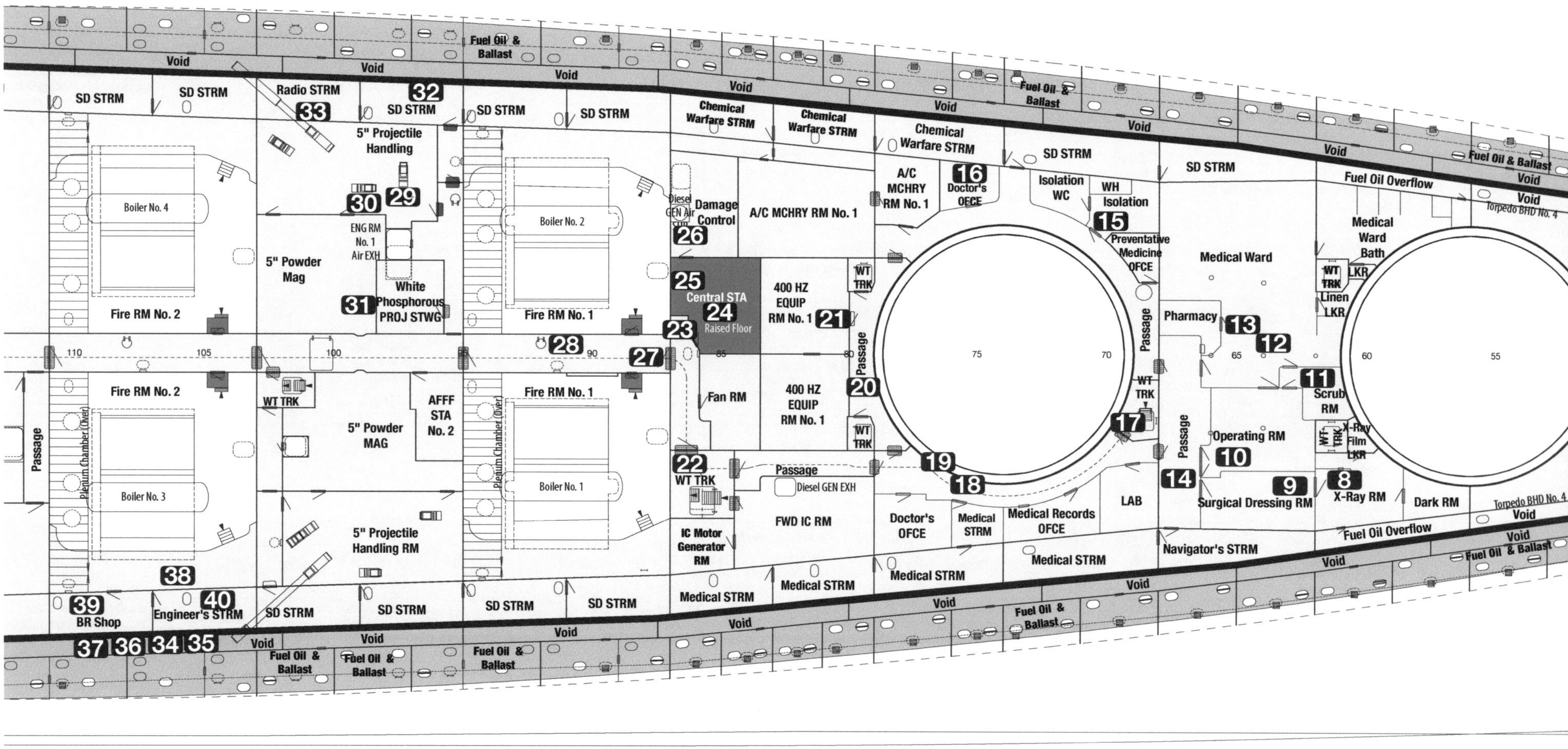
Fuel Oil & Ballast
Void
SD STRM
Radio STRM
5" Projectile Handling
Boiler No. 4
5" Powder Mag
ENG RM No. 1 Air EXH
White Phosphorous PROJ STWG
Fire RM No. 2
Boiler No. 2
Fire RM No. 1
Chemical Warfare STRM
Diesel GEN Air
Damage Control
A/C MCHRY RM No. 1
A/C MCHRY RM No. 1
Doctor's OFCE
Isolation WC
WH Isolation
Preventative Medicine OFCE
Medical Ward
Fuel Oil Overflow
Medical Ward Bath
LKR
Linen LKR
Torpedo BHD No. 4
Central STA
Raised Floor
400 HZ EQUIP RM No. 1
WT TRK
Passage
Pharmacy
Scrub RM
WT TRK
5" Powder MAG
AFFF STA No. 2
Plenum Chamber (Over)
Boiler No. 3
Boiler No. 1
Fan RM
Operating RM
X-Ray Film LKR
Diesel GEN EXH
IC Motor Generator RM
FWD IC RM
Doctor's OFCE
Medical STRM
Medical Records OFCE
LAB
Surgical Dressing RM
X-Ray RM
Dark RM
5" Projectile Handling RM
Navigator's STRM
Medical STRM
BR Shop
Engineer's STRM
110
105
100
90
85
80
75
70
65
60
55
1
2
3
4
5
6
7
8
9
10
11
12
13
14
15
16
17
18
19
20
21
22
23
24
25
26
27
28
29
30
31
32
33
34
35
36
37
38
39
40

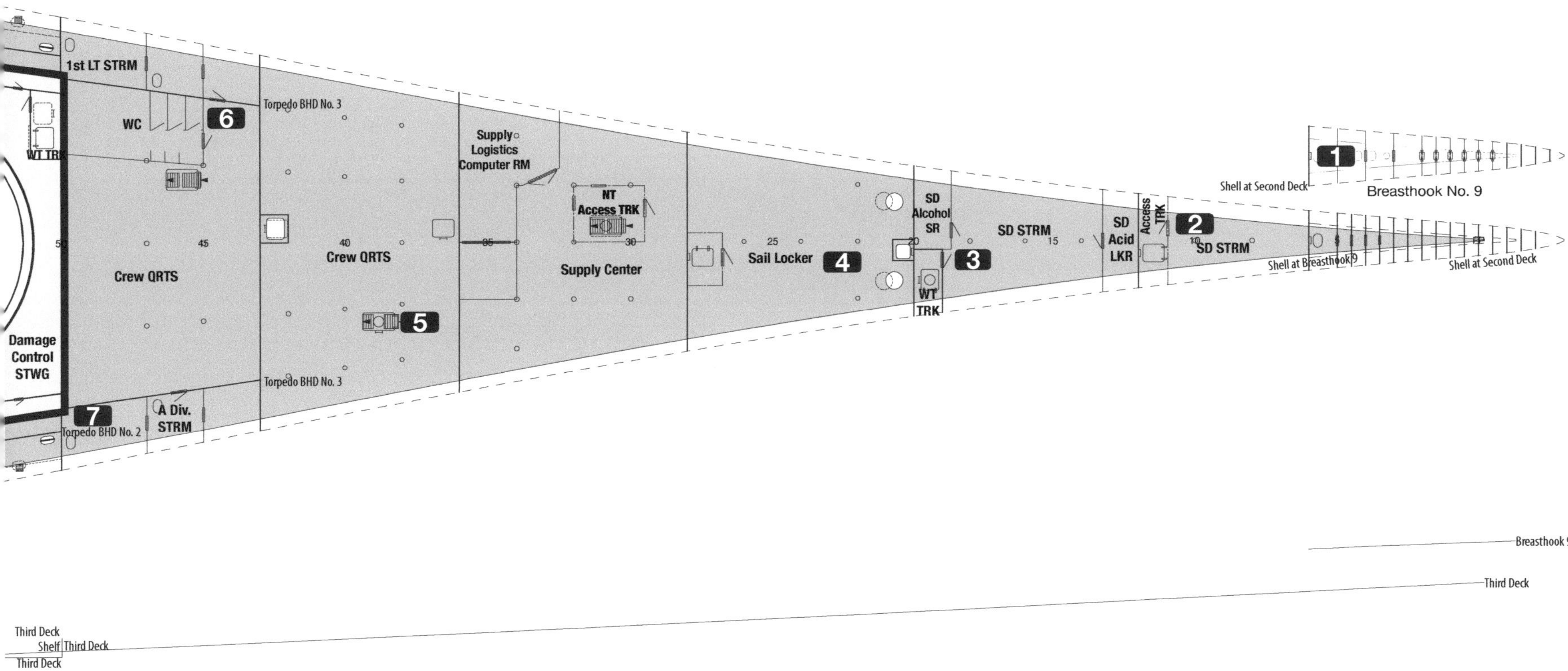

Third Deck
1"=24'

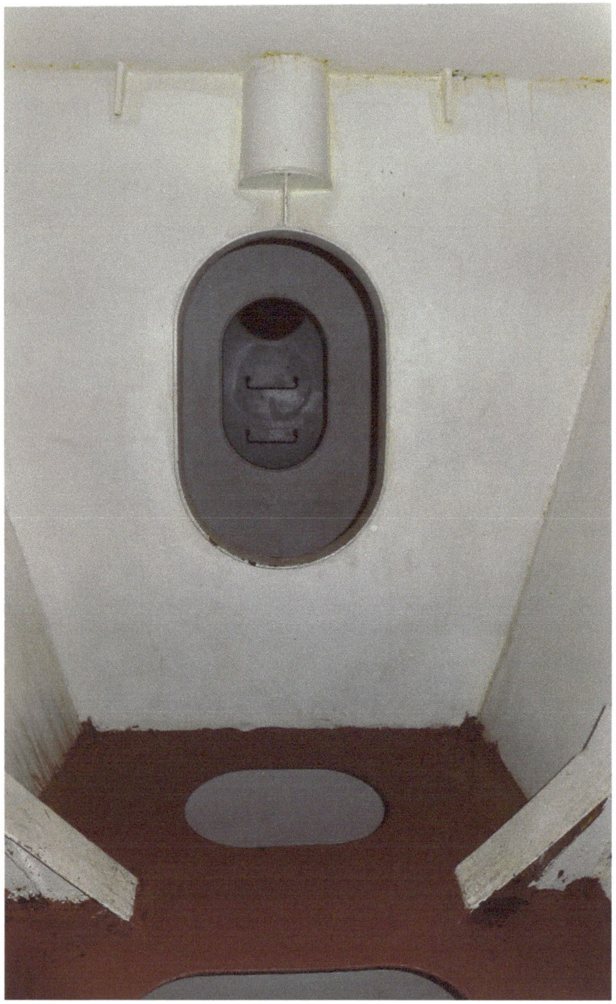

(1) Breasthook no. 9 looking forward. Breasthooks are horizontal triangular members that reinforce the bow of a ship. On *Iowa*-class breasthooks are numbered 1 to 9, bottom to top. Some breasthooks are extensions of a deck while others, including this one, form intermediate levels between decks. All of the breasthooks, except for no. 2, merge with stringers at their aft end. A foundation was cut into the web above the archway to support the stanchion in the bosun's storeroom on second deck.

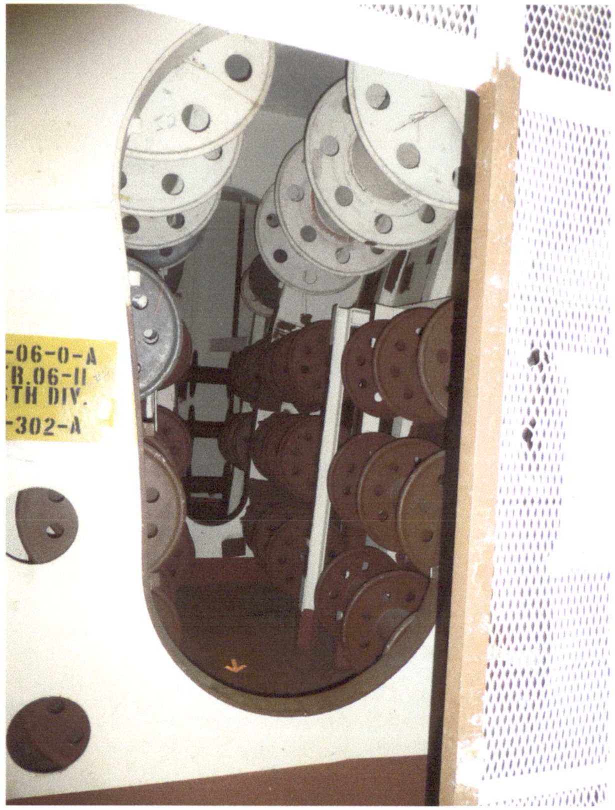

(2) The supply divisions maintained twenty-three storerooms throughout the ship. The narrowness of the hull at the waterline causes the supply division storeroom at the bow to have this unusual shape. This storeroom can only be accessed from the second deck through a watertight trunk.

(3) This supply division storeroom is aft of and adjacent to the one shown in photograph (2). However, they are reached through separate watertight trunks so there is no direct access between the two. The view here is forward from the trunk. In the background (between the two reel racks) is the door to the acid locker. The alcohol storeroom is at the aft end of the compartment. The alcohol stored there was intended for industrial purposes.

(4) The sail locker is where canvas awnings were repaired and stored. The name of this function is a traditional carry over from the days of sailing ships. This view is looking forward and to port towards the anchor chain pipes and the watertight trunk leading to the chain locker. There is a large table to the left of the photograph. The opening visible in the table marks the former location of the large sewing machine used to repair canvas that was removed when the ship went into mothballs. There are storage racks all around the around the compartment.

This is the sail locker during the Korean War showing sailors at work and the equipment missing in photograph (4). (1951 Cruise Book)

(5) There are few hatches that were dedicated to moving supplies. In nearly all cases deck hatches accessing storerooms were also used for personnel access. In order to assist in loading supplies, some inclined ladders have fold out slides so that boxes can move along the ladders. Inclined ladders also could be removed to speed up the process of storing or accessing supplies but, unless a large quantity had to be moved or an item had to be moved vertically, it was easier to use a slide than to remove and replace an inclined ladder. This inclined ladder is located within crew berthing and there are storerooms below.

(6) The space outside torpedo bulkhead no. 3 between Frames 43–50 was used for storage. The torpedo bulkhead is sloped at 19 degrees to match the slope of the belt armor. In order for the doors to the storage areas to function without effort, they are housed in recesses with vertical faces.

(7) The diagonal structure here is the forward end of the belt armor at Frame 50. Angle bars were tap riveted to the front and outer surface of the armor plate to hold it in place. Only the forward outer corner of the armor plate is visible between the angles. The angle bar at the inside (soft) face of the plate is welded to torpedo bulkhead (at the right) along scalloped edges while the angle at the outside (hardened) face is riveted to the bulkhead at Frame 50 to prevent welding heat from cracking the armor. Torpedo bulkhead no. 3 runs between the belt and the transverse armor so these two armor segments do not connect directly.

(8) This is the X-ray room looking forward. The door in the background leads to the darkroom. The X-ray technician's station is located behind the leaded bulkhead at the left. There is a leaded window allowing the technician to see the patient.

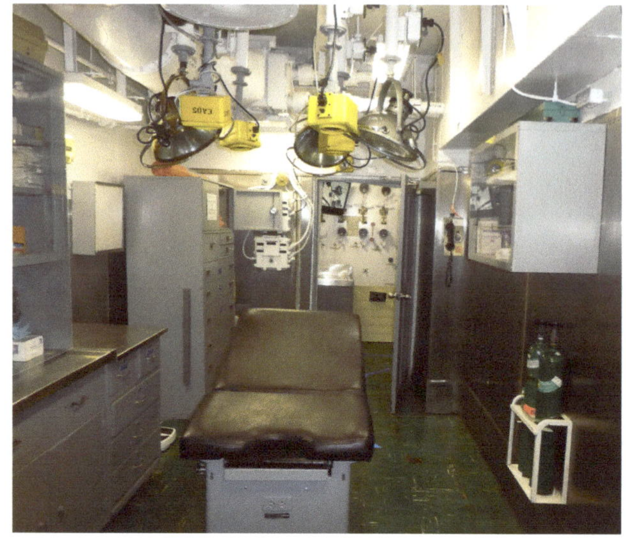

(9) Just aft of the X-ray room is the surgical dressing room. The view shown is looking aft.

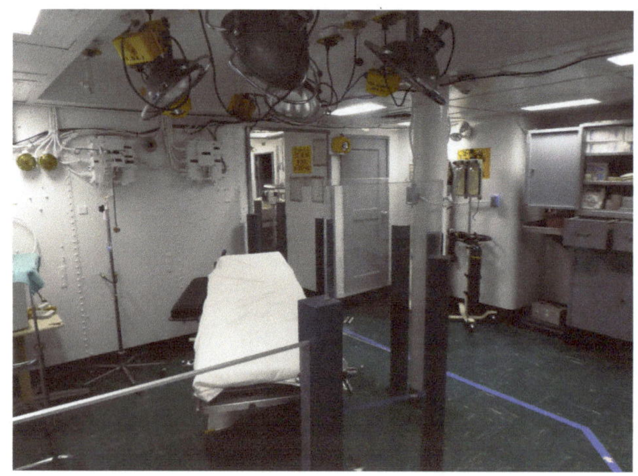

(10) The operating room has been set up for museum display. The open door leads to the medial ward and the closed door leads to the scrub room. USS *New Jersey* provided medical support to her own crew as well as her escorts who did not have the facilities to perform surgery.

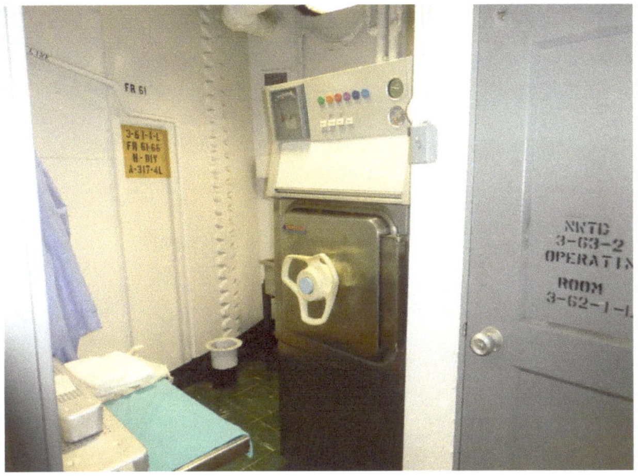

(11) The surgical scrub room. An autoclave is in the center. The lower barbette for turret no. 1 is at the left. The door at the right leads to the operating room.

(12) The medical ward viewed from the operating room door. The pharmacy door is to the left. The barriers are for museum display.

(13) The pharmacy has a work area to the right and dispensing window to the left so that prescriptions can be picked up without entering the medical rooms.

(14) This transverse passage runs through the middle of the medical spaces. The view is inboard with the operating room, dressing room, and medical ward to the right. The pharmacy's dispensing window is at the far end of the passage to the right. Most of the third deck is below the waterline. For damage control, the door thresholds are elevated and there are steps giving access to the door openings.

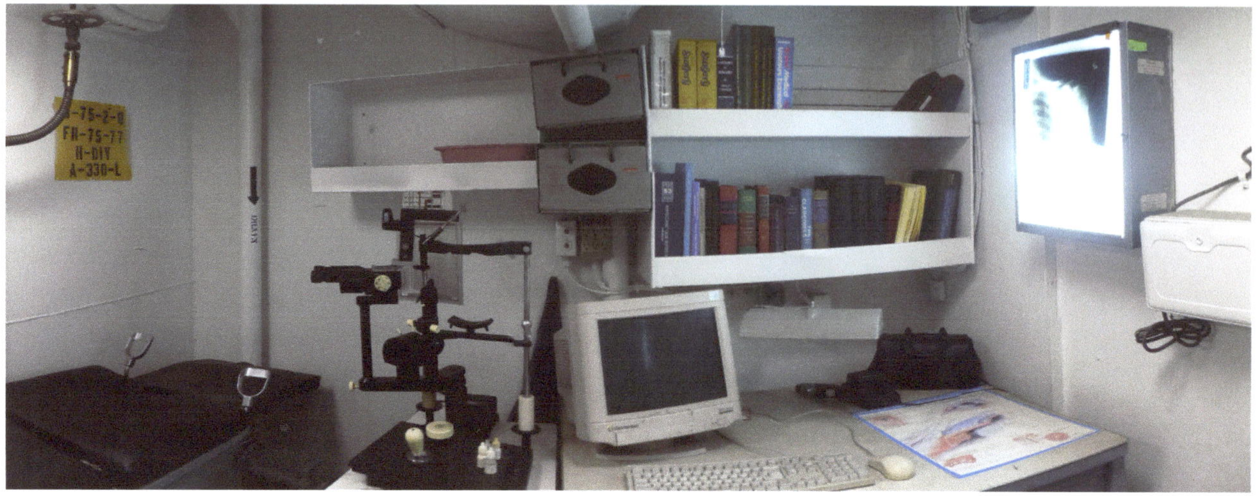

(16) The doctors' offices also served as examination rooms. They are quite compact. There are two empty containers for emergency escape breathing devices (EEBD) in the center of the photograph.

(15) The medical area contains a small isolation ward with its own water closet. The isolation ward was once used as the stateroom for a female officer who was temporarily assigned to USS *New Jersey*.

(17) The longest segment of the overhead trolley system runs from turret no. 3 and ends here at turret no. 2. It was designed to allow ammunition to be moved between the forward aft magazines but it appears to have rarely been used for that purpose (other than in action films). There is a large winch on the bulkhead to move materials between the trolley and lower decks.

This trunk was used to transfer ammunition between the magazines and main deck. The crew could remove the inclined ladder and open the hatches to create a clear path from main deck to first and second platform magazines.

(18) The museum cut this opening into turret no. 2 at the third deck to give visitors access to some of the turret machinery. At the right is the nonrotating turret. This is a $1\frac{1}{2}$" thick circular structure that supports the rotating turret. The nonrotating turret was riveted on USS *Iowa* and USS *New Jersey* and welded on USS *Missouri* and USS *Wisconsin*. The cut goes through a riveted joint so the cover plates give additional thickness. To the left is the 3" barbette support. This supports the weight of the armored barbette above and provides additional protection for the turret.

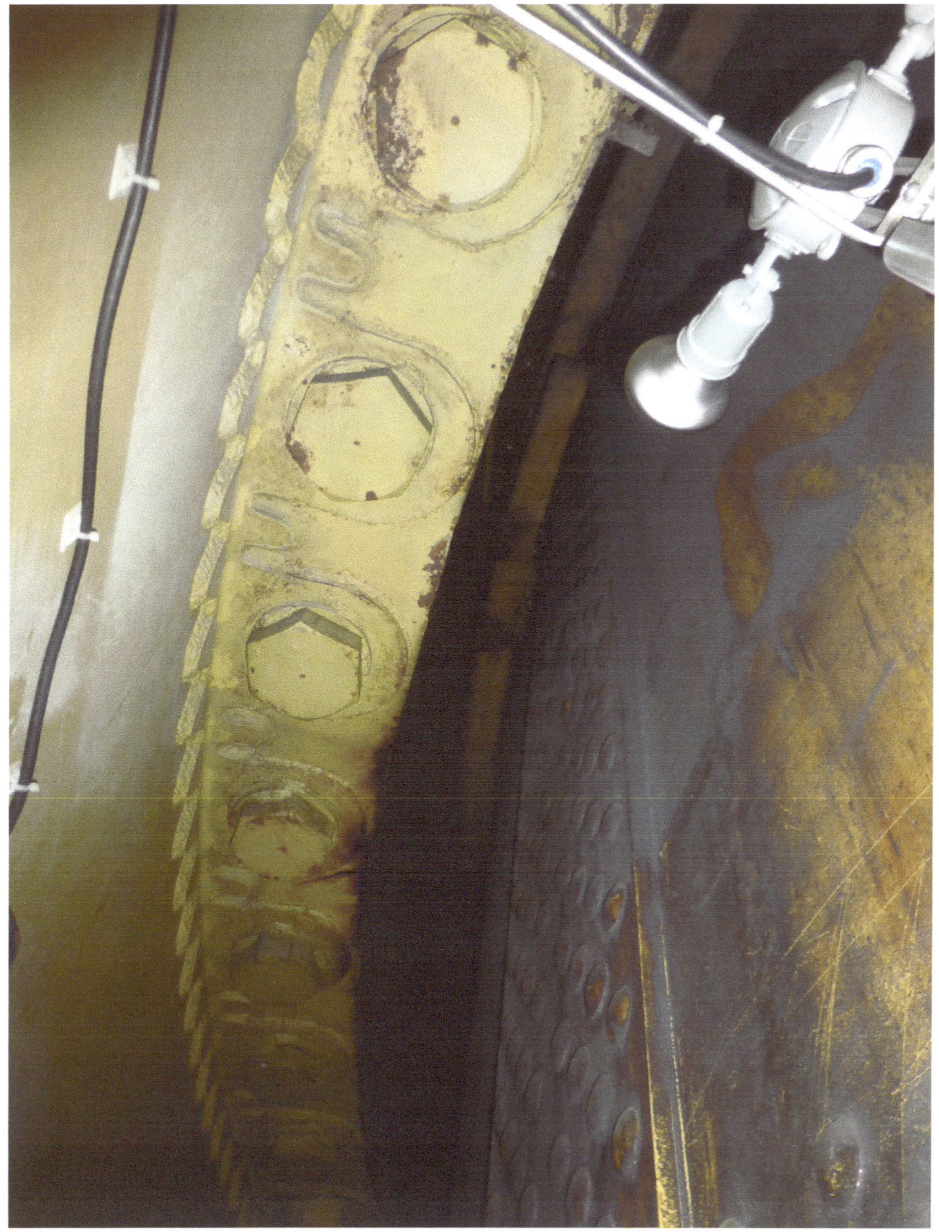

(19) This view upwards from the opening cut in turret no. 2 shows how the barbette armor plates are secured to the ship using bolts at the underside of second deck. The barbette support is to the left and the nonrotating turret is at the right. Scalloped angles were used to weld the barbette support to second deck which extends inward over the barbette support. The large bolts pass through the angle, through second deck, and into the barbette plates to hold them in place. One of the riveted butt plates holding the fixed turret together is visible at the right.

(20) The *Iowa*-class battleships were constructed using (in increasing strength and cost) medium steel, high tensile steel, and special-treatment steel (STS). STS was developed for armor plating. Because of its cost, STS was not used everywhere on the *Iowa*-class but it was used liberally, especially on decks. The Navy did not count steel as armor unless it was at least four inches thick so the *Iowa*-class have much in the way of armor plating that does not show up in tabular data.

Because STS is homogenous, it is a type of Class B armor. At the time of construction, the Bureau of Ordnance had a similar standard alloy for Class B armor. There were no published specifications for the composition of Class A or Class B armor because this information was confidential and the specifications for STS allowed wide variations in composition. One finds in the ship plans that Class B and STS frequently were used interchangeably and for all practicable purposes, STS and Class B were the same.

This photograph shows the outside of the three-inch STS barbette support. STS is not face-hardened so the brackets that connect the circular Carlin beam supporting second deck are welded directly to the barbette support. In this area, second deck is $1\frac{1}{4}$" STS with 4.75" of Class B armor pinned to the top and third deck is 1" STS.

(21) The area aft of turret no. 2 originally held the main radio and decoding rooms. In the 1980's this area was converted to 400 Hz equipment rooms and air conditioning. AC current for electronics in ships and aircraft is usually 115 volt/400 Hz (compared to U.S. household current at 60 hz). The Navy scavenged the equipment from this compartment when the ship was mothballed. The museum has installed historical equipment to return the room to its World War II appearance.

(22) These are the counter balances for the second deck armored hatch in the watertight trunk at Frame $84\frac{1}{2}$. Due to the greater weight of such hatches, the counter balances are much more massive than the balances found at second deck.

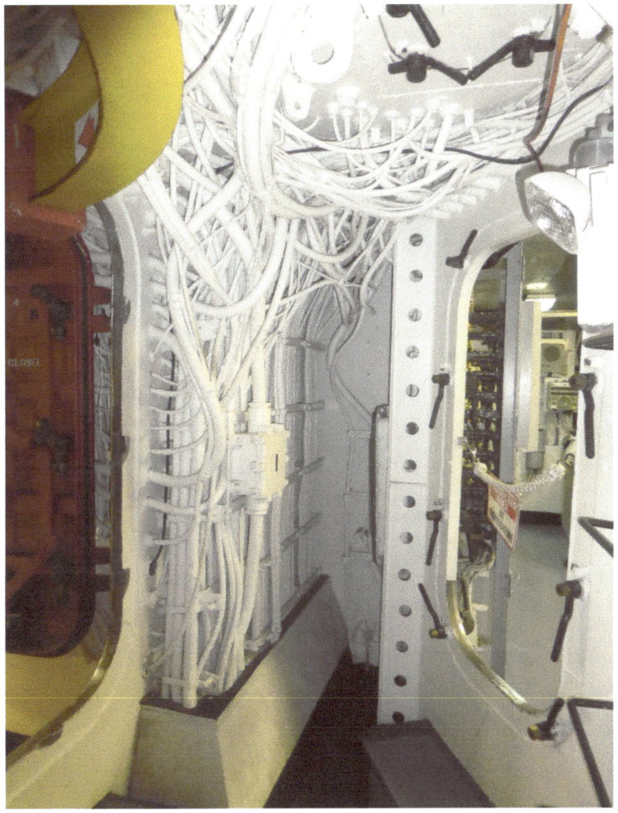

(23) An armored tube connects the conning tower to the citadel. This mass of wiring emerges from the bottom of that tube then flows to the fire control systems and ship control systems in various parts of the citadel. The splinter deck hatch leading to the conning tower tube is visible at the top of the photograph. The forward entrance to Broadway is at the left and the entrance to central station is at the right.

(24) This is the Central Station with the steering stand at the center. Central Station was the battle station for the first lieutenant. A gyroscope is in the foreground. The steering stand has an engine order telegraph to the left and a wheel. There are other wheels on the ship but this wheel rules them all. The switchboards allow the central station to delegate control to the primary conn at the 08 level or the secondary conn at 04 level.

(25) This is a close up view of the switchboard at the aft bulkhead of the Central Station. The switch shown here can be pulled out and rotated to route engine telegraph control to central station, primary conn, or secondary conn in the conning tower.

(26) Damage control is located next to central station. During battle, damage control teams would be dispatched from here. The gray containers on the bulkheads contain damage control diagrams. The intake trunk for the forward auxiliary diesel generator room is to the left.

(27) This is the forward entrance to Broadway. Broadway runs the length of the engineering spaces and is the longest and widest passage on the ship. The engineering spaces consist of four alternating fire and engine rooms. There are watertight doors at each end and six more doors subdivide Broadway into watertight segments.

A trolley track runs through Broadway. The gap in the trolley track allows the door to close. The missing section of track is stowed on the bulkhead at the left. The museum has added protective padding over the end of the track segment.

The engine rooms extend to third deck. However, the four fire rooms go up to splinter deck so Broadway forms a tunnel through each of them. An entrance to fire room no. 1 is at the left. There is another entrance on the opposite bulkhead that is obscured by the door to Broadway. The fire room door is mounted within a removable plate that allows large machinery to be moved from Broadway into the boiler room. Each fire room has an escape trunk that exits within Broadway. The escape hatch from fire room no. 1 is visible in the deck at the center of the photograph.

(28) Much of the bulkhead space within Broadway is devoted to engineering-related controls. These are fuel cutoffs for fire room no. 1. If the crew had to evacuate the fire room, they could shut down the boiler from here.

(29) Five-inch ammunition was semi-fixed meaning that that powder and primer were integrated into a single brass cartridge but the projectile was separate. Projectiles were stored in the handling room. Powder was stored in separate, adjoining magazines.

There were originally six magazines located above engine rooms nos. 1–3. The magazines over engine room no. 2 served one mount each while the other magazines served two mounts. When four five-inch mounts were removed in the 1980's, the magazines over engine room no. 2 were replaced with machinery rooms.

This is the lower end of a five-inch dredger hoist. Each mount has two hoists like this one located within the projectile handling room. The hoists could be used to move either projectiles or propellant cartridges to the gun mount. However, the normal practice was to use one hoist for projectiles and the other for powder. A powder canister is sitting in this hoist. Powder charges were stored in such canisters until they were ready to be fired. Powder cartridges traveled through the hoist to the mounts in their canister.

Between the handling rooms above the main deck and the third deck handling rooms and magazines there were about 450 rounds per gun (5,400–9,000 total). The magazines have different sizes and shapes so they held different amounts of ammunition.

(30) Powder canisters were passed through scuttles in the magazine doors into the projectile handling room where they were loaded into hoists. The scuttles ensured there was no opening into the magazine except when a canister was passing through.

(31) A five-inch powder magazine. While at sea, the ammunition racks shown here would have been completely filled with propellant cartridges stored in canisters. Projectiles were stored in similar racks within the projectile handling room.

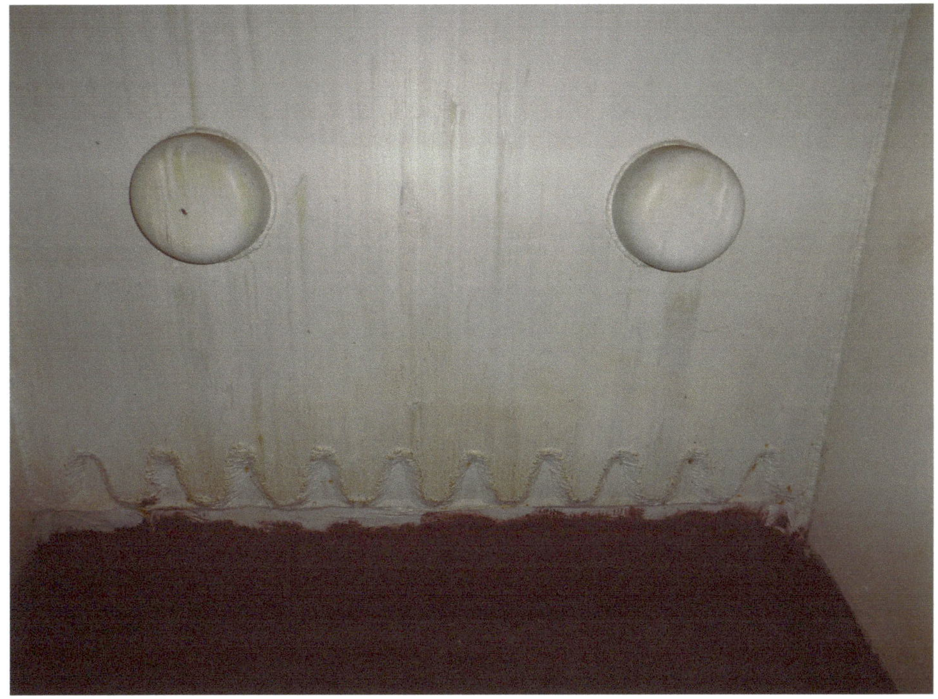

(32) One can catch a glimpse of the lower belt in the third deck storerooms. At the top is the backing bulkhead that forms part of torpedo bulkhead no. 3. The upper belt plates are bolted to the backing bulkhead. The two bumps shown are $\frac{1}{8}$" thick protective covers welded over the bolts. A projectile strike could have caused the connecting nut to pop from the bolt. In such an event, the cover would have prevented the nut from causing damage or injury inside the armor. The plating visible below the scallop and above the deck is the lower belt. The lower belt is welded to the backing plate along the scalloped lower edge of the backing bulkhead. The joint between the upper and lower armor belt is located above the scallop and below the covers but it is obscured by the backing bulkhead. While the upper belt armor is attached to the torpedo bulkhead, the lower belt armor forms part of the torpedo bulkhead.

(33) The five-inch hoists take a variety of paths from the third deck to the mounts and the paths and there is limited symmetry among the port and starboard hoists. This photograph shows how one of the hoists curves upwards through a storeroom.

(34) This is a void space that is part of the ship's underwater protection system looking aft at the starboard side. Torpedo bulkhead no. 2 is at the left and the outer face of the belt armor with Torpedo bulkhead no. 3 is at the right. The belt is sloped inward towards the bottom at nineteen degrees. This made it nearly impossible for a projectile to strike the belt normal to the face. The horizontal division shown is between the Class A upper belt and the STS lower belt. The upper belt is 12.1" and is bolted to torpedo bulkhead no. 3 as shown in photograph (32). The heat treatment used to face-harden the armor plates could produce irregular surfaces that would not fit flush against the backing bulkhead. By design, the upper belt plates are offset $1\frac{1}{4}$" from the backing bulkhead so that any irregularities in the armor will not affect the fit between adjoining plates. The space between the backing bulkhead and the Class A plates is filled with cement. The cement-filled gap creates the overlap visible between the upper and lower belt. The lower belt Class B plates are 12.1" at the top. Then they taper at knuckles to 1" at the bottom. In this arrangement, the upper armor belt is bolted to torpedo bulkhead no. 3 while the lower armor belt is part of torpedo bulkhead no. 3. The plates of the lower belt were connected by welding scalloped plates over both sides of their joints. Part of such a plate is visible at the far right.

(35) Each armor plate used in the *Iowa*-class has an identifier. The Midvale Company painted the identifier UB-8-S (Upper Belt, Eighth Plate, Starboard) on this plate before it was delivered. This plate was never painted after that. This specific plate weighs 15,015 lbs. Creating armor plates for battleships strained the capacity of the U.S. steel industry so the armor plates were farmed out to multiple manufacturers. The Midvale Company also made this plate for USS *Missouri*, Bethlehem Steel. Co. made it for USS *Illinois*, and Carnegie Steel made this plate for USS *Iowa*, USS *Wisconsin*, and USS *Kentucky*.

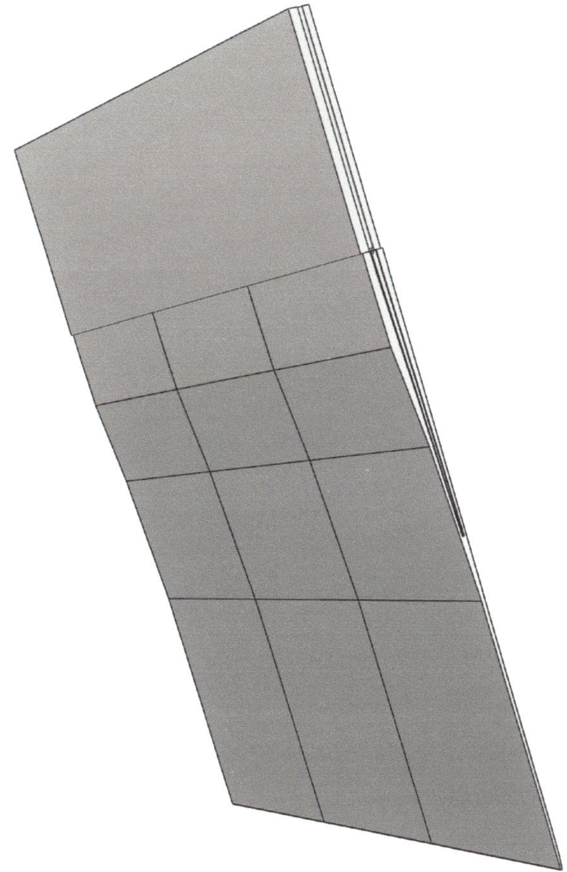

This rendering depicts a group of belt armor plates with one Class A upper belt plate over three Class B lower belt plates looking inboard. The outer faces of Class B plates taper at knuckles with a final taper to one inch at the bottom on the inside face. The knuckles and plate thicknesses on the actual ship are not mathematically precise as shown in this illustration. The slots at the edges of the plates are keyways for alignment. Scalloped plates were welded over the lower belt joints on both sides after the armor was installed.

(36) An opening in third deck shelf allows one to look downward and see how the lower belt plates taper towards their bottoms. The lower belt plates taper at knuckles from 12.1" at the top to 1"at the bottom. The plates were grinded from slab of STS and grinding marks are clearly visible in the surface. Knuckles are also visible in the plate but they are irregular as are the plate thickness. A scalloped butt plate is visible along the bulkhead at the left.

The belt plates were ordered oversized and the shipyards had to grind the edges to make them fit on the ships. On USS *Iowa* the sides of the plates were trimmed so the joints were normal to the centerline. On the later ships, the amount of grinding was reduced by trimming the plate edges normal to the face (except at facet angles which were made normal to the centerline). The Midvale Company supplied the lower belt plates for the New Jersey.

(37) This is another view from the opening in third deck shelf (looking downward and aft) showing the transverse frames support the lower belt from the outside. The supports for the lower belt alternate between the two sides at two-foot intervals.

(38) The machinery areas do not have decks above the hold. Instead, there are platforms at various levels that provide access to the machinery. Within the fire rooms, third deck ends at Broadway then resumes just inboard of torpedo bulkhead no. 4. This is the resumption of the third deck within fire room no. 2 at the starboard side looking aft. Torpedo bulkhead no. 4 is at the left. The right side of the deck is lined with storage racks. In the background is the plenum chamber that supplies air to the no. 3 boiler. The lower opening provides access to a manhole above through which one can enter the plenum chamber. The opening above that is the ventilation exhaust for the fire room. Rather than sending the ventilation air out of the ship, a fan pulls it into the plenum chamber where it joins the boiler intake air and then it flows into the boiler.

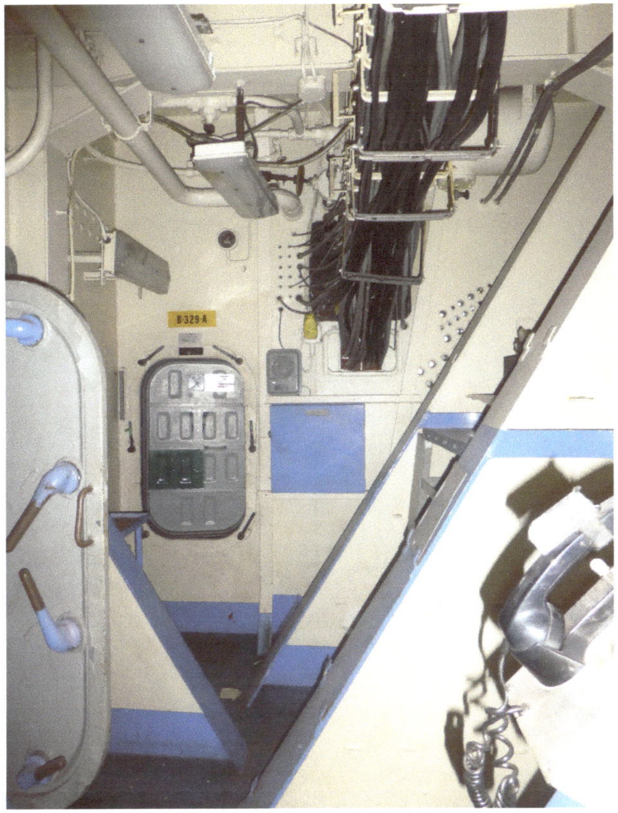

(39) A door through torpedo bulkhead no. 4 in boiler room no. 2 leads to the boiler room shop. The view is forward with torpedo bulkhead no. 4 (the holding bulkhead) to the left and torpedo bulkhead no. 3 to the right. This was originally a storeroom but had been converted in the 1980's into a shop. A bundle of control cabling runs overhead. Below third deck, torpedo bulkhead no. 4 slopes outboard toward the top. It shifts to vertical between third deck and second deck. On the incomplete USS *Illinois* and USS *Kentucky* this bulkhead remained sloped to second deck

(40) Store rooms fill the spaces between torpedo bulkhead no. 3 and no. 4. This storeroom is forward of the boiler room shop shown in photograph (39). The view is forward so that torpedo bulkhead no. 3 is to the right with the armored belt behind it. torpedo bulkhead no. 4 (the holding bulkhead) is to the left. The support frames for the bulkheads staggered. The supports for the armor (right) are partial frames that just run along the bulkhead and are positioned at half frame locations. The supports for the holding bulkhead (left) are part of complete frames that run the full beam. These are at integral frame locations. The purpose of this arrangement is to minimize the impact of an explosion against the belt armor. The inward force from such a blast would be transferred to the partial frames but not the full frames. This could possibly keep the blast from breaching the holding bulkhead and would prevent the deformation of the armor from distorting the hull framing (and shell form) beyond the blast area.

(41) Each engine room has a large hatch that opens into Broadway. A trolley car can be positioned directly over the hatch so that equipment can be lowered into or raised out of the engine room.

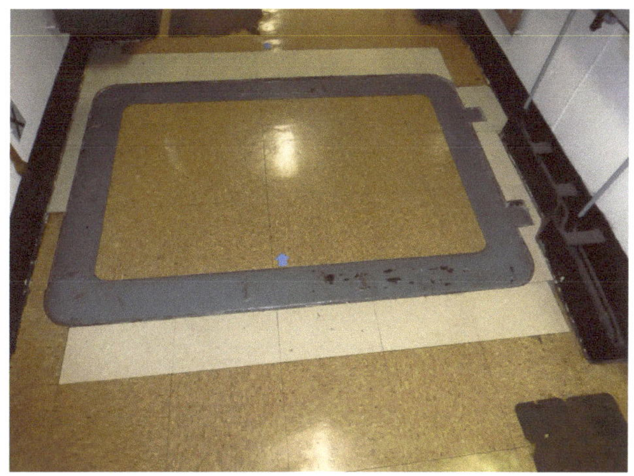

(42) This is one of two damage control repair lockers located off of Broadway. The repair lockers house tools and material for damage control. Damage control diagrams are hanging from the bulkhead at the right.

(43) The radio transmitter room at the starboard side of Broadway looking inboard. The circular deck fixture at the lower left is the escape hatch from engine room no. 2. The Navy cleared the equipment out of this compartment when the ship was mothballed. All of the equipment shown are replacements the museum has acquired to restore the space. While the equipment functions, it does not reflect what was installed in the ship while in service.

(45) When four five-inch mounts were removed in the 1980's to make room for missile platforms, their hoists were removed. However, all but one of the armored hoist trunks remained in place. Some of those hoist trunks were repurposed as armored wiring trunks. The box at the overhead is the end of one such trunk that was reused to protect wiring for Harpoon missiles. Once the wiring reaches the armored citadel, it can run in the open. From here, the wiring shown passes through torpedo bulkhead no. 4 where joins a mass of fire control cables running forward shown in photographs (39) and (40). Two functioning hoists remain in this shell handling room. The racks in the back and to the right are for storing five-inch projectiles.

(44) The tank level indicator room is located in an alcove at the starboard side of Broadway. This room contains the gas gauges for the ship. The large number of gauges reflects the number of tanks on the ship.

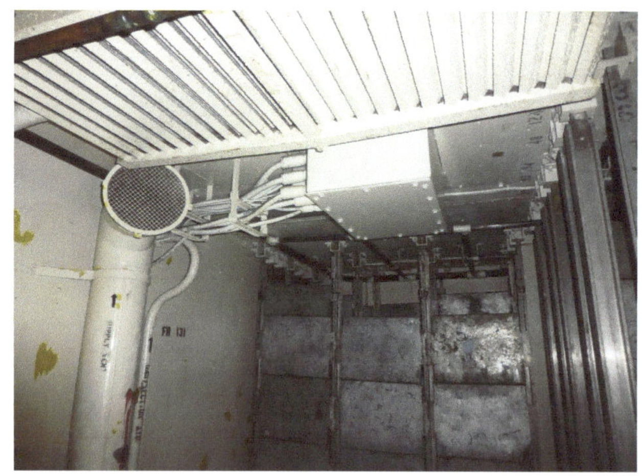

(46) Third deck splits towards the aft end of Broadway. The outer areas of the deck slope up while Broadway and the areas near it remain level. This is the aft most port side passage off Broadway. The step up at this point illustrates the difference in elevation between Broadway and the outer part of the third deck.

(47) The telephone exchange room has a 1980's era AT&T PBX (Private Branch Exchange) that is still functioning. It is likely that the World War II-era telephone equipment completely filled the room.

(48) The five-inch guns were normally controlled from one of two plotting room. This is the secondary battery aft plotting room looking to starboard. There is a nearly identical forward plotting room on the first platform. The description of that room goes into more detail about the operation of the Gun Fire Control System Mk.37 whose functions are duplicated here. The large devices in the center of the photograph are the Computer Mk.1A. The rectangular units mounted on top of the computers are the Starshell Computer Mk.1. The two smaller devices in the foreground and between the computers are the Stable Element Mk.6. This is a gyroscopic device that supplies ship pitch and roll angle to the computers. The firing keys for the guns were normally mounted on the stable element but they have been removed from these units. The linkages among the guns and the fire control system components are configured using the switches on the fire control switchboards at the left and in the background

(49) This is a close up view of the fire control switchboard. The switches lock into detents. Pulling the switch out slightly allows it to be rotated. The gunnery department would establish named configurations and define the switch setting for each configuration. Each input and output function had to be switched individually on the switchboard. Connecting a computer to a director, for example, required making multiple switch settings. The crew has marked the normal settings with stars. Computer no. 4 was normally linked to director no. 4 (Sky 4).

(50) Like the five-inch guns, the sixteen-inch guns were normally controlled from a plotting room. This is the aft primary battery plotting room. There is another primary battery plotting room forward at the first platform. The description of that room goes into more detail about the operation of the Gun Fire Control System Mk.38 whose functions are duplicated here. The device in the center of the photograph is a Rangekeeper Mk.8. The rangekeeper calculated how to aim the guns. The unit in the foreground is a Stable Vertical Mk.41. The stable vertical provided the ship pitch and roll angle to the rangekeeper and was where the gun firing keys were mounted. A Radar Mk.13 repeater is behind the rangekeeper. In the 1960's a Computer Mk.48 that was designed to simplify aiming at shore targets was added but it was scavenged when the ship was mothballed. The fire control switchboards are along the bulkheads.

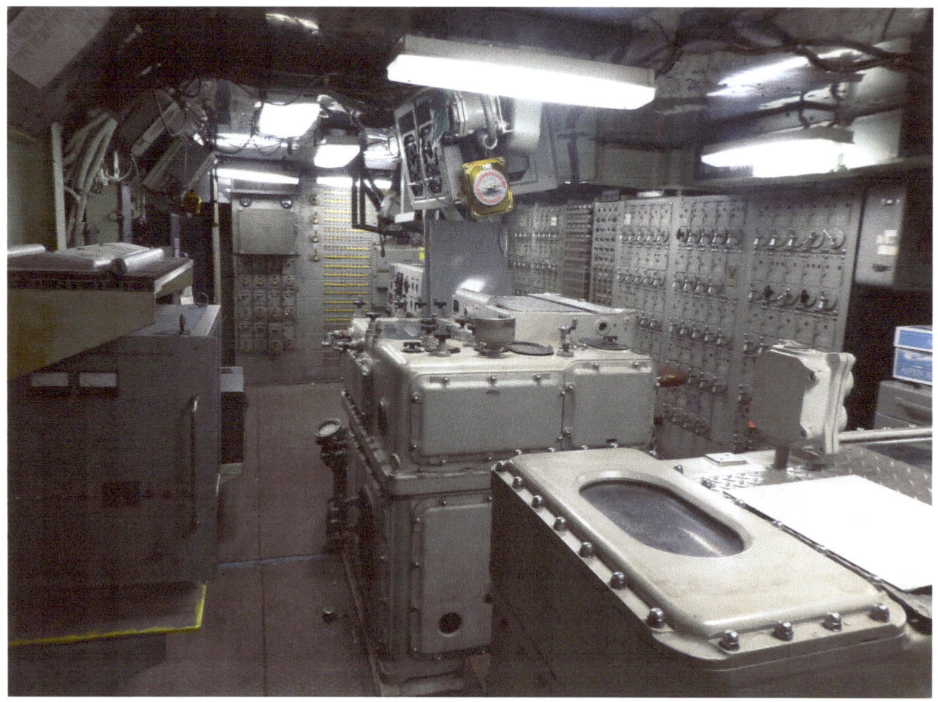

(51) The aft entrance to Broadway looking forward. The door to the right leads to Broadway. The hatch to the left leads to a watertight trunk that goes directly to the aft emergency diesel generator room. The mass of cables runs into the aft primary battery plot behind the bulkhead to the right.

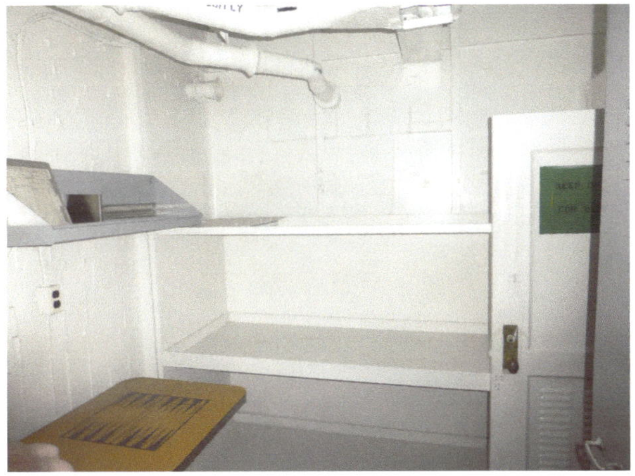

(52) Except during the 1960's, USS *New Jersey* carried a detachment of marines while in service. In World War II there were about 85 marines with three officers. In the 1980's there were 30–40 marines with one or two officers. The marines provided a security force, a potential landing force, and they traditionally manned weapon mounts at general quarters. Most of the marine berthing was on third deck. These are the marine sergeant's quarters located within this area. There are three bunks but the mattresses are not in place. This had originally been the secondary combat information center.

(53) This is a view of the underside of second deck looking to port. The aft transverse armored bulkhead is 11.3" of Class A armor is to the left and the barbette support for turret no. 3 (3" of STS) is at the right. The untreated (soft) inside face of the Class A armor could be carefully welded without cracking the hardened face. Scalloped plates were welded to the second deck on either side of the armor to hold it in position. Then a scalloped angle was welded on the inside face to the first scalloped plate and the armor plate. This created the double row of scallops shown. The transverse armor was bolted to third deck from the underside. The transverse armor is an example where Class A armor was used structurally. The armor supports the second deck and forms a watertight bulkhead.

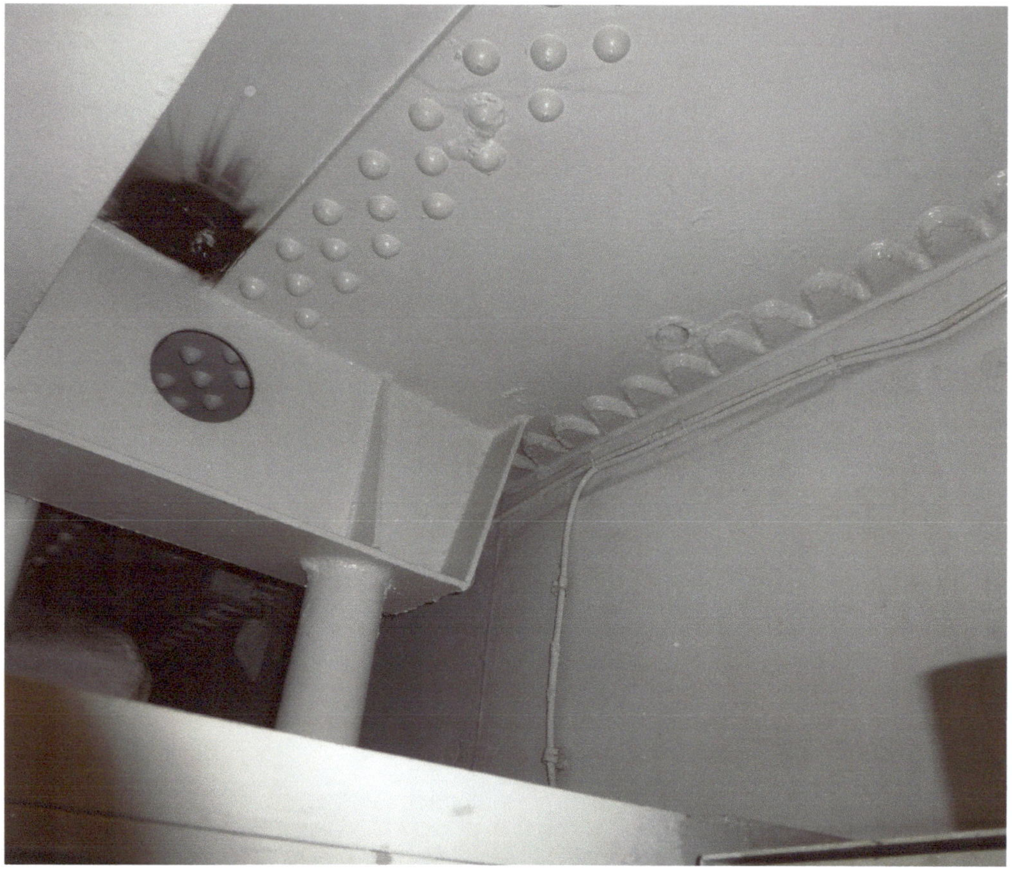

(54) Here is aft (hardened) face of the transverse armor at Frame 166 looking forward. An angle scalloped on one edge was welded to the second deck. The other edge is covered by insulation but it was not welded to the armor. Note the girder is not connected directly to the armor.

(55) Haircuts were officially free for enlisted sailors. However, it was customary to tip the barber to influence the quality of the cut. The end of third deck armor passes through this photograph but it is not visible because of the deck tile. The upper face of third deck armor, unlike that of second deck, is flush with the surrounding deck. There is no place on third deck where the edge of the armor is visible.

(56) The detention center (brig) contains three individual cells and one large detention cell. During World War II there was a fourth cell in the space next to and aft of the current cells. The walls of the cells bulkheads are made of expanded metal. During the 1980's the brig was used as a storeroom and was gutted to house a weight room paid for from the crew activities fund. The museum has restored the brig.

(57) The dry cleaning room contains dry cleaning and pressing machines.

(59) Industrial washing machines in the laundry. Each machine could wash 100 lbs. of clothes in three separate loads simultaneously. The ship's laundry only handled uniforms. Crew activities funds were used to purchase household washers and dryers for the crew to clean civilian clothes on their own.

(58) Clothes were brought to the laundry receiving room (left). Sheer raises third deck above the waterline at the ends of the ship. In such areas the door thresholds are low and lack a step. The clothes were collected from receptacles in the berthing areas (right).

First Platform

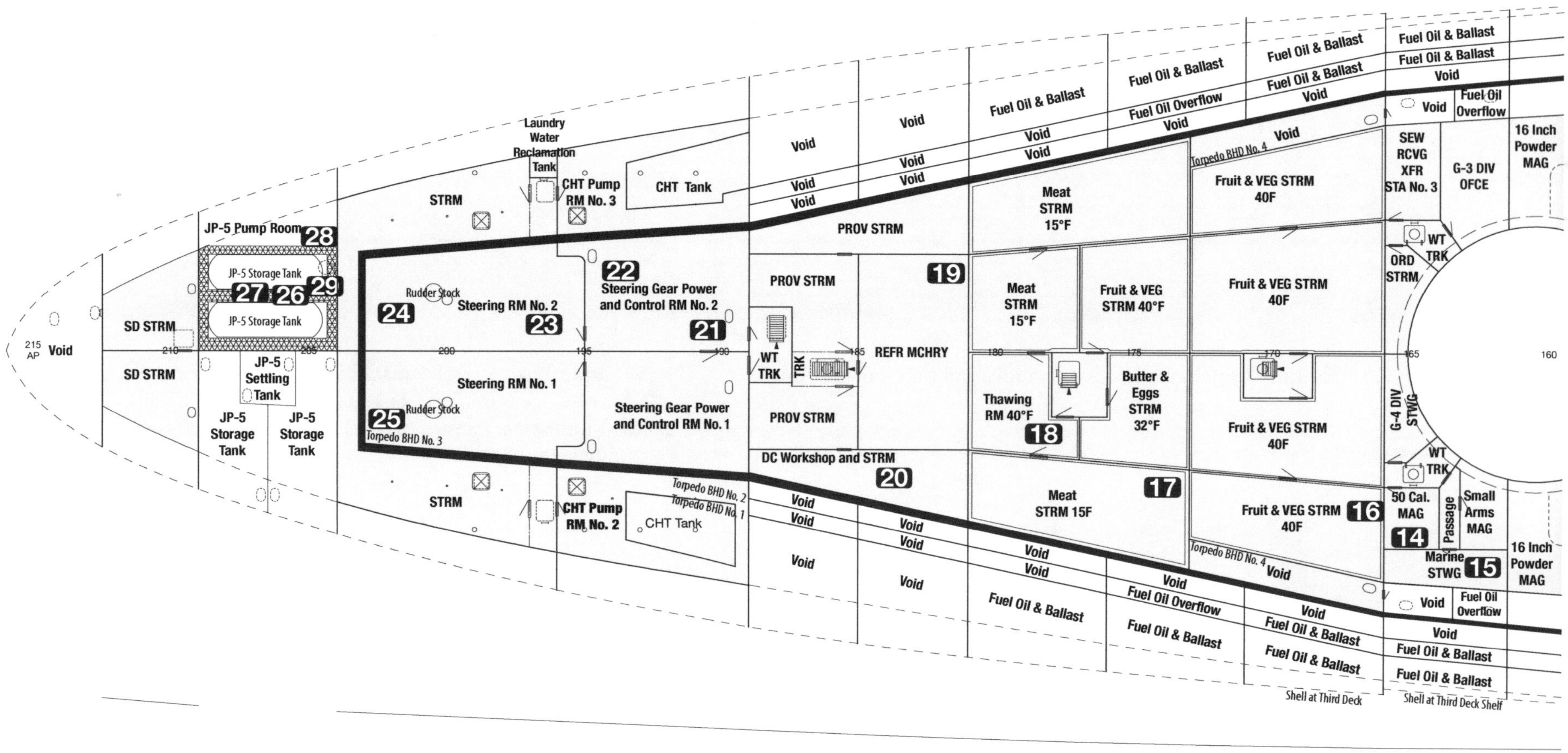

First Platform
1"=24'

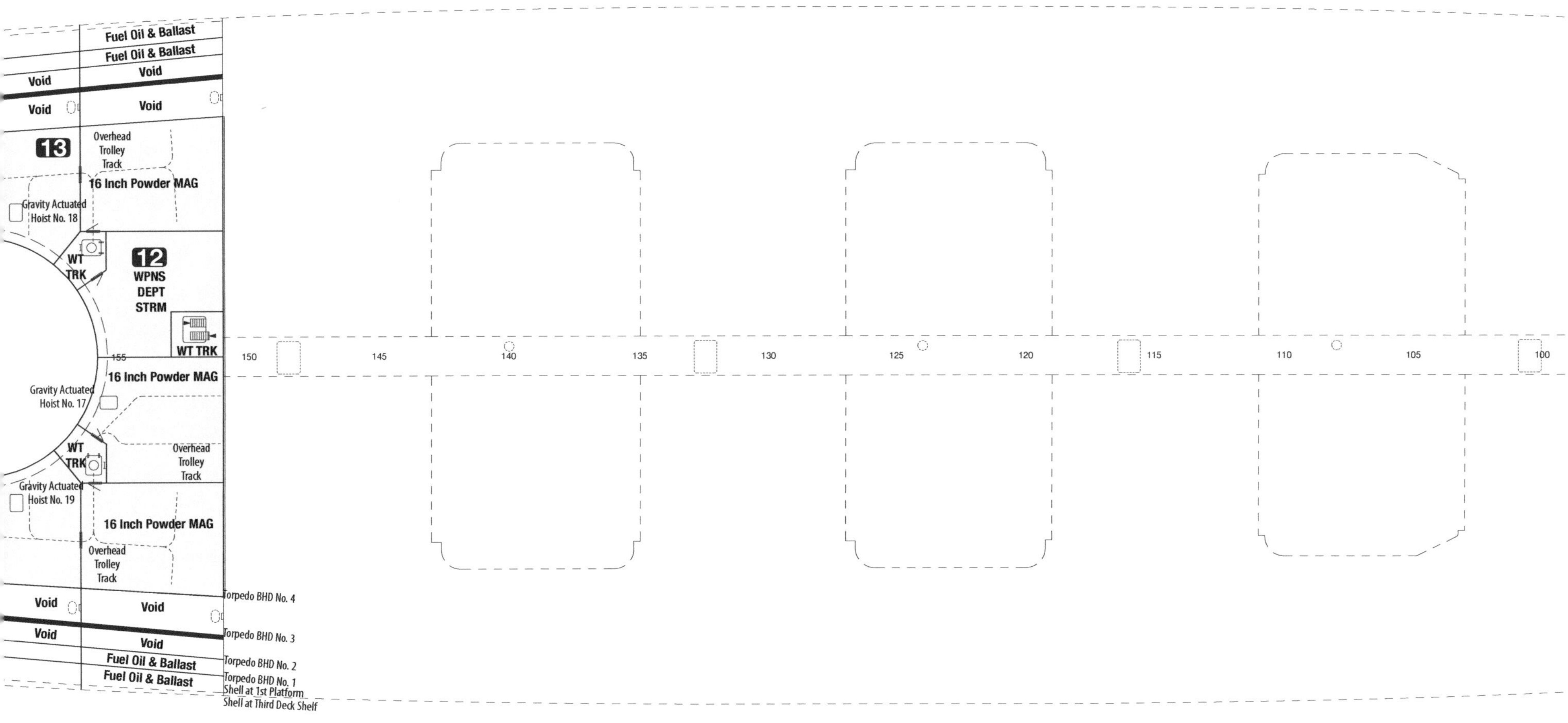
Fuel Oil & Ballast
Fuel Oil & Ballast
Void
Void
Void
Void
13
Overhead Trolley Track
16 Inch Powder MAG
Gravity Actuated Hoist No. 18
WT TRK
12
WPNS DEPT STRM
WT TRK
155
16 Inch Powder MAG
Gravity Actuated Hoist No. 17
WT TRK
Overhead Trolley Track
Gravity Actuated Hoist No. 19
16 Inch Powder MAG
Overhead Trolley Track
Void
Void
Void
Void
Fuel Oil & Ballast
Fuel Oil & Ballast
Torpedo BHD No. 4
Torpedo BHD No. 3
Torpedo BHD No. 2
Torpedo BHD No. 1
Shell at 1st Platform
Shell at Third Deck Shelf
150
145
140
135
130
125
120
115
110
105
100

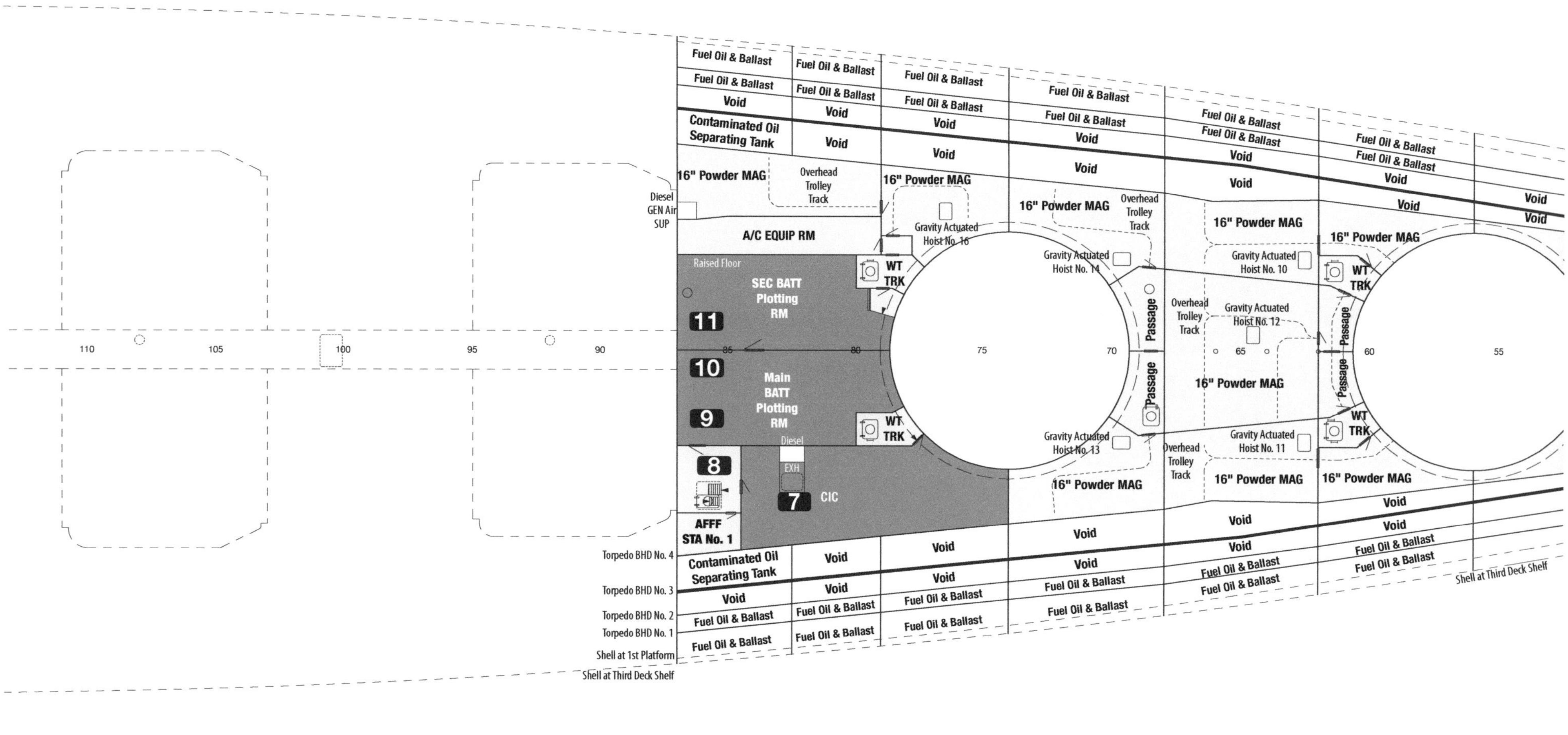
Fuel Oil & Ballast
Void
Contaminated Oil Separating Tank
16" Powder MAG
Overhead Trolley Track
Diesel GEN Air SUP
A/C EQUIP RM
Gravity Actuated Hoist No. 16
Gravity Actuated Hoist No. 14
Gravity Actuated Hoist No. 10
Gravity Actuated Hoist No. 12
Gravity Actuated Hoist No. 13
Gravity Actuated Hoist No. 11
Raised Floor
SEC BATT Plotting RM
Main BATT Plotting RM
WT TRK
Passage
11
10
9
8
7
CIC
Diesel
EXH
AFFF STA No. 1
110
105
100
95
90
85
80
75
70
65
60
55
Torpedo BHD No. 4
Torpedo BHD No. 3
Torpedo BHD No. 2
Torpedo BHD No. 1
Shell at 1st Platform
Shell at Third Deck Shelf

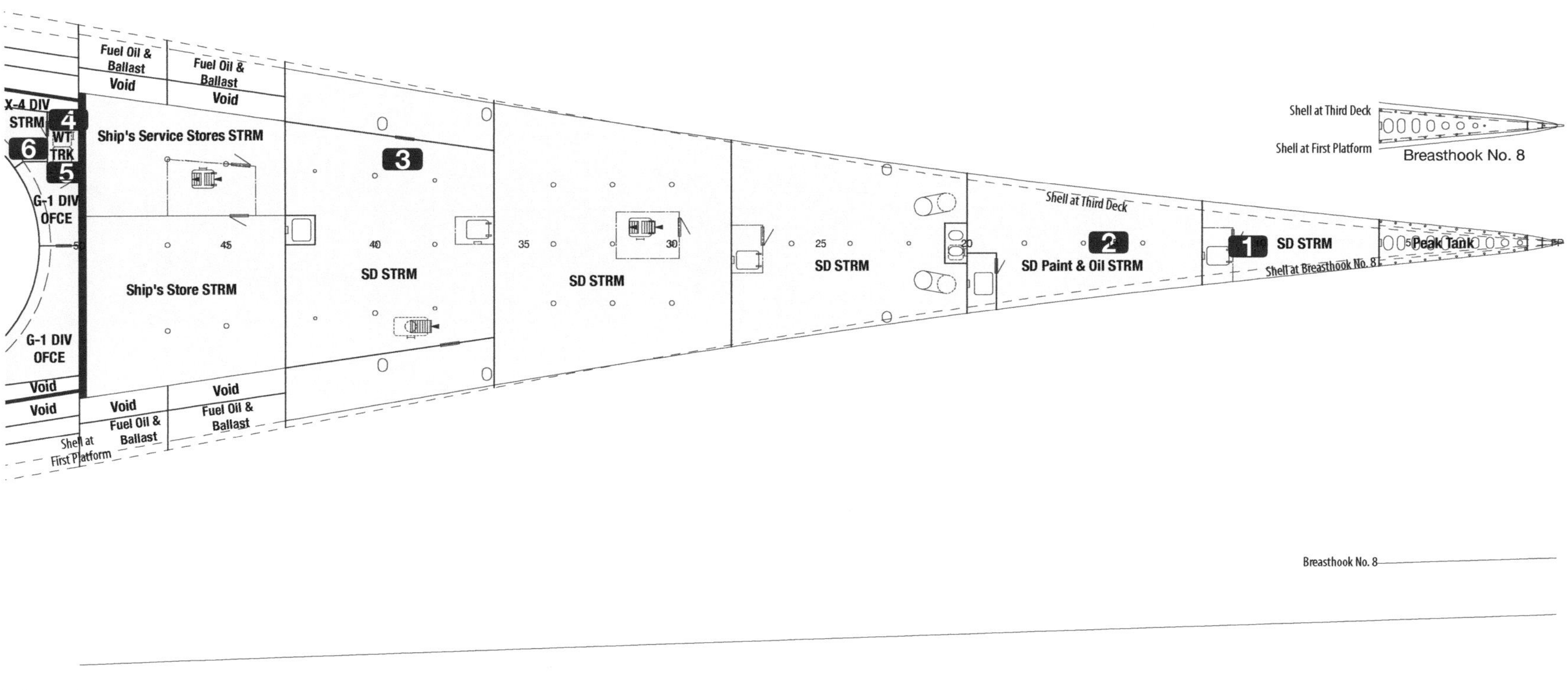

First Platform
1"=24'

(1) The framing in the forward supply division storeroom illustrates the shape of the hull at the bow. The hull at the bow narrows at the waterline and widens to form a bulbous bow near the keel. Some of the larger frames form arches in the storeroom. This view is looking forward. The curve of the frames shows this location is below the waterline.

(2) The paints and oils storeroom looking aft. The storage racks are sized to hold paint cans. The watertight trunk in the background provides the only access to this compartment.

(3) Torpedo bulkhead no. 3 extends to third deck from Frame 36–43. At first platform this bulkhead is structural but not watertight. The storeroom in the area extends through archways through the torpedo bulkhead here.

(4) Third deck is welded to a perforated plate welded to the aft face of the transverse armor plate. The perforations create additional welding surface. The inside face of the transverse armor is not face hardened so it would be welded. This is the underside of third deck viewed from first platform. The plans do not explain why the deck was not welded directly to the transverse armor. The perforated plate is also visible at third deck.

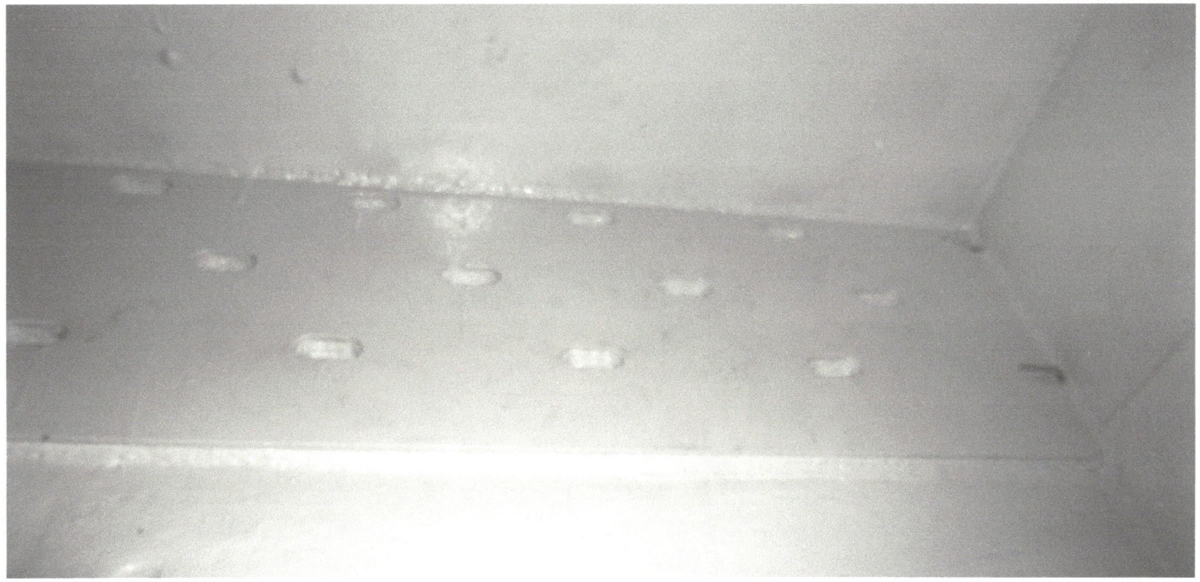

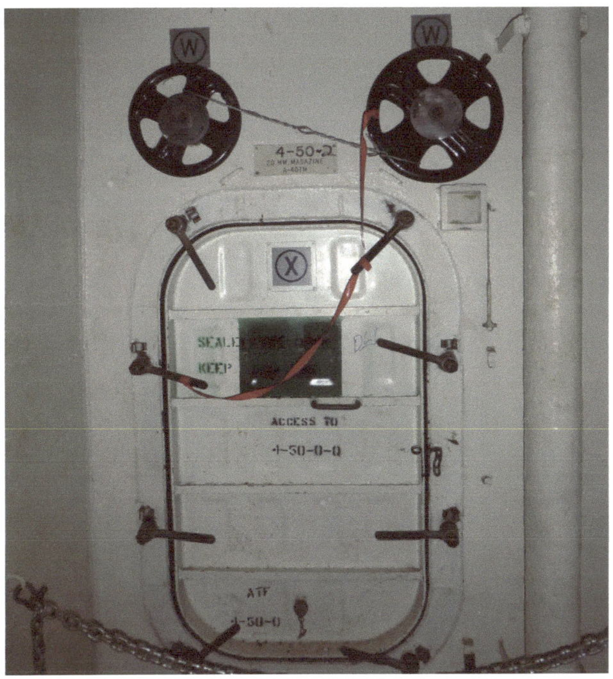

(5) This door leads from the watertight trunk to the G-1 division office. A the sign above the door indicates, that compartment was originally a 20mm magazine. Even though the 20mm mounts were removed in 1952, the sign remained in place.

(6) This is the X-4 (administration) Division storeroom looking aft. The G-1 (turret no. 1) Division office at the port side looks similar. This photograph shows a key structural feature of the turrets. The circular bulkhead is the fixed turret no. 1. The rotating turret rests on this support. The circular Carlin beam at the overhead (third deck) transfers the weight of the barbette support and barbette to the fixed turret.

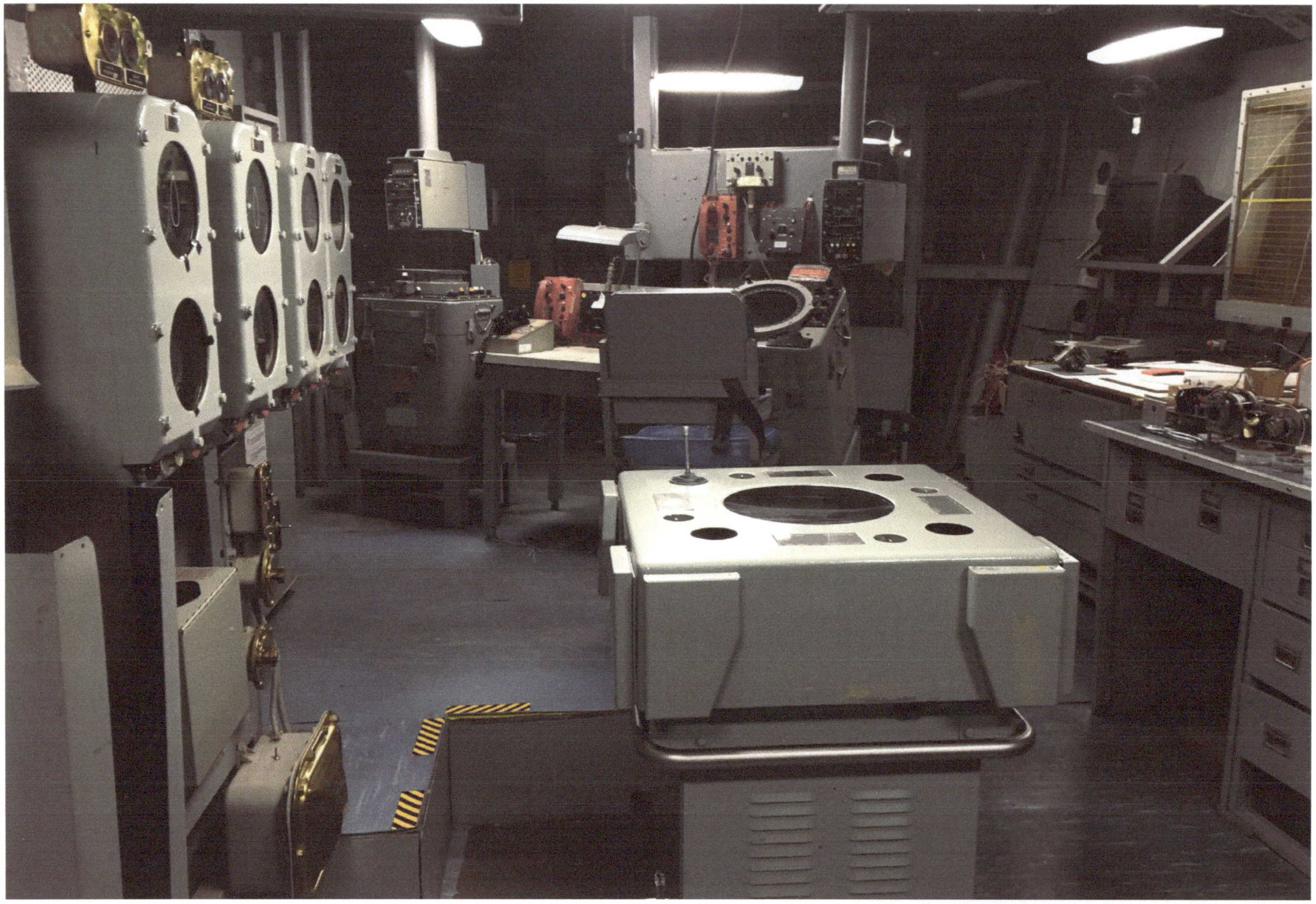

(7) The Combat Information Center (CIC) looking forward. This was the original tactical control center for the ship. Targets were tracked using radar and the crew here maintained the ship's situational awareness. The forward end of this compartment was originally a 40mm magazine. The dividing bulkhead was removed and CIC was extended forward in 1968 when the 40mm mounts were removed. The CIC assumed a backup role in the 1980s when the Combat Engagement Center (CEC) was created at the 02 level and took over the functions of the CIC.

(8) This ladder leads from first platform to the forward emergency diesel generator room. The deck hatch is mounted within a removable plate. That plate could be removed to allow larger machinery to be moved in or out of the diesel generator room.

(9) The forward main battery plot looking forward. There is an additional main battery plot on third deck aft. The Gun Fire Control System (GFCS) Mk.38 was the system that handled fire control for the 16-inch main battery. The GFCS incorporated components within the turrets, the Directors Mk.38 at the tops of the fire control towers, and radar. The heart of the system was the main battery plotting rooms. These rooms are where the 16-inch guns were normally aimed and fired. The two plotting rooms had the same capabilities so one could take over if the other were put out of action. The ship tended to switch between the two plot rooms to keep both crews in practice. In theory, the 16-inch turrets could be divided between the plot rooms. For example, if only the two forward guns could be brought to bear on the primary target in a naval engagement, turret no. 3 could have been directed to a secondary target by the other plot room. However, this situation never arose and the same plot room normally controlled all the turrets.

The units at the center of the room (from left to right are) Stable Vertical Mk.41, Rangekeeper Mk.8, Mk. 13 radar repeater, and Computer Mk.48. Fire control switchboards for configuring the system run along the bulkhead. The stable element is a gyroscopic instrument that supplied ship's pitch and roll measurements to the rangekeeper. The firing keys for the guns are on the stable element. The left key activates the salvo alarm letting the ship know the guns are going to be fired; the center key tells the system to fire the guns when it detects all is ready; and the right key fires the gun immediately. In operation, the key with the bumps would have been installed in the center position.

The museum replaced a door in the bulkhead at the left with the arch to make movement easier for visitors. An EM Log for measuring the ship's speed is on display at the right rear.

(10) This is the forward main battery plot viewed from the inboard side. The Stable Vertical Mk. 41 is a the lower right and a repeater for the SPS-10 surface search radar added in the 1980's is at the lower left. The Rangekeeper Mk.8 is at the center. The large hand crank on top could power the rangekeeper in case of a power failure. The other dials are for entering parameters. Next to the range-keeper is the repeater for the radar Mk.13 in the fire control towers. Next to that is the Computer Mk. 48.

The purpose of any fire control system is to have ordnance and the target arrive at the same point at the same time. In a naval engagement, both the source and the target are moving constantly. At short ranges, gunners could estimate the amount of lead to hit a target but the aiming problem (known as *rangekeeping*) became more difficult as engagement ranges grew. In the Spanish–American War of 1898, battleships fought targets at 2,000 yards. At the World War I Battle of Jutland in 1916, battleships were fighting at 15,000 yards. Navies responded to increasing ranges by developing mechanical computers to aim guns in the early 20th century.

The U.S. Navy adopted the first of the Rangekeeper series of mechanical computers developed by the Ford Instrument Company in 1916. The Rangekeeper tracked the movement of the target and its own ship independently. Consequently, changes in its own ship's course or speed (which were fed into the rangekeeper automatically from the ship's log and compass) had no effect on accuracy. Through various mechanical devices, the Rangekeepers made aiming calculations using arithmetic, trigonometry, and calculus. As the Rangekeepers grew in complexity, they also grew in size. *Iowa*-class battleships used the Rangekeeper Mk.8 in the plotting rooms and the smaller (and older and less capable) Mk.3 in the turrets.

The gun directors supplied the target range, bearing, and speed to the rangekeeper. The GFCS Mk.38 was designed before radar became available so optical range finders were the original source of this information. However, radar was quickly incorporated into the system. The main- and secondary-battery directors on *Iowa*-class battleships had radar from the beginning so the optical systems generally served as backups. Once an initial target fix had been entered, the rangekeeper would order the directors where to look for the target. This feature would have been more important in the absence of radar where the target could become temporarily obscured causing the director to lose track of it.

The main function of the rangekeeper was to calculate where the guns had to be aimed in order to hit the target. The aiming point was continuously calculated. Changes to input parameters entered into the system (such as wind speed) immediately updated the aiming point. The gun aim was transmitted to the guns in the form of elevation and traverse orders. After the guns were fired, the rangekeeper would activate a buzzer in the director a few seconds before the projectile arrived. The director would observe the fall of the salvo (both visually and by radar) and send aiming corrections to the plot room that would be entered into the rangekeeper.

The Rangekeeper Mk.8 was designed to aim at moving targets at sea level. Nearly all of the targets *Iowa*-class battleships engaged were on land. While the Rangekeeper could handle land targets it was difficult to use for that role. At the same time, mechanical computers could not be reprogrammed like a digital computer can be. The Computer Mk.48 was developed after World War II for indirect fire against land targets and could be used in place of the Rangekeeper. The Computer Mk.48 is an electronic analog computer where values are represented as voltages. Rather than tracking the target directly, the computer tracked the location of a fixed, known visible point on shore. The target was designated using its distance and bearing from that fixed location. The computer calculated the aiming point from these two distances and bearings. USS *Iowa*, USS *New Jersey*, and USS. *Wisconsin* had a Computer Mk.48 installed in their forward primary plot during the 1950's. USS *New Jersey* had a second Mk.48 added to her aft plot in in the 1960's. All the *Iowa*-class battleships had two Mk.48 computers installed during their 1980's upgrades.

(11) The Gun Fire Control System (GFCS) Mk.37 handled fire control for the 5-inch guns. Similar to the GFCS MK.38, the Mk.37 incorporated radar and directors Mk.37, with the heart of the system being the plotting rooms. The GFCS Mk.37 could also control the 16-inch guns in an emergency. However, it was limited in range and this appears to have only been done during testing.

This is forward secondary battery plot. This room duplicates the aft secondary battery plot at third deck. The secondary battery plots normally controlled the 5-inch guns.

The unit in the foreground is a Stable Vertical Mk.6. This unit is similar to the Stable Element Mk.41 in the primary plot rooms. It supplied the ship's pitch and roll angle to the fire control system and mount the firing keys for the guns. The left firing key sounded the salvo alarm on the ship; the center key (shown correctly with the bumpy trigger) fired the guns when the fire control system determined all was ready; and the right trigger fired the guns immediately.

Behind that is a Computer Mk.1A with a Starshell Computer Mk.1 mounted on top. There is a second Computer Mk.1A in the background and a second Stable Element Mk.6 is behind but it is obscured. Both computers could operate simultaneously. The Computer Mk.1A could address both surface and air targets. The level of complexity of this machine led the navy to call it a *computer* rather than a *rangekeeper*. However, the manufacturer's plaques on these units calls them rangekeepers.

One of the functions of the Computer Mk.1A not present in the Rangekeeper is the ability to calculate fuze settings and transmit those to the fuze setters in the projectile hoists. When firing against aircraft, 5-inch projectiles were unlikely to hit their targets directly due to the distance and speed involved. Instead, the projectiles were set to explode near their targets. Similarly, starshells had detonate above their target. The fuze setter wound a clockwork timer in the fuze to adjust the time between firing and detonation. The Navy introduced VT (proximity) fuzes in 1943 that exploded when the projectile was near the target that did not need to be set.

When attacking surface targets at night, it was normal to first illuminate the target with a starshell. The starshell computer took the input values from the Computer Mk.1A and adjusted the aim and fuze timing so that the projectile would explode above the target and lower a flare on a parachute.

Unlike the 16-inch turrets that generally operated in unison, the 5-inch mounts were usually divided among plot rooms and computers within the plot rooms. The system was capable of engaging four targets at once. In fact, the siting of the 5-inch mounts made it impossible for them to all fire at the same target.

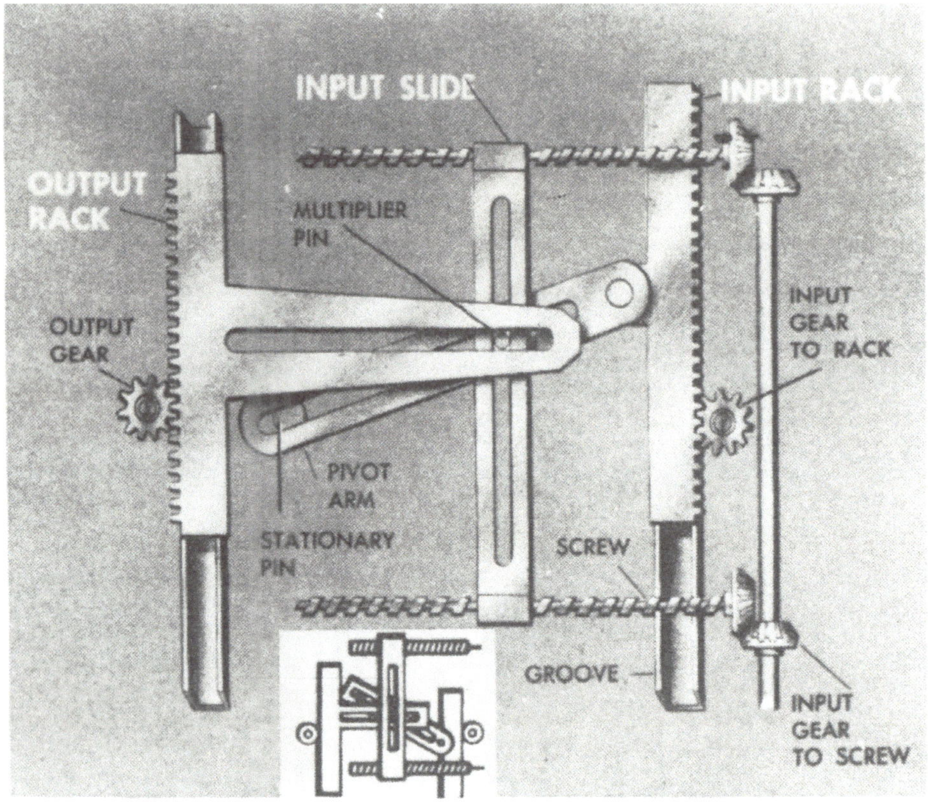

Electro-mechanical analog computers, like the Rangekeeper Mk.8 and Computer Mk.1A, use the position of rotating rods to represent values. This is similar to the way a heat-setting dial on a stove represents temperature. Rotating a rod in one direction increases its value and rotating it the other way decreases the value. The computers combined various mechanical devices that implemented arithmetic operations, trigonometric functions, and even calculus.

This diagram shows one way of implementing multiplication in a mechanical computer, as in X×Y=Z. In this diagram, position of *Input Gear to Screw* represents the X value, *Input Gear to Rack* represents the Y value, and *Output Gear* is the Z value. As the X and Y input values change, the Z value updates with them. The precision with which such a device is manufactured determines the precision of its calculations.

The *Stationary Pin* marks the location of zero. If either the *Input Slide* (connected to X) or *Pivot Arm* (connected to Y) is aligned with that pin, the *Output Rack* is also aligned with the pin, reflecting the fact that anything multiplied by zero is zero.

In the arrangement shown, Y and Z can have positive or negative values while X can only have a positive value because the *Input Slide* cannot move past the *Stationary Pin.* If this arrangement were changed so the *Screws* and *Pivot Arm* were extended to the left of the *Stationary Pin*, X could have negative values as well.

Complex calculations were implemented by connecting the output rods to the input rods of another operations. In an analog computer, each operation added to the size and weight of the machine. The Computer Mk.1A weighed over 3,000 lbs.

(12) A weapons department storeroom. When the ship was constructed, the plans often indicated that openings were to be cut in bulkheads so that machinery could be moved through. These openings were welded closed after the machinery was installed. The bulkhead at the center shows an example of this. When USS. *New Jersey* was built, welds were not always ground flush so the former opening locations are visible. The bulkheads to the left are the watertight trunk leading from third deck to the aft emergency diesel generator room. At the right is the Carlin beam that transfers the weight of the barbettes to the turret support for turret no. 3.

(14) During 1980's and 1990's the marine detachment was equipped with 0.50 caliber M-2 machine guns. These guns could be mounted at positions along the main deck to defend against small targets. This was the magazine for 0.50 ammunition. The ammunition was stored in boxes holding linked belts. This rack could hold a significant number of ammunition boxes. Originally, this was a magazine for 40mm ammunition.

(13) An explosion in a power magazine was the greatest threat a battleship faced. During World War II, USS *Arizona*, HMS *Barham*, HMS *Hood*, *Roma* and *Yamato* all suffered magazine explosions resulting in large losses of life. *Iowa*-class battleships were designed with multiple powder magazines at first and second platforms. The powder bags were stored in sealed canisters that, in the event of a magazine fire, were expected to delay ignition and give time to flood the magazine to put out the fire. By having multiple magazines serving each turret, the loss of one would not put the turret out of action.

This is the port first platform magazine for turret no. 3. When the ship was in service the aluminum racks would have been stacked from deck to overhead with powder canisters holding three bags. The only access to the turret is at second platform. Powder bags would be removed from canisters and slid along the brass tray to the gravity-operated hoist at the center. The weight of the powder bag would send the hoist gently down to second platform where it could be moved into the turret. Brass trays were used because they would not cause sparks.

The overhead trolley tracks were used to move the powder canisters because a full canister was too heavy for a person to move. The powder trolley tracks are narrower than the projectile and machinery tracks. The bend at upper center of the photograph is a removable track section. It could be replaced with a straight section that would connect with the track at the right.

(15) This was a marine's storeroom after the 1980's. It appears that clothing was stored hanging on the rack shown. Originally, this was part of the aircraft bomb magazine. During the 1940's, *Iowa*-class battleships carried seaplanes (Vought O2SU Kingfisher and later Curtis SC-1 Seahawk). These aircraft were capable of carrying bombs.

(16) Refrigerated food storerooms are located between Frame 166–181. The storerooms were kept at different temperatures, optimized for their contents. This is the starboard fruits and vegetables storeroom that was kept at 40 degrees F. Cooling coils cover the bulkheads. In service, the deck was covered with grates to allow cool air to flow under the stored food. The grates have been stacked along the bulkhead to right. The end of torpedo bulkhead no. 4 is to the left.

(18) Meat was stored at 15 degrees. Its temperature had to be raised above freezing before it could be butchered. Meat was transferred from storage into this thawing room kept at 40 degrees before being brought to the butcher shop. This view is inboard with the access trunk to the right.

(17) The starboard meat storeroom. This may look a lot like photograph (16) but there is a major structural difference. The belt armor forms the bulkhead to the left. Welding was rather new to ship building at the time *Iowa*-class battleships were built. The hulls of *Iowa*-class battleships were constructed using an odd mixture of welding and riveting, with riveting dominating. In many cases the approach to welding mirrored the method used in riveting. Rather than butt weld armor plates together directly, cover plates were placed over the joints on both sides and welded to the armor. The cover plates have scalloped edges to create a larger welding surface. This photograph shows scalloped cover plates used to weld the armor plates together and scalloped angles used to weld the armor to the deck. The armored plates here are 13.5" at third deck with the outer face tapering at knuckles. The thickness at the bottom depends upon the location of the plate. The belt armor here also serves as torpedo bulkhead no. 3.

(19) The refrigeration machinery room. The machinery sits on an elevated platform. Originally, this compartment held ice making machinery and the ice was used to cool the refrigerated storage compartments.

(20) The damage control workshop. All of the equipment was removed when the ship was retired.

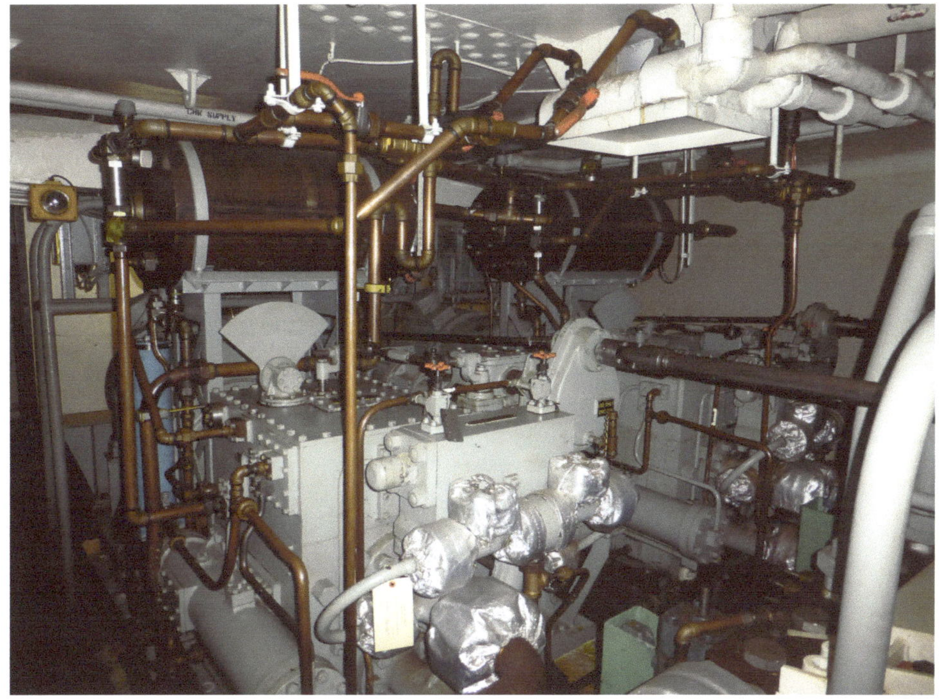

(22) These are the port side A-end pumps for the port rudder. The large tanks are hydraulic fluid reservoirs.

(21) The port rudder control. This is one of the four places where the ship could be steered. There are duplicate systems for powering the rudders. Only one could be operating at a time with the other serving as a backup. There is a control wheel for each system. Such controls that were directly connected to the steering motors were known as *trick wheels*. The electric motors that drive the hydraulic pumps are immediately behind the wheels. This equipment is duplicated on the starboard side for that rudder. From these locations the two rudders could be operated independently so that they could be turned in different directions for braking in an emergency.

(23) The port rudder B-end machinery looking aft. Each rudder has two pistons that can move fore and aft. Each piston has two cylinders; one at either end of the piston. Rods that connect to the rudder are attached to the top and bottom of each piston in the exposed area between the two cylinders. The two forward cylinders are cross connected.

To turn the rudder, the A-end supplies pressure to one of the aft cylinders. The cross connection creates pressure at the forward cylinder at the other side. This causes the pistons to move in opposite directions.

(24) This is the port rudder crosshead. It is at the very top of the rudder structure. The rudder extends 26'-6 $\frac{1}{2}$" below the hull at this point. The pistons on both sides have two rods that apply torque to opposite ends of the crosshead to turn the rudder. The two pipes extending from the crosshead were installed during the mothball process to immobilize the rudder.

(25) This is the aft end of the armored citadel. The belt armor to the left is 13.5" Class A, the transverse armor to the right is 11.3" of Class A, and the horizontal armor above is 6.2" of Class B. This is an area where Class A armor was used structurally. It forms the watertight compartment around the steering machinery and the end of torpedo bulkhead no. 3. The Class A armor here is thick enough that it could be welded in the inside faces using scalloped plates without cracking the outside face. The Class B deck armor is bolted to the belt and transverse armor. The curved plates at the top were welded along the joint to make the joint watertight. The pole at the bottom of the photograph immobilizes the rudder.

(26) When USS *New Jersey* was constructed it was equipped with four 2,260 gallon fuel tanks for aviation gasoline. The fuel lines ran up a trunk to a filling station at main deck. When the USS *New Jersey* was reactivated in 1968, the helipad was erected over the trunk opening. The fuel pipes were rerouted to the port stern tub that formerly held a 40mm mount. The fuel system was modified to handle the JP-5 jet fuel used by turbine-powered helicopters.

The vertical keel divided the fuel tank compartment in half. There were two tanks on each side and there were arches in the keel to allow access between the two sides. In the 1980's the starboard fuel tanks were removed, the arches were sealed, and the entire starboard side of the fuel tank compartment became fuel tanks. This more than doubled the ship's aviation fuel capacity. These are the two remaining original fuel tanks.

(28) The pumps for the fuel system are at the port side of the compartment. Their location here precluded converting the port side of the compartment into a tank as well. There are three pumps. The one farthest to the left is a service pump added during the 1980's reactivation to send fuel to the main deck. The pump at the center is a stripping pump for unloading fuel and the one at the right is a fuel transfer pump.

(27) There is no deck in the aviation fuel tank compartment. The tanks are mounted on supports and walkways run around the tanks. One can see the ship's outer hull in this compartment.

(29) This is a view up the aviation-fueling trunk. Forward is to the right. The main deck hatch has been welded shut and a battle lantern has been attached to it. The trunk widens at third deck to create a small control space.

Second Platform

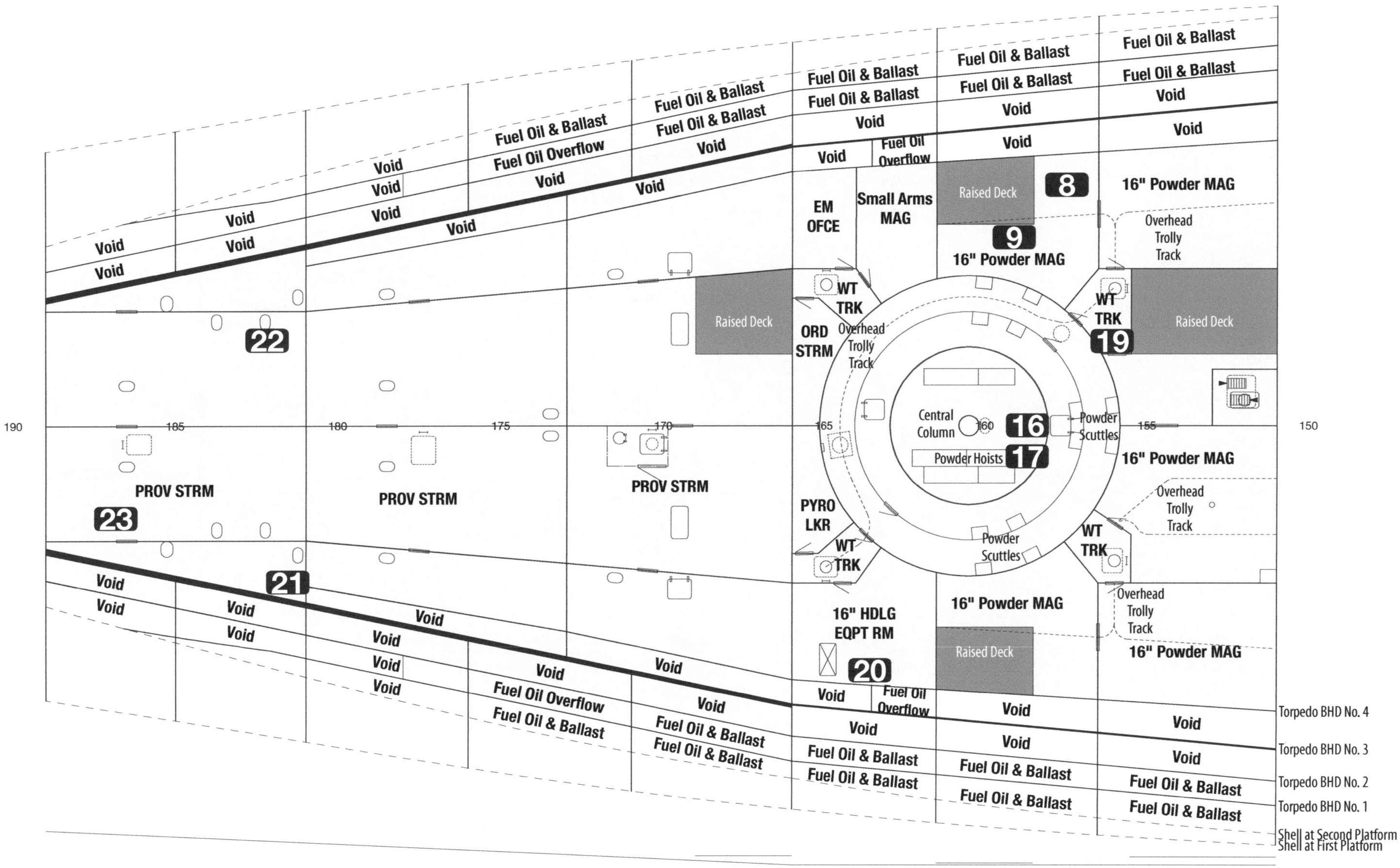

Second Platform
1"=18'

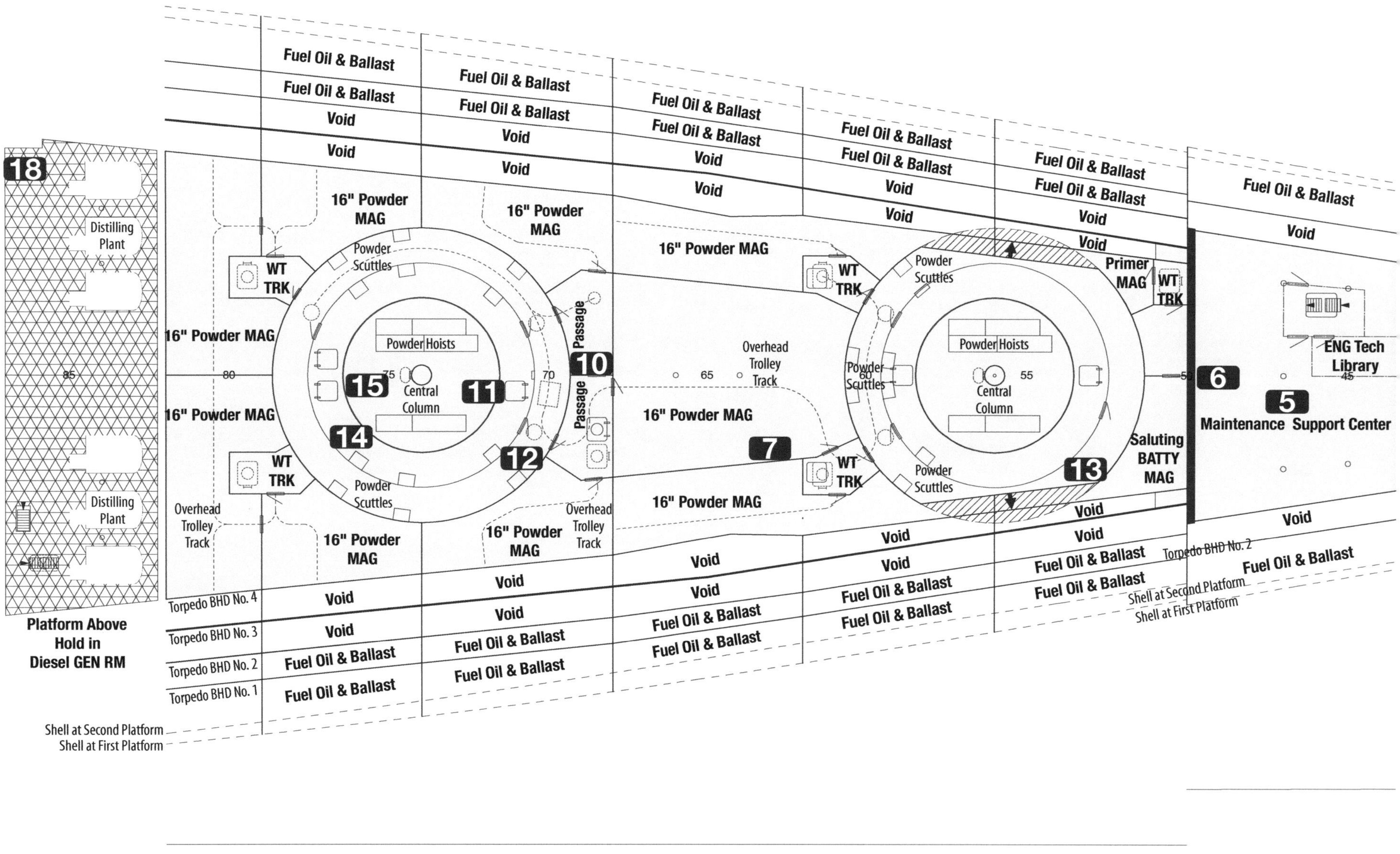
18
Distilling Plant
Distilling Plant
Platform Above Hold in Diesel GEN RM
Fuel Oil & Ballast
Void
16" Powder MAG
WT TRK
Powder Scuttles
Powder Hoists
Central Column
15
11
14
12
10
Passage
Overhead Trolley Track
7
13
Primer MAG
Saluting BATTY MAG
6
5
Maintenance Support Center
ENG Tech Library
Torpedo BHD No. 4
Torpedo BHD No. 3
Torpedo BHD No. 2
Torpedo BHD No. 1
Shell at Second Platform
Shell at First Platform

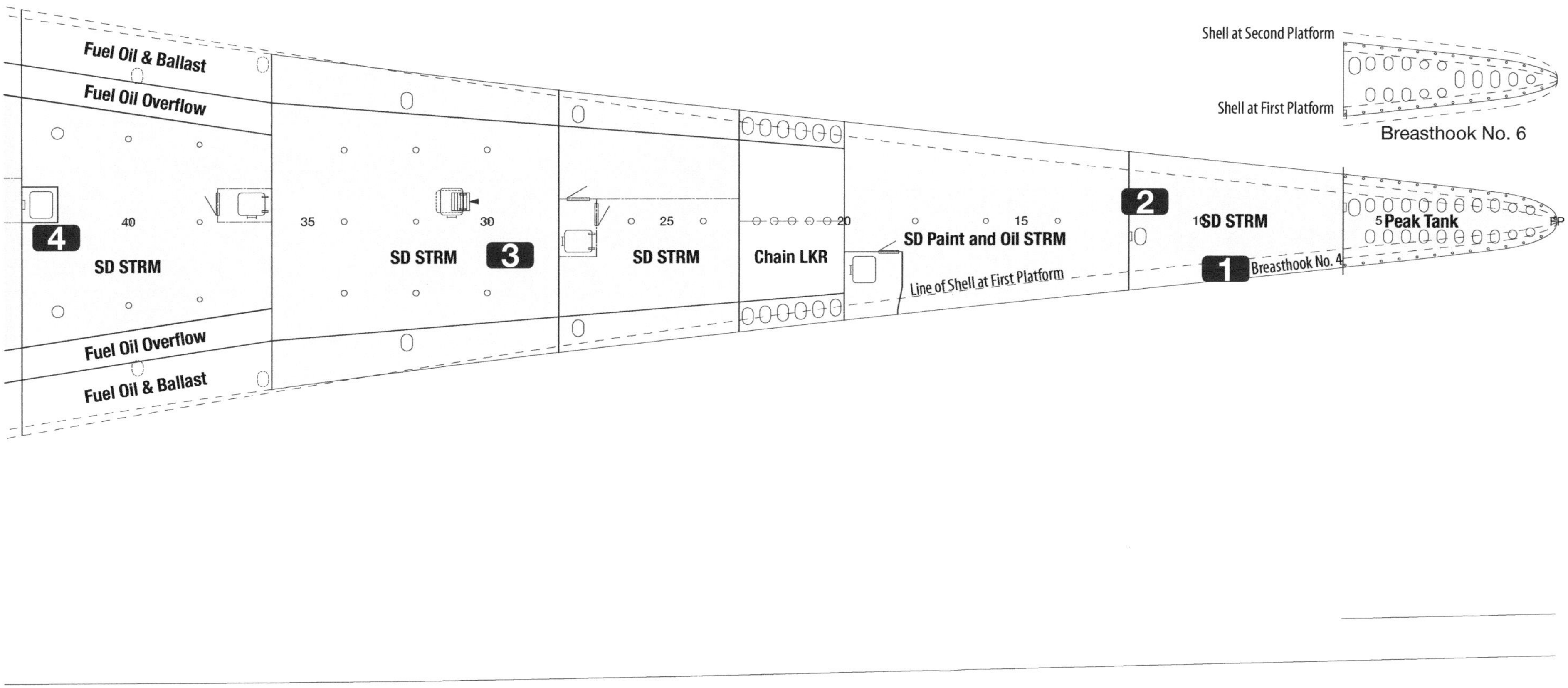

Second Platform
1"=18'

(1) The forward supply division storeroom looking aft. The curved framing follows the hull form. At the bow the hull narrows at the waterline and widens below to form a bulbous bow. The deck here is below the waterline where the hull widens. At the right are removable aluminum stowage racks. Such racks are the most common storage arrangement in storerooms. They are found in storerooms that held large items.

(2) The storage racks in the various storerooms reflect the type of material stored within them. This is the paint storage room. The stowage racks are sized to accommodate paint cans.

(3) This is another supply division storage room. This storeroom was used to hold small parts in numbered drawers.

(4) This storeroom extends between Frames 36–43. The large aluminum racks indicate that it was used to hold large items. The unusual detail of this storeroom is this access trunk that runs between Frames 42–43. This trunk connects second deck with the pump room. It just passes through the intermediate decks without any access so the trunk just forms a bump at the end of the storeroom.

(5) This area was a storeroom until the 1980s when it was converted into the S-1 (supply) division's maintenance support center. The furniture moved into this compartment was apparently taken from whatever source was available. The various desks date from the 1940's to the 1980's. A rotary dial telephone hangs on the stanchion at the right.

(6) The armored transverse bulkhead at Frame 50 closes off the aft end of the Maintenance Support Center. The forward face shown is Class A (face-hardened) so it could not be welded without risking cracking. The brackets supporting the first platform above are wedged into slots machined into the plate joints. Similar brackets support second platform from below. A scalloped plate that is wedged against the armor plate is welded to the second platform. This plate just uses friction with the armor to resist pulling up. The transverse armor plates start to taper towards the bottom just above the scalloped plate.

Many sources incorrectly state that Class A armor could not be use for structural purposes and that, unlike Class B, it is all dead weight. This is an example of where Class A armor was used structurally. It was difficult to use face-hardened armor structurally and doing so was generally—but not completely—avoided.

(7) One of three powder magazines for turret no. 1 at the second platform looking forward. The powder bags were shipped and stored in sealed canisters. Some such canisters are on display here. When the ship was in service, the canisters were stacked to the overhead. Each canister holds three 110 lb. powder bags. A full canister was too heavy for a man to lift. The overhead trolley tracks could be used to assist with moving the canisters.

When needed for firing, powder bags were removed from the canisters and placed on the brass trays shown. Brass was used because it would not cause sparks as the powder bags were slid towards the scuttles in the turret. The scuttle openings have both a watertight door and a flame tight rotating scuttle. One of the two watertight doors is open and the rotating scuttle is visible within the opening.

A gravity hoist is in the foreground. There is no access from first platform to the turrets. Therefore, gravity hoists were used to transfer powder bags from the first platform magazines to the second platform where they could be moved to the turrets through the scuttles.

The powder bags on display are inert training bags.

(8) This is the port powder magazine for turret no. 3 looking aft. Part of the deck is raised to accommodate shaft a bearing located below at the hold. The racks for powder canisters are empty. At the left there is a gravity hoist and trays for moving powder bags.

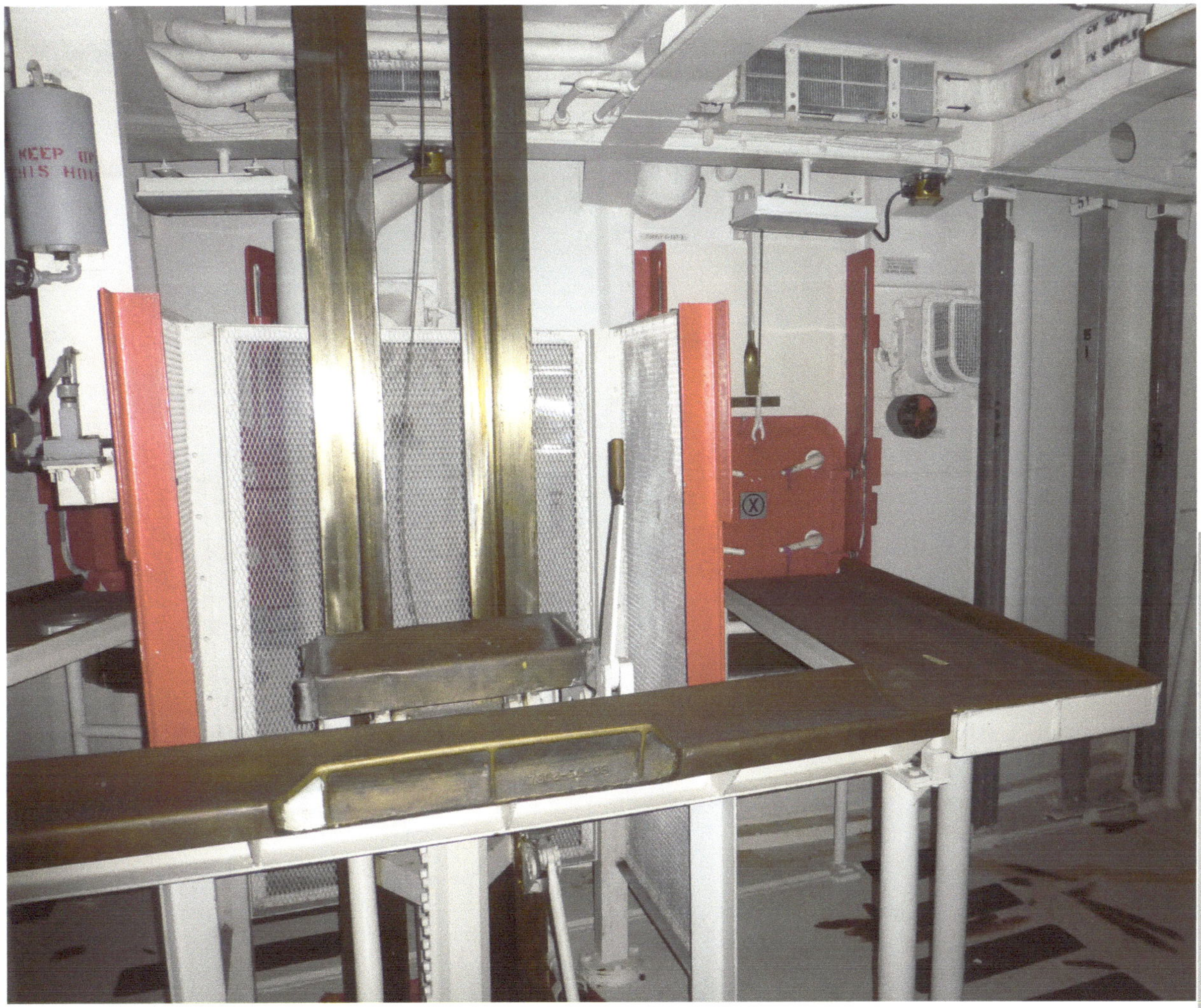

(9) This is a gravity hoist in the down position. The weight of the powder bag moved the powder downwards and slight resistance in the hoist provided a gentle ride. This hoist leads to a brass tray with paths to powder passing scuttles on either side of the hoist. This is the same magazine shown in photograph (8).

(10) This cylinder is the nonrotating turret support of turret no. 2. This bears the weight of the rotating turret and the barbettes. The butt plate shown joins $1\frac{1}{2}$" plates of the turret support. The turret supports on USS *Iowa* and USS *New Jersey* were riveted as shown. They were welded on USS *Missouri* and USS *Wisconsin*. The nonrotating turret rests on the second platform. The turret foundation is immediately below the second platform. The door leads to the powder passing ring.

(11) The forward end of the powder handling flat of turret no. 2 viewed to port from the shell storage platform. The powder handling flat is the only turret level aligned with a deck. The powder handling flat is divided into the three sections.

First, between the bulkhead visible here and the turret support there is a powder passing area. This area is isolated from the powder flat by flame tight doors and scuttles and is part of the nonrotating turret. The only way to access the turret from within the hull is through the doors to the powder-passing area.

Second, just inside the bulkhead shown there is another ring that is part of the nonrotating turret. In this photograph one powder-passing scuttle is visible. To the right is an access door. Next to the door there is a water immersion tank that could be used to render a broken powder bag inert.

Third, at the center there is a circular area with a radius of circle 9'-9" that is part of the rotating turret. This area holds the powder hoists. The division between the rotating turret and nonrotating turret is visible in the deck and there is no barrier between these two zones.

Each of the three hoists contains a car that moves up and down on a cable, similar to an elevator except that the path is sloped. The powder car has two levels that each holds three powder bags. The powder handlers moved powder bags from the powder scuttle and placed them as shown on shelves outside the hoist door. When the powder car arrived and the door opened, someone had to roll the six bags into the car. The navy allotted just three seconds in the loading cycle for the powder car to be at the bottom of the hoist so the powder bags had to be quickly rolled into the car when it arrived.

(12) The powder-passing area forms a ring around the rest of the powder flat. It surrounded by flame tight bulkheads that are designed to isolate the turrets from the powder magazine. This is the powder-passing ring of turret no. 2. High pressure air flasks are at the center. To the left is a powder scuttle leading to a magazine. Powder bags would pass through the scuttle one at a time. The operator would remove the powder bag from the scuttle then put it in a matching scuttle to the right that leads to the main area of the powder handling flat. The scuttles can only be open to one side so there is never a clear path from one side to the other. This arrangement places two bulkheads between the working areas of the turret and the magazines.

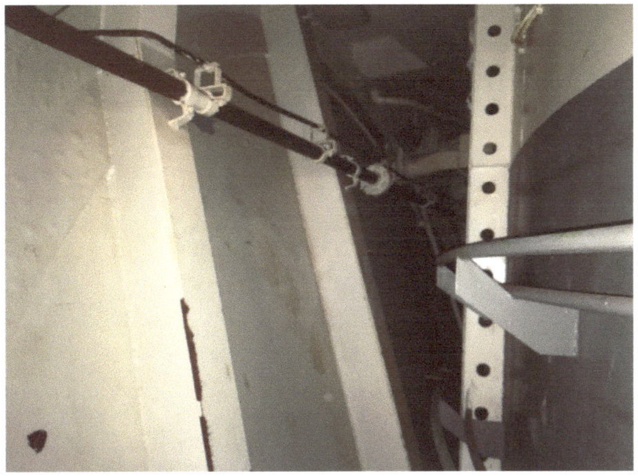

(13) The hull is extremely narrow around turret no. 1. In order for the turret to fit within the torpedo bulkheads the side of the fixed turret slopes inward as shown here (starboard side looking aft), dividing the powder-passing ring into forward and aft areas. The slope prevents one from walking between the forward and aft segment of the powder-passing ring but one could climb over the narrowest part.

(14) These are two powder passing scuttles in the powder handling flat of turret no. 2. The scuttles contain a rotating cylinder that has about a third of its side open. The crank at the side rotates the cylinder so that the open side can either face the powder passing ring or the powder handling flat. The scuttle at the left is open to the powder handling flat. The scuttle at the right is open to the powder-passing ring. When the scuttle opened, the powder bag rolled into the fold out tray. Powder men would grab the powder bag and take it to one of the trays at the powder hoists. Under the official manning requirements there would be nine powder men per turret. Each one would have to move a 110 lb. powder bag every fifteen seconds to maintain the official firing rate of once every thirty seconds.

(15) The rear of the powder-handling flat of turret no. 2. At the center are the booths for the powder hoist door operator for the center and left powder hoists. The door operator opened the powder hoist door when the powder car arrived at the bottom of the hoist and closed it when the car was loaded and ready to be sent up to the gun. To the right is the central column that supports the rotating part of the powder flat and holds the wiring connections between the turret and the rest of the ship. A complex wiring connection was not required because the turret cannot revolve a full circle. The wiring bundle entering from below just needed to be able to twist from side to side.

The platform at the top provides maintenance access to the gypsy head drive gears for the lower shell flat. This platform is unique to turret no. 2 where the lower shell flat is two levels above the powder handing flat. The stairs at the left are a museum addition.

(16) This the central column for turret no. 3. Turret no. 1 is nearly identical. The powder handling flats in these turrets are not as spacious as in turret no. 2 because they lack the powder storage platform. The hatch at the top of the ladder opens into the center of the lower shell flat.

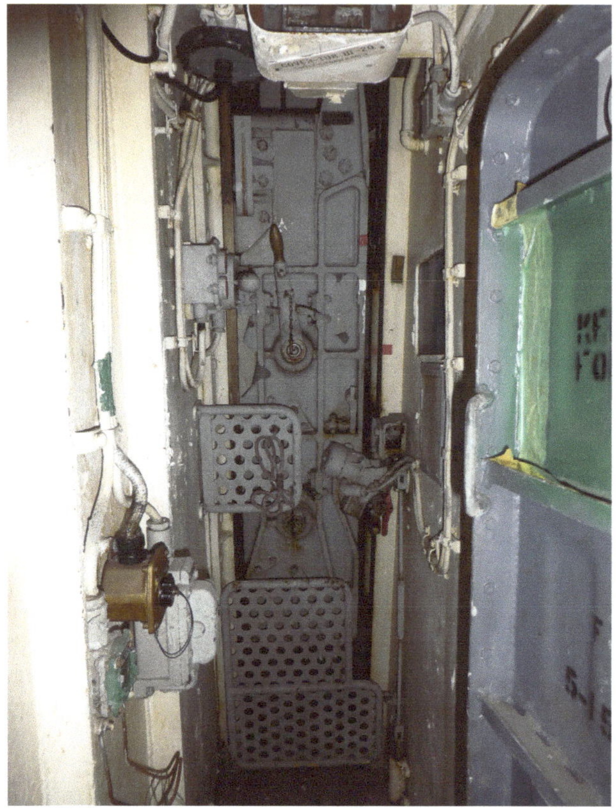

(17) The lower powder door operator's position for the left powder hoist of turret no. 3. The powder car is at the bottom position. The two vertical levers on the powder car (lower obscured) are for dumping the powder bags. They are operated by the powder hoist operator at the top of the hoist when the gun crew is ready to load powder. A cam unlocks the hoist doors when the powder car reaches the bottom and a foot pedal locks the doors before the car can travel upward.

The powder bags are about eighteen inches long, eighteen inches in diameter and weighed about 110 lbs. However, the term *powder bag* is misleading. The propellant within the bags was composed of grains that were two inches long and one inch in diameter. The grains were carefully packed in stacks within the powder bag. The only *powder* was a pad of black powder quilted to one end whose purpose was to ignite the bag. The quilting kept the powder from shifting to the bottom of the pad. This photograph shows powder grains being arranged in stacks and placed into bags. (Nat'l Archives)

(18) The main distilling plants are located within the forward emergency diesel generator room. This compartment is at the hold. However, distilling plants are mounted on an elevated platform that is roughly at the level of the second platform. This view is to starboard. There are three-stage distillers on both the port and starboard sides of the platform. The system used steam from the propulsion plant assisted by low pressure to boil sea water. Water that remained after the third stage was concentrated brine that was returned to the ocean. The steam condensed to become liquid water for the crew and propulsion plant.

(19) One of watertight trunks outside turret no. 3. All of the turrets have similar trunks. These trunks create vertical paths that were used to move ammunition between the magazines and the main deck. A bracket for an overhead trolley track is visible at the top of the photograph. The track segment has been removed and stowed on the bulkhead at the center to allow the turret door to be closed.

(20) The 16-inch Handling Equipment room looks like it has been untouched since the ship left service. At the sides there are hand trucks that were used to move projectiles and powder canisters on the main deck to the trolleys. Stacked in the center are shell carriers that wrapped around 16-inch projectiles and created hookup points for winches and trolleys. To the left of the aluminum racks there are boxes of trolley cars with their integrated winches.

(21) This is the bottom of the belt armor. Between Frame 155–189 the belt extends from second platform to third deck. The vertical scalloped plate is welded over a joint between belt armor plates. The inside face of the belt is flat. In this area the outside face of the belt tapers at knuckles. The armor in the area of this photograph tapers between 13.5" and about 5". The belt is welded to the deck using scalloped plates. There is a double layer of scalloped plates at the base of the armor. The bottom layer is a flat plate used to align the armor. The upper layer is a scalloped angle that is welded to the flat scallop and the inside face of the armor.

(22) Manholes give access to the twin keels that support the inboard propeller shafts. This is a view into the twin keel on USS *Iowa* from second platform. The tube for shaft no. 3 is at the center of the photograph. (Mike Getscher)

(23) The keel is the main longitudinal support of a ship. However, the beam of an *Iowa*-class battleship (about 108 feet) is too wide for the keel to provide sufficient longitudinal support on its own. As is common with large ships, *Iowa*-class battleships have additional longitudinal members that run alongside the keel. These are longitudinals at the start of the second platform where the hull shell rises at the stern. The longitudinals have to step down from the first platform to the second platform here. Without the slopes here, there would be stress points in the longitudinals at the step. The bulkhead at the right is located directly over and forms an extension of the vertical keel.

Third Platform

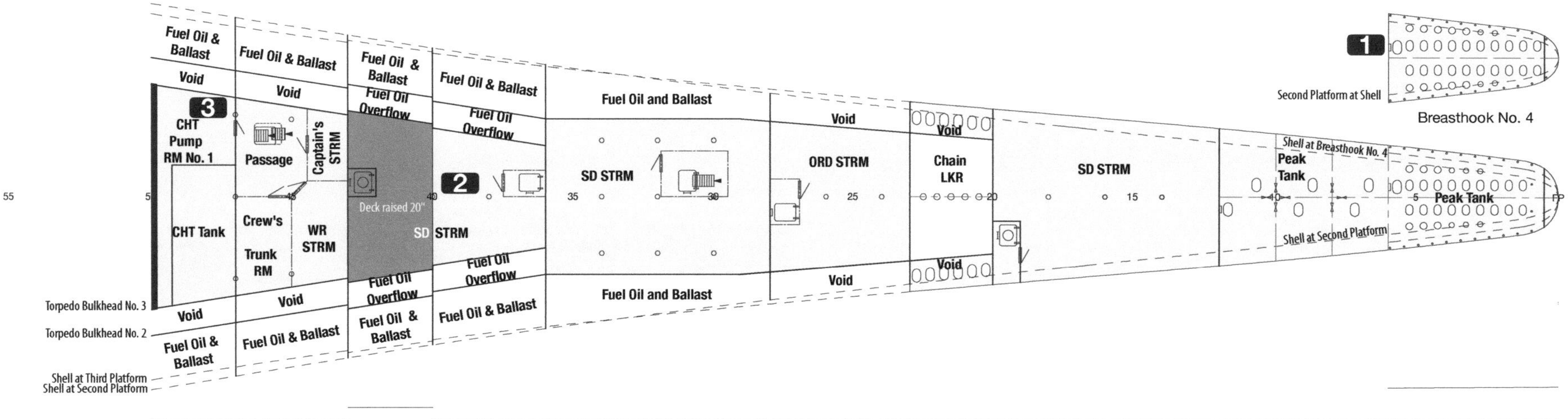

Third Platform
1"=24'

(1) Water could be pumped into or out of the peak tank to trim the ship. This is the port side of the peak tank at breasthook no. 4. A swash bulkhead divides the peak tank along the centerline to slow the side-to-side movement of water in the tank.

(2) This storeroom is located Frames 36–43. The overhead of pump room at the hold below divides the storeroom by raising the deck 20 inches. The watertight trunk that connects the pump room with second deck is in the background center. The trunk has no intermediate entrances. There is a hatch in the trunk at the level of the raised platform with a quick acting scuttle. The circular structure on the trunk is a recess for the hatch's opening wheel. Without the recess the hatch would be able to open fully.

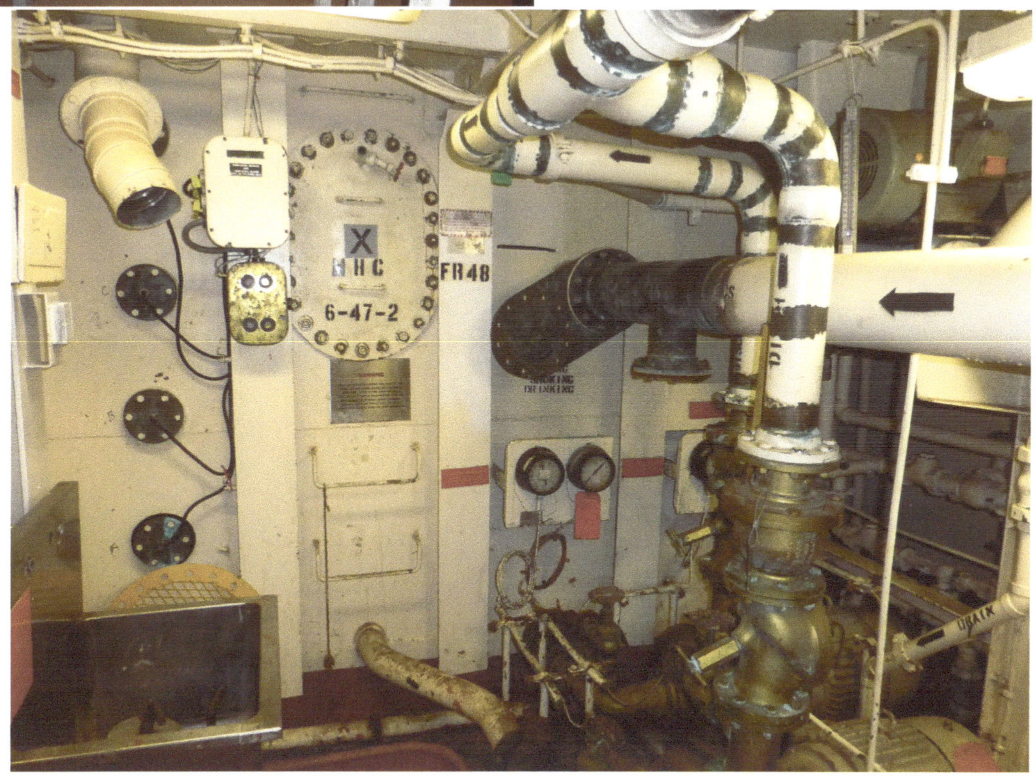

(3) Between the time USS *New Jersey* was built in the 1940's and its final reactivation in the 1980's there was a dramatic change in environmental laws. By the 1980's it was no longer legal for ships to send their sewage overboard while in port. One of the changes to USS *New Jersey* was the addition of collection and holding tanks (CHT) to store sewage until it could be discharged in deep water. This is one of the CHT tanks and some of its pumping equipment.

Hold

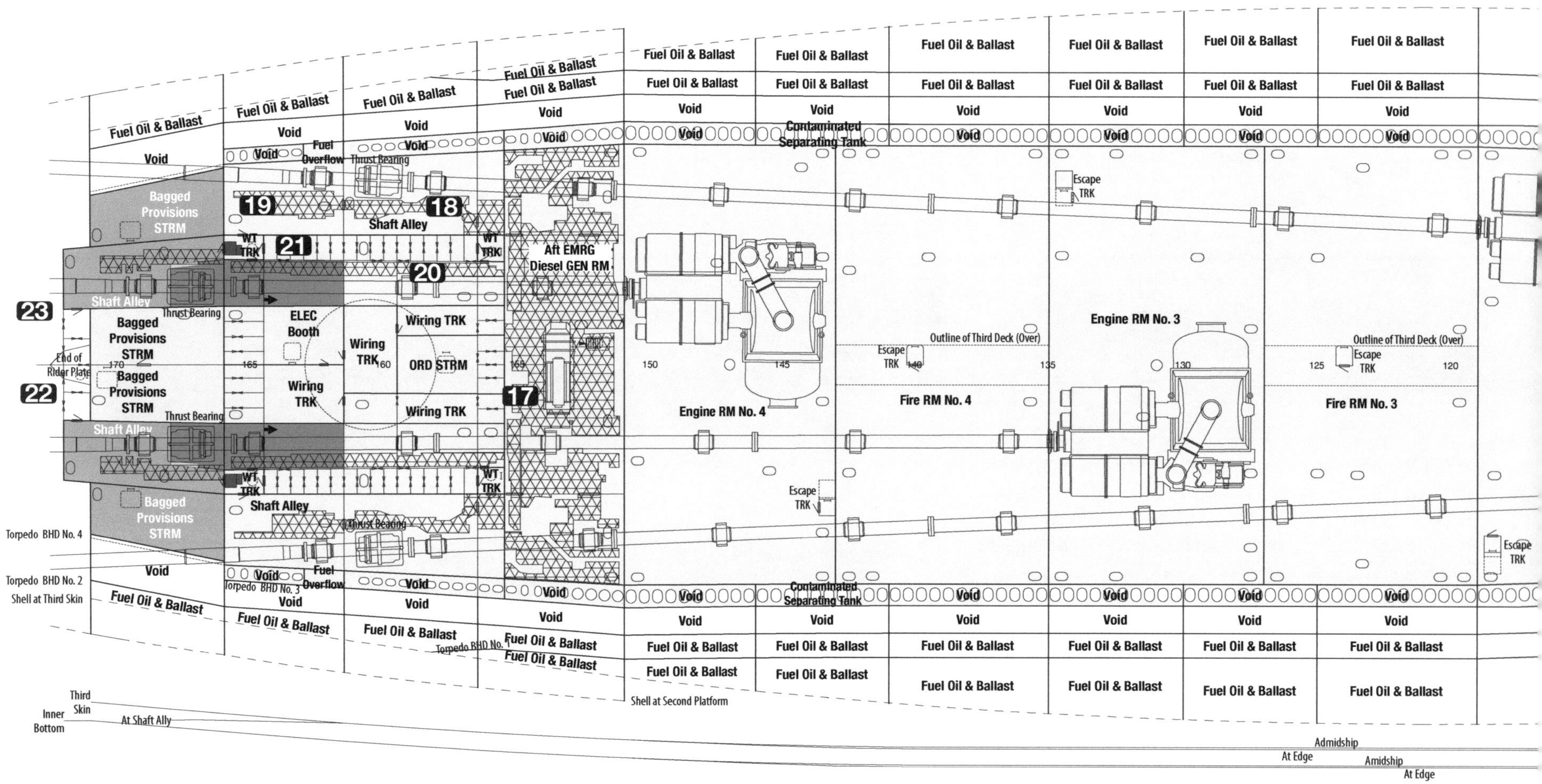

Hold
1"=24'

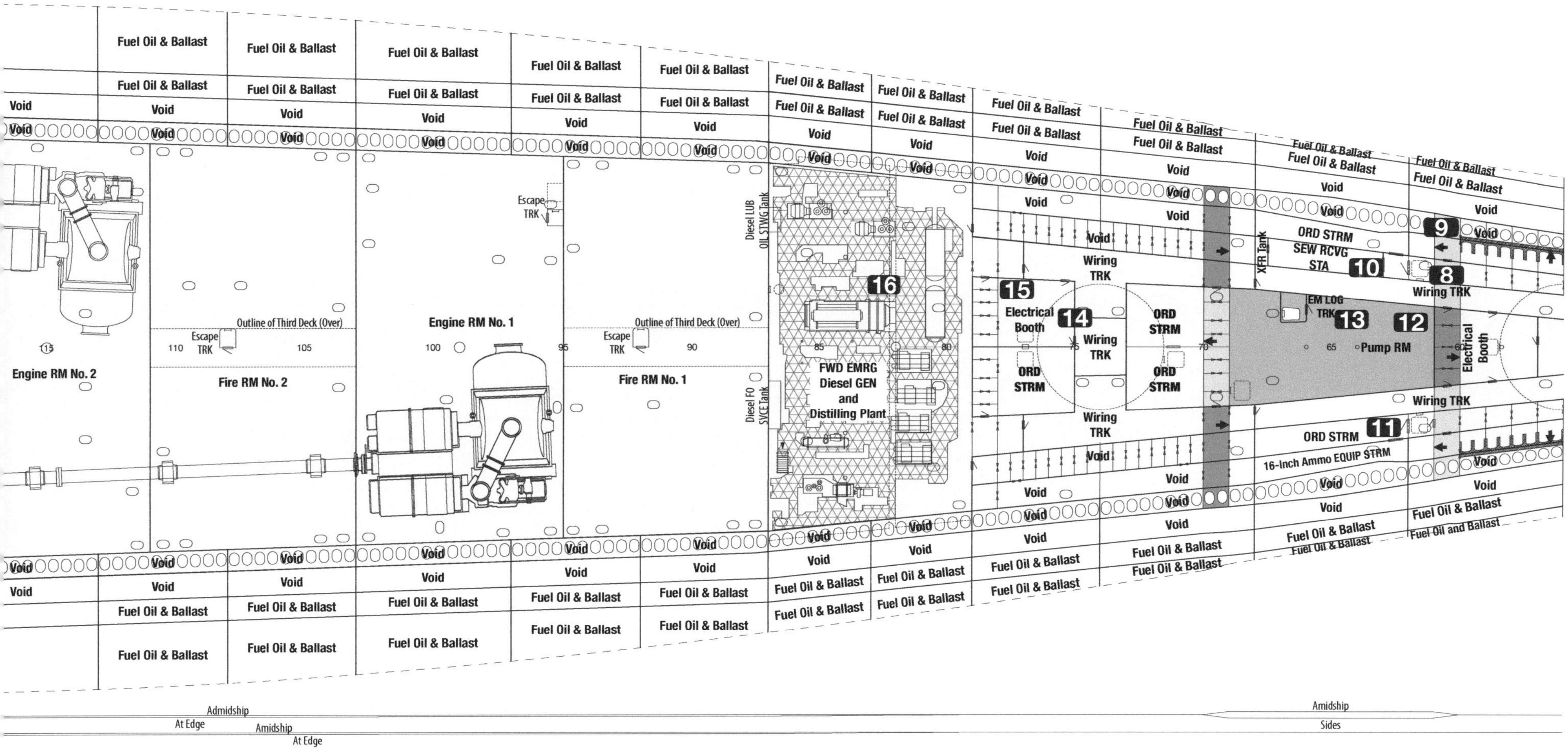
Fuel Oil & Ballast
Void
Engine RM No. 2
Fire RM No. 2
Outline of Third Deck (Over)
Escape TRK
Engine RM No. 1
Fire RM No. 1
Diesel LUB OIL STWG Tank
Diesel FO SVCE Tank
FWD EMRG Diesel GEN and Distilling Plant
Electrical Booth
Wiring TRK
ORD STRM
XFR Tank
ORD STRM SEW RCVG STA
EM LOG TRK
Pump RM
16-Inch Ammo EQUIP STRM
Fuel Oil and Ballast
16
15
14
13
12
11
10
9
8
Admidship
At Edge
Amidship
At Edge
Amidship
Sides

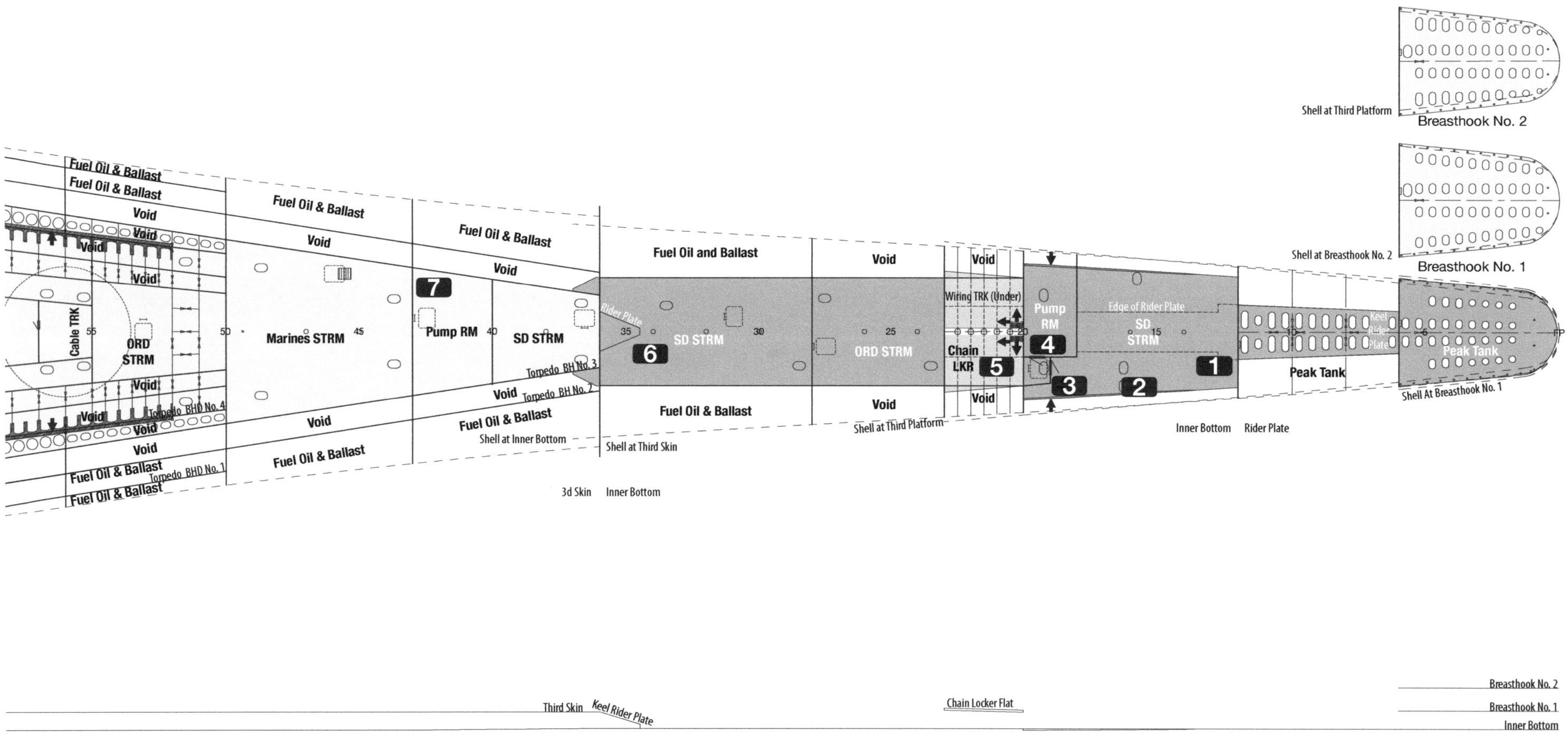

Breasthook No. 2
Breasthook No. 1
Shell at Third Platform
Shell at Breasthook No. 2
Fuel Oil & Ballast
Void
Cable TRK
ORD STRM
Marines STRM
Pump RM
SD STRM
Fuel Oil and Ballast
ORD STRM
Wiring TRK (Under)
Chain LKR
Edge of Rider Plate
Peak Tank
Keel Rider Plate
Torpedo BHD No. 4
Torpedo BHD No. 1
Torpedo BH No. 3
Torpedo BH No. 2
Shell at Inner Bottom
Shell at Third Skin
Shell at Third Platform
Shell At Breasthook No. 1
Inner Bottom
Rider Plate
3d Skin
Third Skin
Keel Rider Plate
Chain Locker Flat

Hold
1"=24'

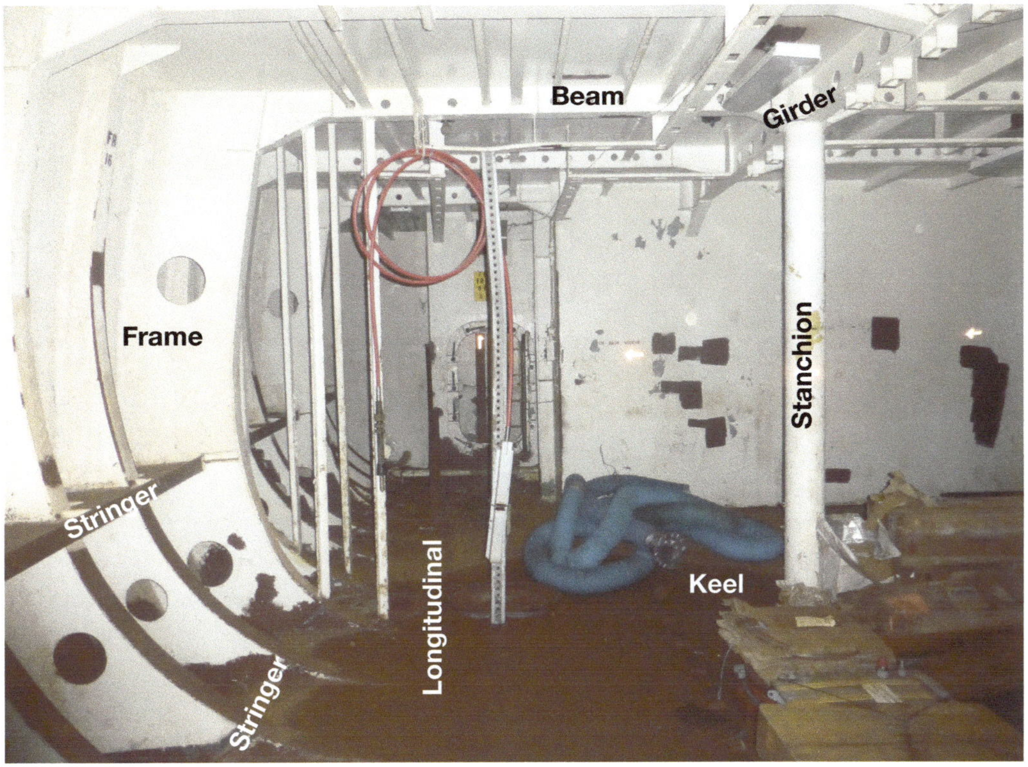

(1) The forward supply division storeroom illustrates several of the structural features of the hull. The view is aft with the watertight access trunk at the center and bulkhead separating the pump room at the right.

The supports for the hull shell are oriented in three directions. Framing provides the structural support for the shell parallel to the transverse axis. The framing forms loops where the sides are connected with floors running below the hold and beams running below the other decks. Beams are visible in this photograph. Frames are spaced every two feet in this area of the ship.

Along the lengthwise axis of the ship, the keel provides the main support to the shell. Longitudinals sit on either side of the keel to provide additional support in this direction. Longitudinal no. 8 runs through the center of this photograph with its outline faintly visible in the deck. In the background next to the door, this longitudinal rises 5'-1" above the deck to merge with the start of torpedo bulkhead no. 4 at the left (starboard) side of the access trunk. Girders provide longitudinal support to the decks. Stanchions support girders and beam mid-span. Finally, stringers connect the framing at the sides. Stringer no. 9 runs prominently along the left side of photograph. Stringer no. 10 runs along the edge of the deck.

(2) *Iowa*-class battleships have a triple bottom for most of their length. That means there are three layers at the bottom: (from outside to inside) the hull shell, the inner bottom, and the third skin. The three layers provide underwater protection, additional strength, and create storage tanks for fuel and water. At the bow the hull only has a double bottom with no third skin. In this area the hold deck corresponds to the inner bottom as at the forward storeroom. An unusual feature of the forward storeroom is that the deck edge merges with stringer no. 10 so the outer edge of the deck slopes downward.

(3) This is the view upwards in the watertight trunk at Frame 20. This trunk leads all the way to main deck. The netting was added in the 1980's as a safety measure to make it less likely that one could fall all the way down the trunk. The netting could be removed for transferring supplies through the trunk.

(4) Pump room no. 1. The machinery is positioned on raised platforms so that it can operate with water in the hold.

(5) This is starboard side of the chain locker looking upward. The chain pipe leads from main deck to the overhead at third deck. The ends of the chains are secured to the aft bulkhead. A partial, non-tight bulkhead divides the chain locker in half and separates the two anchor chains. Torpedo bulkhead no. 2 forms the outboard bulkhead and is at the left. The chain locker is on an elevated platform that is sloped towards a drain forward that extends to either side of the dividing bulkhead.

(6) The double bottom becomes a triple bottom at Frame 36. This is within another supply division storeroom. The structure at the center is the keel rider plate. For most of its length, the keel has three main components. First, there is the flat keel that runs along the bottom of the hull. The flat keel is six feet wide for most of its length but it narrows at the bow and stern. Second, the vertical keel rises from the center of the flat keel. Finally, the rider plate rests on top of the vertical keel. The rider plate is six feet wide for most of its length but it narrows at the stern and widens at the bow. Those three elements combine to form a giant I-beam. The rider plate follows the level of the hold. At the double bottom the rider plate is 3'-3" above the baseline of the hull. When the hull becomes a triple bottom, the rider plate rises to 5'-9". That is the transition shown here. The triangular extension of the upper rider plate overlaps the lower level of the rider plate here. The keel is visible underneath.

(7) Pump room no. 2. The view is forward and to port. The only access to this pump room is through a watertight trunk from the second deck. This trunk has no other openings.

(8) The closely spaced webs shown here are part of the port foundation for turret no. 1. The areas under the turrets are known as "the catacombs." (USS *New Jersey*, 1988 Cruise Book)

(9) The sides of turret no. 1 are truncated at the bottom with flat angled surfaces so it can fit within the hull. These angled sides are supported from below by a row of sloped arches at both sides. These are the arches at the port side looking forward from the only entrance to this space. The opening in the web in the background leads to another group of arches and then the passage comes to a dead end forward. Torpedo bulkhead no. 4 is to the left.

(10) During the 1980's reactivation this port side ordnance storeroom was converted into a sewage receiving station. The bolted plate manhole in the bulkhead in the background leads to a sewage transfer tank.

(11) This is the starboard ordnance storeroom. The sewage receiving station was originally a mirror of this compartment.

(12) In the 1980's the trunk shown was added to the pump room in order to house an electromagnetic (EM) log. An EM log measures the ship's speed. The log extends from the bottom of the hull and creates an electromagnetic field. As the ship passes through salt water, the field creates a potential difference. As the ship moves faster the difference in potential increases. A voltmeter measures the difference and that is translated into speed.

(13) This is the frame used to lower the EM log through the bottom of the hull. The EM log could be raised for cleaning or for protection in shallow water.

(15) The electrical booth under turret no. 2. The rotating turret extends below second platform, creating a very low overhead at the hold. The door shown leads to the wiring trunk shown in photograph (14). The open hatch at the upper right leads to the powder handing flat of the turret.

(14) This wiring trunk is at the bottom of the rotating section of turret no. 2. The mass of wiring shown runs up the central column of the turret. Because the turret cannot make a complete revolution, there is no need for a complex wiring connection between the ship and the turret. The wiring just needs to be able to move back and forth. The wiring is protected against chaffing and the edge of the opening is rounded. The gas ejection air supply runs through a swivel pipe that goes through the center of the column.

(16) The forward emergency diesel generator room looking aft with the generator to the left. The machinery in this compartment is surrounded by access platforms. A ladder to the upper platform mounting the distilling plants is in the background center.

(17) The aft emergency diesel generator room looking to port. The 250 K.W. emergency generator is at the center of the photograph. A gauge board is to the left. This is the only compartment that all four propeller shafts pass through. However, there is so much equipment that there is no place where one can view all the shafts at once.

(18) This is propeller shaft no. 4 looking aft. The large unit to the left is the thrust bearing. This is where the thrust from the propeller is transferred to the ship. At either end of the thrust bearing there are shaft bearings that support the shaft. One is located behind the bulkhead in the background and the other is at the right.

(20) This is shaft no. 3 looking forward. The shafts are too long to be made in one piece so they are made in sections. Shaft no. 1 has eight segments, shaft no. 2 has six, shaft no. 3 has five, and shaft no. 4 has seven. The segments vary in length but most are about forty-eight feet long and are about 23" in diameter. This photograph shows a joint between shaft segments. The bar across the bottom was added during the mothballing process to keep the propeller from windmilling.

(19) Where the shaft passes through a watertight bulkhead, packing glands are required to maintain a watertight seal. Here shaft no. 4 passes through that bulkhead at Frame 166 with a packing gland.

(21) This is the port side of the foundation for turret no. 3. The closely spaced webs support the weight of the turret above. Second platform is higher above the baseline at turret no. 3 than it is at the other turrets so the foundation webs are taller. The arches here are 42" while they are only 23" for the forward turrets (see photograph (8)). Each turret has four such foundation passages at the sides, fore and aft.

(22) Aft of Frame 171 the triple bottom becomes a single bottom. Here at Frame 172, the rider plate ends in a triangle. From there to the stern, the flat keel and vertical keel form an inverted T that meets the lowest deck above. The vertical keel here rises to the second platform and blocks the view of the port side of the rider plate where it is mirrored.

(23) Twin keels support the inboard propeller shafts. The shafts run through watertight shaft tubes. This photograph is looking over the opening into the port twin keel. Propeller shaft no. 3 runs above.

01 Level

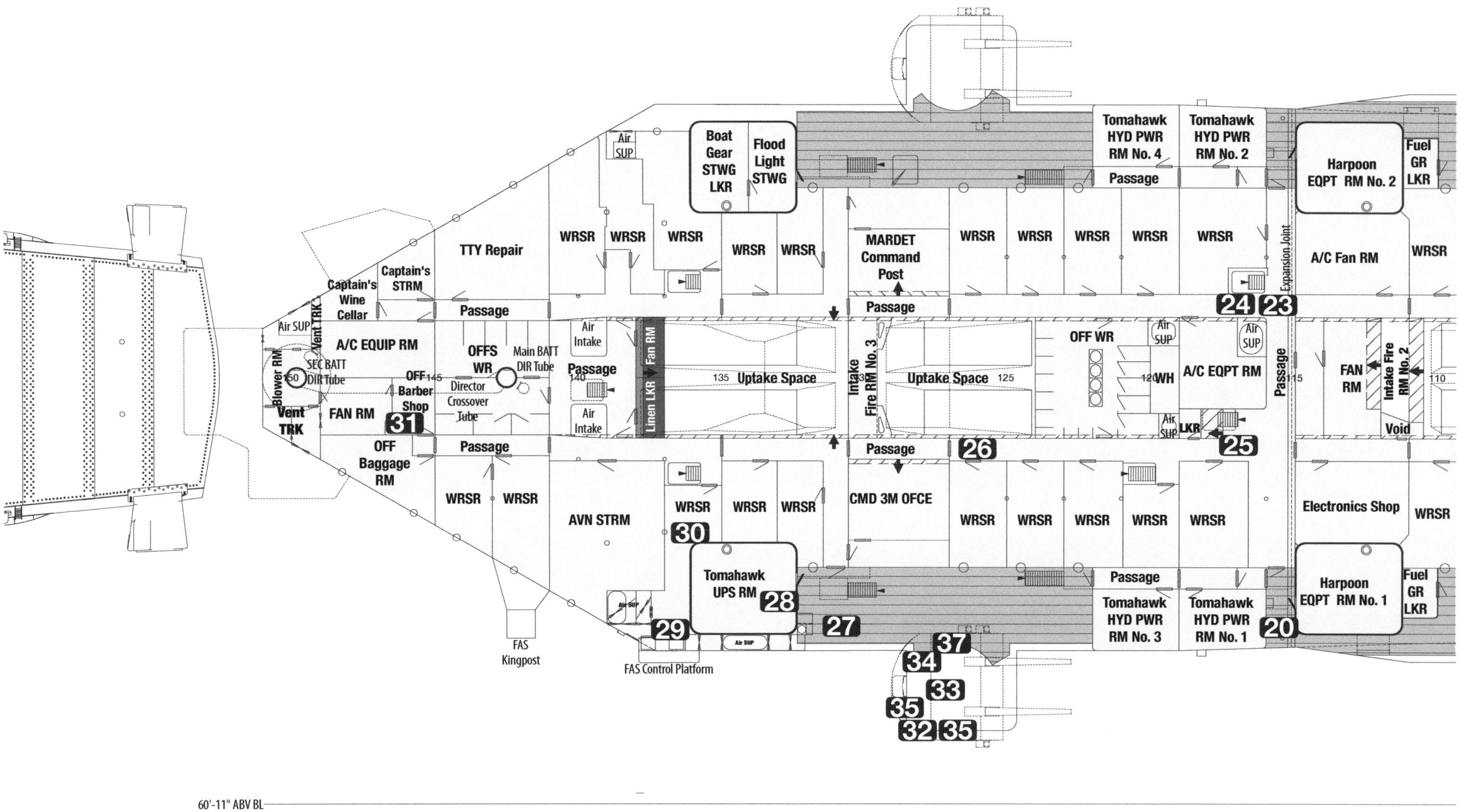

01 Level
1"=20'

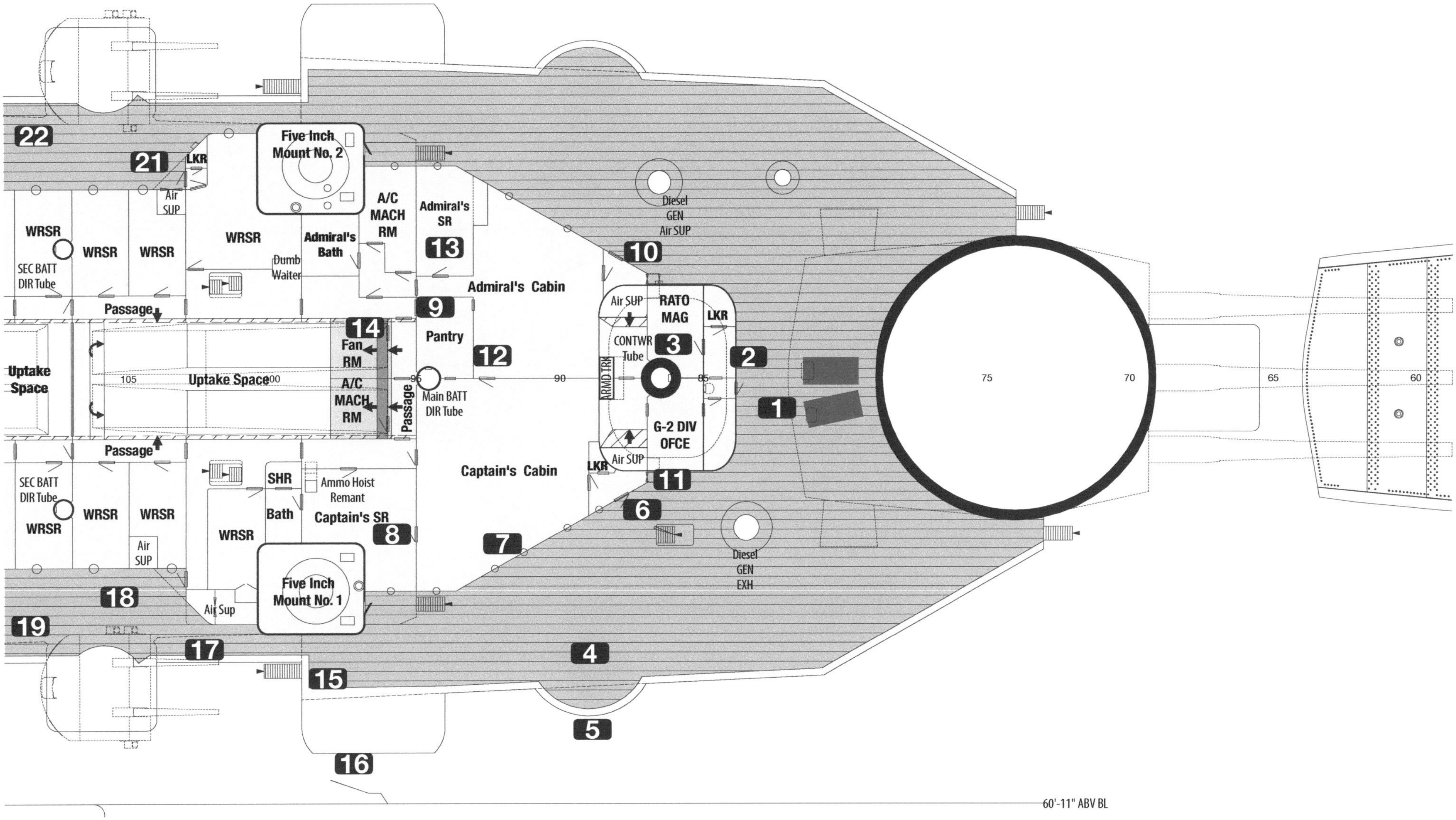
Five Inch Mount No. 2
LKR
Air SUP
WRSR
WRSR
WRSR
WRSR
SEC BATT DIR Tube
Dumb Waiter
Admiral's Bath
A/C MACH RM
Admiral's SR
Diesel GEN Air SUP
Admiral's Cabin
Passage
Uptake Space
Uptake Space
Fan RM
A/C MACH RM
Passage
Pantry
Main BATT DIR Tube
Air SUP
RATO MAG
LKR
CONTWR Tube
ARMD TRK
G-2 DIV OFCE
Air SUP
LKR
Captain's Cabin
Passage
SEC BATT DIR Tube
WRSR
WRSR
WRSR
Air SUP
WRSR
SHR
Bath
Ammo Hoist Remant
Captain's SR
Air Sup
Five Inch Mount No. 1
Diesel GEN EXH
105
100
95
90
85
75
70
65
60
60'-11" ABV BL
1
2
3
4
5
6
7
8
9
10
11
12
13
14
15
16
17
18
19
21
22

(1) The gunhouse entrances to turret no. 2 are at the 01 level. The main hatch is located close to the centerline of the turret. The emergency hatch is offset to the right. Because turret rotation can position these hatches over the main deck, turret no. 2 has platforms suspended below the hatches to reduce the drop. The museum repositioned the main hatch platform to put in place the ladder for visitors shown. The hatches on turret no. 2 swing rearward while those on the other turrets swing forward. A ventilation trunk is to the right of the emergency hatch. The four-inch thick shelf plate at the bottom of the gunhouse has four clusters of holes for intake air. These are covered by four trunks with screens to keep out bugs.

(3) Only the 01 level of the conning tower support has usable space. The usage changed over time. Originally the starboard side was a magazine for 20mm ammunition magazine and the port side was for film storage. After the 20mm guns were removed, both sides became offices. The final use of the port side was as the rocket assisted takeoff (RATO) magazine for rockets used to launch Pioneer remotely piloted vehicles (RPV). The RATO racks that remain in the magazine.

(2) This hatch in the forward face provides the only practicable access to the interior of the conning tower support.

(4) This is the former location of the port swimming pool that was created from a 40mm tub in the 1960s. Only the outer part of the tub remains as a support for frames holding inflatable life rafts. The wood deck is under repair revealing the outlines of the former 40mm mount. The outline of the ammunition-handling cupola is visible at the bottom of the photograph. A gun director was mounted on top of this cupola. The swimming pools were only used briefly because there was no means of keeping the water clean. Their legend is much greater than their use.

(6) A pair of doors keeps the outside air from rushing into the captain's cabin. There is a linen locker between these doors that could be restocked without disturbing the captain

(5) An underside view of the swimming pool showing how the 40mm tub extended over the side of the superstructure. The frames for inflatable life rafts were added in the 1980's.

(7) The captain's cabin provided a place for the captain to take meals, entertain, and hold meetings. The captain did not eat in the wardroom with the officers, except when invited for special occasions. The captain could invite officers to dine with him in his cabin. Contrary to the practice of some other navies, the captain's cabin is the largest on the ship and is even larger than either of the two admiral's cabins that existed at various times. The captain typically would have a larger staff and needed a larger meeting space than an admiral would. The door in the background was created in 1989 to connect to the new admiral's cabin. To the left is the food-passing window from the pantry. The circular structure in the corner is the main battery director tube that connects the forward director Mk.38 (Spot 1) with the citadel.

(8) The captain's suite includes a pantry. The pantry has a refrigerator but lacks an oven and stove top. Food would have to be cooked in the officer's galley and sent up to the pantry through the dumbwaiter.

(9) The captain's stateroom is adjacent to his cabin. It is not ornate but the captain has a real bed. The trunk in the background was for a 40mm ammunition hoist that runs to the 05 level. In the 1980's it was repurposed as an armored wiring trunk. The captain's bath is through the door to the left of the hoist. For most of the ship's service the captain's bath had a shower. However, during the 1960's the captain's bath had a bathtub. The stateroom was only used while the ship was in port. The captain normally slept in the at sea cabin on the 04 level outside the bridge. During the 1940's there was an additional captain's sea cabin at the 08 level. It was rarely, if ever, used and was converted into the meteorological facility.

(10) During their 1980's reactivations, the admiral's cabins were removed from *Iowa*-class battleships. USS *New Jersey* served as a flagship for her final cruise in 1989–90. To prepare for that cruise, the crew created an admiral's suite from several officers' cabins. That conversion included adding this door to give the admiral direct access to the weather deck. The ladder leads up a shroud added to the conning tower support during World War II to keep water out of the air intakes for fire room no. 1. The intakes were at the sides of the conning tower support.

(11) Here is a view up the conning tower support shroud. There is not much to see here because the intakes were sealed when the ship was mothballed. However, the outline of the intake is still visible. The triangular shape at the top of the photograph is the overhang of the admiral's bridge. USS *Wisconsin* had her shrouds here removed during the 1980's reactivation, a distinctive feature of that ship.

(12) This is the admiral's cabin the ship's crew added to USS *New Jersey* in 1989 looking to port. The door at the right leads to the weather deck (see Photograph (10)). The television partially obscures a passing window cut into the captain's pantry to allow it to be shared with the admiral. The passage with curtain visible above the television leads to the admiral's stateroom and bath.

(13) The admiral's stateroom added in 1990. The wardrobe at the right extends into the admiral's cabin that is visible in the background of photograph (12).

(14) The upward slope in the deck of the fan room is necessary to accommodate the uptakes for fire room no. 1. The uptake spaces are open areas with no decks between the 01 and 04 levels.

(15) The weapon in the foreground is a Mk.38 25mm Bushmaster chain gun. A chain gun is externally powered. A motor drives a chain that has a linkage with the bolt along a rectangular path. As the chain moves, it pulls the bolt forward and backward causing spent cases to be ejected and fresh ammunition to be loaded. Mounting points for two of these guns were installed on all of *Iowa*-class battleships in the late 1980's. However, the actual guns were only installed on USS *Missouri* and USS *Wisconsin*. USS *New Jersey* received its pair of 25mm guns after it became a memorial. To the left there is a 40mm saluting gun. The only function of a saluting gun is to make noise and smoke. The box above the saluting gun is the ship's side light.

(16) A view of the underside of the saluting gun platform. This gives a better view of the side light enclosure. This platform held two 20mm mounts in World War II. During the Korean War this became the saluting gun platform.

(17) 5-inch mount 53 looking aft. The large door at the lower left is for maintenance. Above it are the hoods for the pointer (left) and checker (right) telescopes. There is only one hood on the right side of the turret for the trainer. The oval door at the rear is one of two access openings. These doors are mirrored left and right. The mount commander's hood is at the top rear of turret. The counter-balance for the mount commander's top hatch is oriented vertically along the turret rear.

(18) This view is forward with 5-inch mount 53 to the right. This photograph illustrates the overhang in the superstructure created by the air intake for engine room no. 1.

(19) The starboard boat davit viewed from behind 5-inch mount 53. The door in the foreground is to the starboard fueling gear locker. The remnant of mount 55 is behind that. Mount 55 was removed from *Iowa*-class battleships in the 1980's. A kingpost was added to the front face of mount 55 to provide a mounting point for the forward davit pulley. This created a tunnel next to the mount. While in service, cables ran from the davit arm at the left to a pulley on the mounting at the left that were used to raise and lower the davit. There is a similar pulley at the aft end of the davit mounted directly to the superstructure. The horizontal bar locks the davit in the up position. Below that bar is the spring buffer that limits the davit's upward travel.

(21) The grating shown is the air intake for boiler room no. 3. The port side of the deckhouse here does not have an extension with an overhang as on the starboard side (see photograph (18)).

(20) This ladder provides access to the boat platform beneath the davit. The superstructure expansion joint is visible in the deck at the right.

(22) This is the port fueling gear locker. The port boat davit is longer than the starboard side davit. Consequently, the mounting for the davit pulley is farther forward and mounted to the fueling gear locker rather than to the remnant of the five-inch mount as on starboard side.

(23) An expansion joint runs along Frame 115. This view is to starboard. The superstructure has welded construction while the hull is largely riveted. This difference in construction results in the superstructure being much more rigid than the hull. The expansion join allows the superstructure to move as the hull flexes.

(24) This is the port side passage of the 01 level at Frame 121 looking aft. The doorway to the water closet illustrates how the bulkhead is sloped to align with the bulkheads at 02 level.

(25) This linen locker at the starboard side passage at Frame 119 was designed to conform to the inclined ladder by having a sloped bulkhead forward.

(26) The starboard side passage of the 01 level at Frame 131 $\frac{1}{2}$ looking aft. The slope of the longitudinal bulkheads is apparent here. The arch in the background at Frame 135 was originally an oval like the one at Frame 131. It was expanded in the 1980's to make it easier to move large objects to the aviation storeroom.

(27) The air intakes for engine room no. 3 wrap around the remnant of 5-inch mount 59. This mount was removed in the 1980's to make room for the missile platform above. The handling room was converted into the Tomahawk UPS room. This smaller intake is on the front face of the mount. The larger intake at the left runs alone the outer face.

(28) The interior of the Tomahawk UPS room (left). This originally was a handling room for 5-inch mount 59 that was removed in the 1980's. The turret ring is visible at the top of the photograph. One of the two armored trunks for the projectile hoists to this mount was repurposed as a protected path for missile control cables. The same was done on the opposite side handling room where mount 60 was converted to storage.

(29) The Fueling at Sea (FAS) control platform. U.S. battleships had long been used to refuel their escorts. During the 1980's the refueling facilities were upgraded with the addition of a FAS kingpost that allowed the cables and hoses used for refueling to be raised so they would be less likely to hit the water during refueling. The controls here operate the winches used for fueling at sea. There were originally two seats at the control panel but they have been removed.

(30) This is an oddly shaped two-berth officers cabin (right). This had been used as a storeroom but it was one of several rooms converted to officers' cabins when cabins were lost to the admiral's cabin. Originally this area had been part of a junior officers' bunkroom.

(31) The officers' barbershop has only one chair. An armored trunk for director cables is visible along the overhead.

(32) A panorama view of the gunhouse of a 5-inch/38-caliber twin mount. This view is from the right side of the mount looking to the left. During World War II there were three basic configurations of the 5-inch/38 gun: open single mounts, single enclosed mounts, and twin enclosed mounts. There were additional variations on these configurations. *Iowa*-class battleships used the 5-inch Mount Mk.28 mod 2. This mount has a 2-inch armored shield and weighs 155,000 lbs.

A 5-inch mount could operate in either automatic mode or under local control. In automatic mode, the gun was aimed and fired by the fire control system. Under local control, the mount did its own aiming and firing.

The official size of the mount crew was thirteen. However, a mount could operate with fewer, especially when under automatic control for shore bombardment where rapid fire was not necessary.

At the left is mount captain's platform. That elevated platform allows the mount captain to view outwards through the top hatch of the mount. When under local control, the mount captain viewed the fall of shot and gave correction orders to the sight setter.

At the right are the two guns. A dummy cartridge is visible in the loading tray for the left gun. To the left of the cartridge is the spade rammer that moves the ammunition into the gun.

(33) The centerline of the mount looking forward. The projectile hoist is at the center with dual projectile paths for both guns. The foot pedals at the base of the hoist release the projectiles from the hoist. Immediately behind the hoist to the right is the fuze setter's station. This station was in standby mode when the fuzes were set automatically by the fire control system. The sight setter's station is behind that. This position had to be manned for both local and automatic mount control. In automatic control, the sight setter received the elevation and traverse orders from the fire control system and matched the mount to those orders. In local mode, the sight setter took corrections from the gun captain and adjusted the sights to compensate for the corrections.

(34) This is a view up the loading tray of the left gun. Normally two men loaded each gun. First the powder man placed a cartridge at the end of the loading tray. Next the projectile man placed the projectile between the cartridge and the breech opening. Finally, the projectile man operated the large lever at the upper right corner of the gun that caused the spade rammer to push the ammunition into the breech and then close the breech. The gun could be loaded at any angle of elevation.

The 5-inch/38 gun uses a simple sliding breechblock that moves downward to open. The brass cartridges expand to seal the breech when the gun is fired so there is no need for a complex breech mechanism as there is on the 16-inch guns. The breach block is at the end of the tray and is in the closed position. The breech had to be manually opened to load the first round. When the gun fired, the breech opened and the cartridge was ejected automatically. When the breech is open the curved upper face of the breechblock acts as a guide for projectiles to enter the barrel.

(35) When the guns were fired, the cartridge cases are automatically ejected and they fell into ejection chutes at the rear of the mount where they were ejected through this door at the rear. The chute for this gun has been removed for visitor access. At high angles of elevation the spent cartridges would have to be manually removed from the loading tray and placed in the chute.

(36) The trainer's position at the right side of the mount. When the mount is in local control the trainer rotates the mount by turning the linked brass wheels at the center of the photograph. Normally the mount was powered by electric-hydraulic drive. If power were lost, the guns could be elevated and the mount could be rotated by direct, manual cranking through reduction gears. The trainer has a periscope sight that extends into the hood at the right side of the mount. When the mount was in automatic mode, the trainer position was in standby mode. The large lever at the upper left is for manually opening the gun breech for loading the first round.

(37) The pointer's and checker's positions are at the left side of the mount. Their view is through periscope hoods mounted outside of the mount. The pointer sits at the seat at the front of the mount. Under local control, the pointer turned the brass wheels at the center to elevate the guns. When the mount was under local control the pointer fired the guns. The propellant cartridges contained an integrated case combination primer that could be fired electrically or mechanically. The firing key for electrical firing was attached to the right elevation wheel. However, that trigger has been removed from this mount. The guns also could be fired mechanically using a pedal. When in automatic mode, the pointer position was in standby mode.

The checker's position is immediately behind the pointer. This position was only used for training. The person at the checker's position could verify the performance of the trainer and pointer.

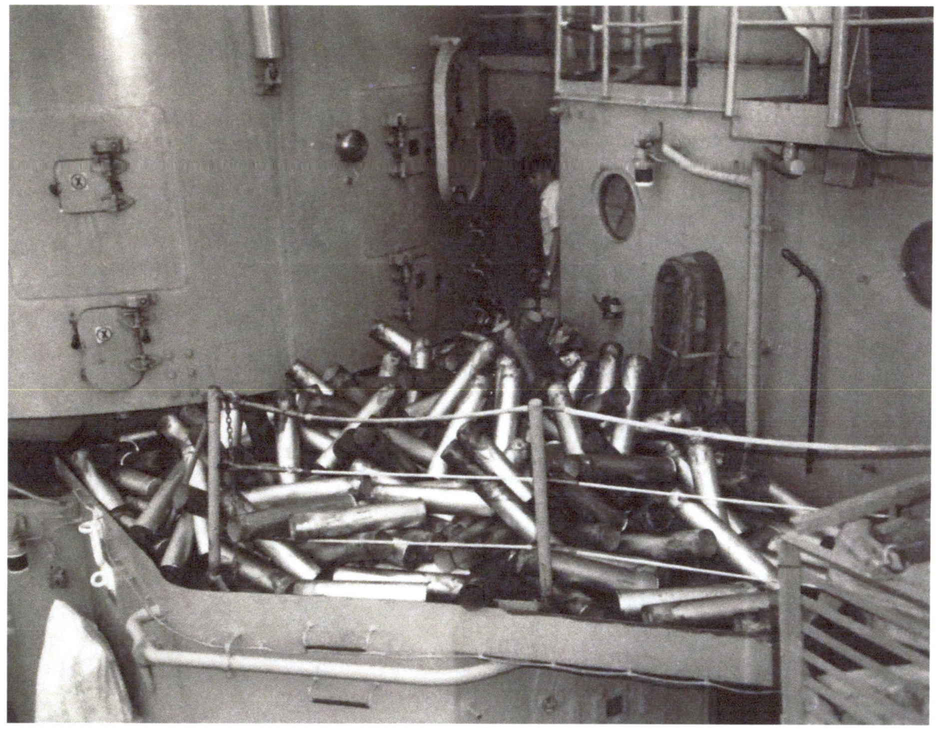

During the night of February 22, 1969, USS *New Jersey* fired over 1,600 5-inch rounds.

02 Level

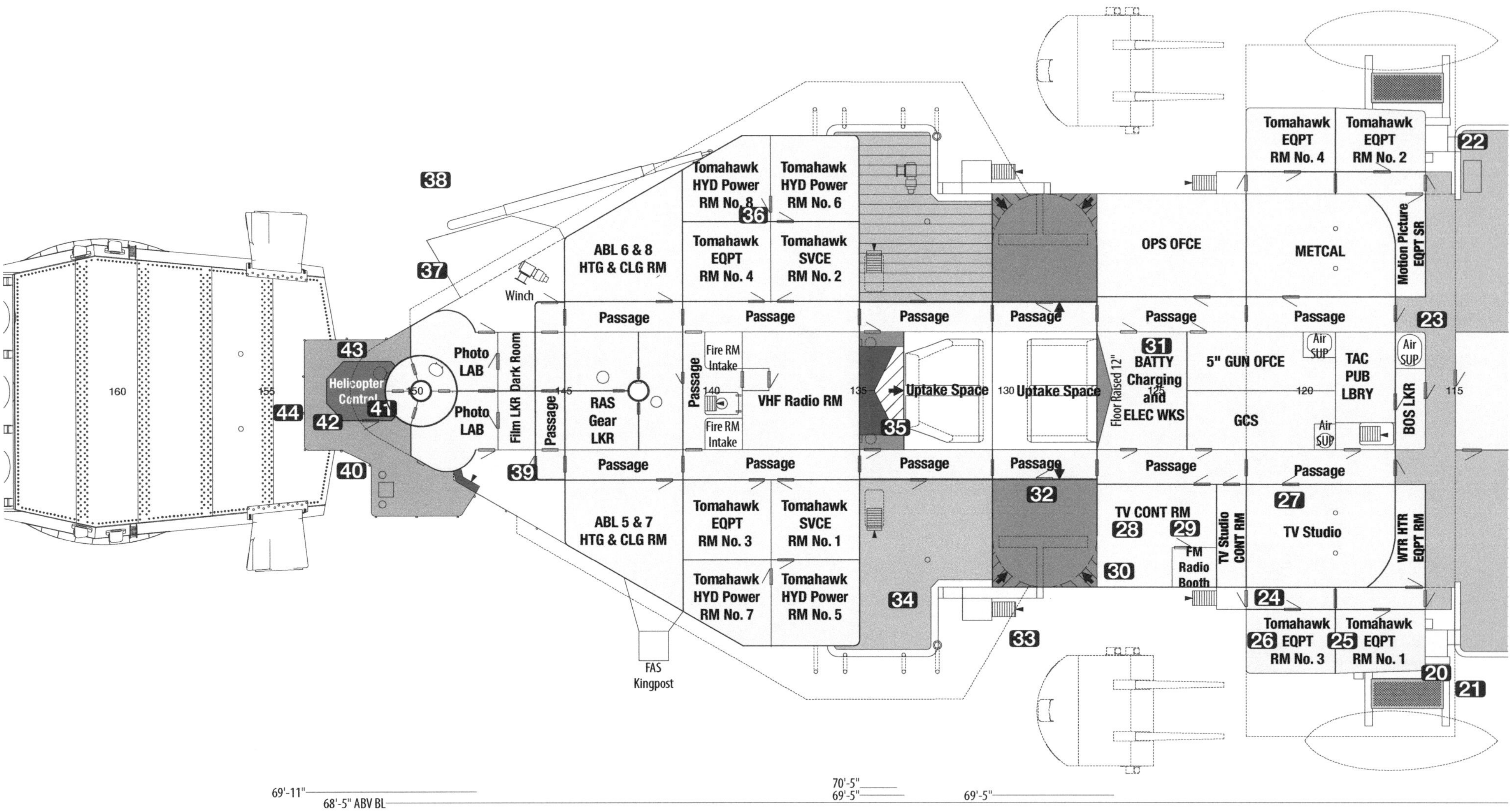

02 Level
1"=20'

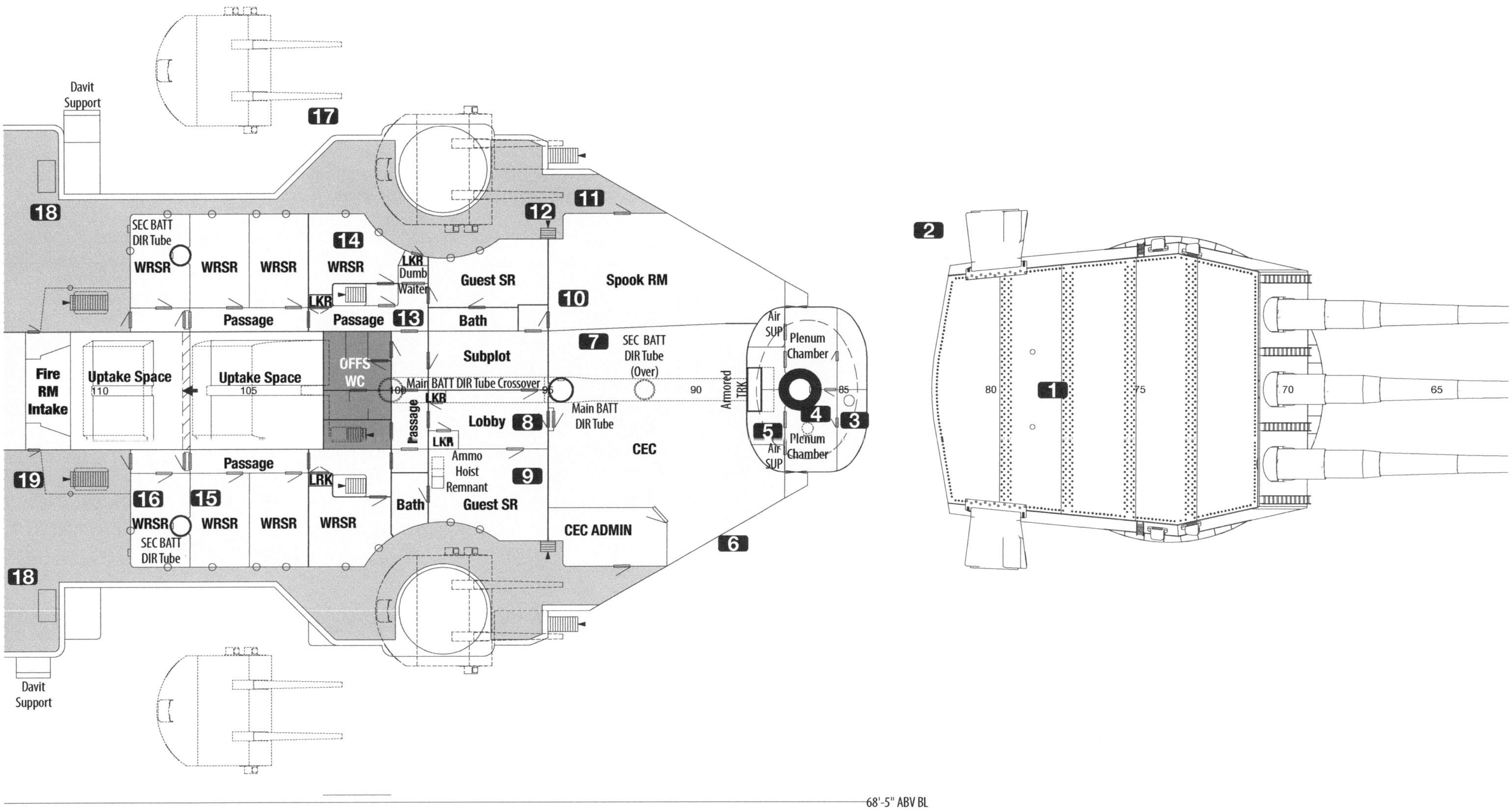
Davit Support
17
18
SEC BATT DIR Tube
WRSR
WRSR
WRSR
14
WRSR
LKR
Dumb Waiter
LKR
Passage
Passage
13
Guest SR
12
11
Spook RM
10
Bath
7
SEC BATT DIR Tube (Over)
Air SUP
Plenum Chamber
Fire RM Intake
Uptake Space
110
Uptake Space
105
OFFS WC
100
Main BATT DIR Tube Crossover
95
90
Armored TRK
85
Subplot
LKR
Passage
Lobby
8
Main BATT DIR Tube
CEC
4
3
5
Air SUP
Plenum Chamber
LKR
Ammo Hoist Remnant
9
Passage
19
LRK
16
15
WRSR
WRSR
WRSR
WRSR
Bath
Guest SR
SEC BATT DIR Tube
CEC ADMIN
6
18
Davit Support
2
80
1
75
70
65
68'-5" ABV BL

(1) The roof of turret no. 2. The dome at the front edge is a radar that was added in 1983 to measure the velocity of the projectiles The age of the powder caused the projectile velocity to deviate from the World War II-era data tables. The radar allowed the actual projectile velocities to be entered into the fire control computers. At the right is a tripod that could be erected at either side of the turret for replenishment at sea (RAS). The supply ship would send a span wire to USS *New Jersey* that would be attached to the tripod. Supplies would be transferred along that wire. The turret's sides have half rounds at the upper edge to minimize cable chaffing.

(2) This is the left rangefinder hood for turret no. 2. The hood is a casting 2.5" thick. The casting process produced an irregular surface. A plate at the side is removable and allowed the rangefinder to be installed after the hood was in place.

(3) Access to the 02 level of the conning tower is through this hatch. This area is a plenum chamber for the fire room no. 1 air supply so it is an empty space.

(4) The air intakes at the conning tower sides were plated over when the ship was mothballed. The opening was originally covered by an externally mounted grating. The ladder leads to the 03 level of the conning tower.

(5) This is a view down the starboard air intake in the conning tower. It is a large open pit. The main deck armored grate is visible at the bottom.

(6) A combat engagement center (CEC) for managing the ship's weapons systems was created in the 1980's at the 02 level. The CEC served essentially the same role as the CIC on the first platform. However, the navy concluded that it would be too difficult to install the modern equipment and its cabling into the existing CIC so the CEC was created in the deckhouse. In service both the CEC and CIC were manned, with the latter largely serving as a backup. This photograph shows the 1$\frac{1}{2}$" HY-80 armor added to the side of deckhouse to protect the CEC. The door below is to the captain's cabin at the 01 level.

(7) This is a panorama view of the CEC (below). The CEC was originally the admiral's cabin. The colored lights at the right show the train angle for the directors and main turrets. The CEC is filled with radar displays and plotting boards. The chair in the center was for the tactical action officer. The positions at the left are for the SPS-10 and SPS-49 radars. Those in the background center are for Tomahawk missiles. The close up photograph (right) shows a Tomahawk console. There is a keyhole (circled) for enabling nuclear weapons at the upper right of the control panel.

(9) Prior to the 1980's, this was the admiral's stateroom. After the admiral's cabin was replace by the CEC, it became a guest cabin. This cabin has its own bathroom. An armored trunk for a 40mm ammunition hoist runs through the cabin.

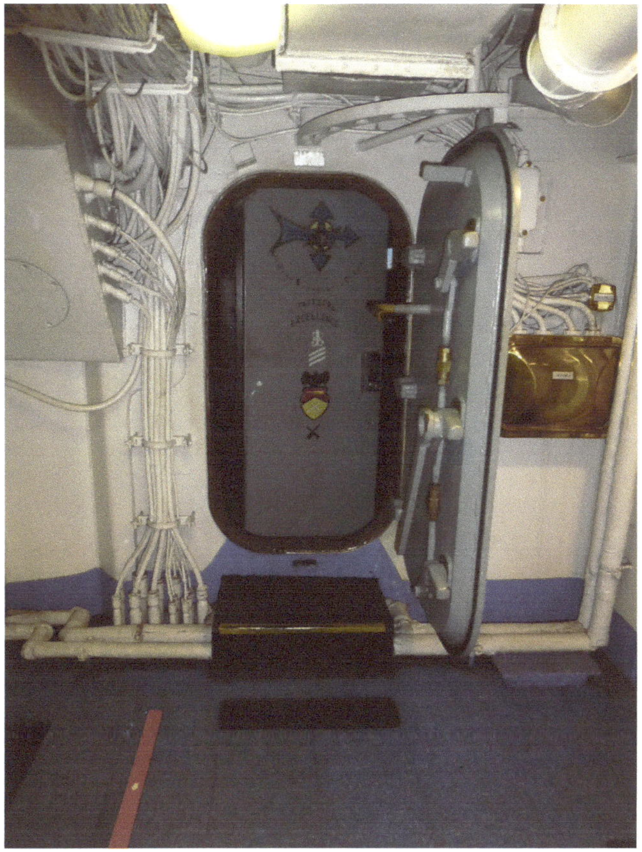

(8) The entrance to the CEC is an armored door. The door gives an idea of the armor protection of the CEC. To the left is the armored director wiring trunk.

(10) This room was originally part of the admiral's suite. It was divided into a stateroom and bath forward and a cabin aft that was used by the admiral's chief of staff. In the 1980's this was converted in to the top secret Spook Room where intelligence gathering was conducted.

(11) The Spook Room has direct access to the weather deck. Oddly, for a room that was top secret, the door is designed to so that it can be kept open.

(12) Originally, there was a vertical ladder between the 02 and 03 levels. By the 1980's this had been replaced by an inclined ladder. The bulkhead on the 02 and bulwark at the 03 levels do not quite line up here. This is due to installation of 2" of HY-80 armor at the 03 Level during the 1980's.

(13) The dumbwaiter from the officers' galley ends at the 02 level near the former admiral's pantry.

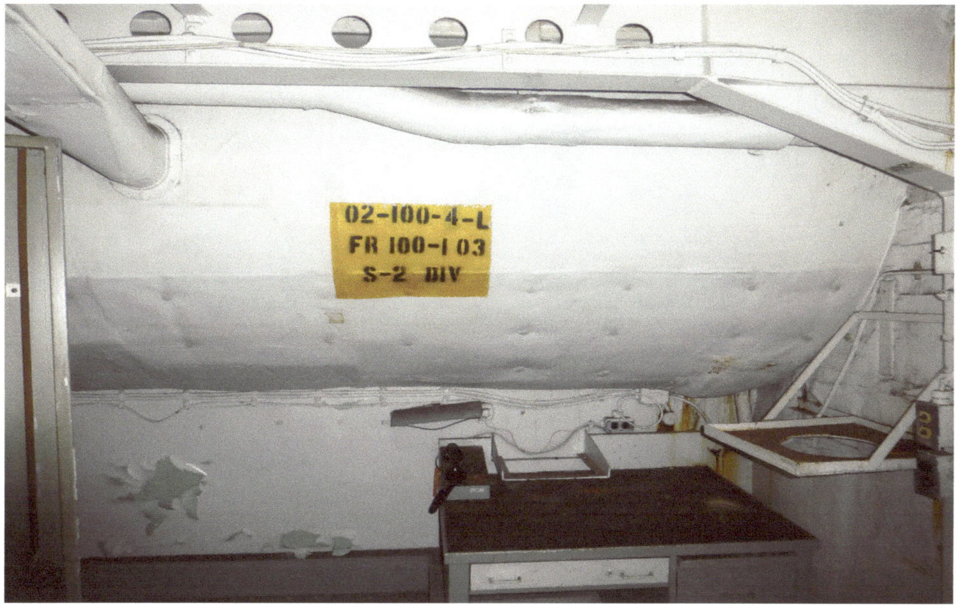

(14) Officer's cabins on each side of the ship are directly below the signal flag boxes on the 03 level. The drains from the flag box pass through the cabin. These were notorious for leaking.

(15) This is a typical stateroom for two officers on 02 level. Each has a desk and the storage space is much greater than allotted to the enlisted crew.

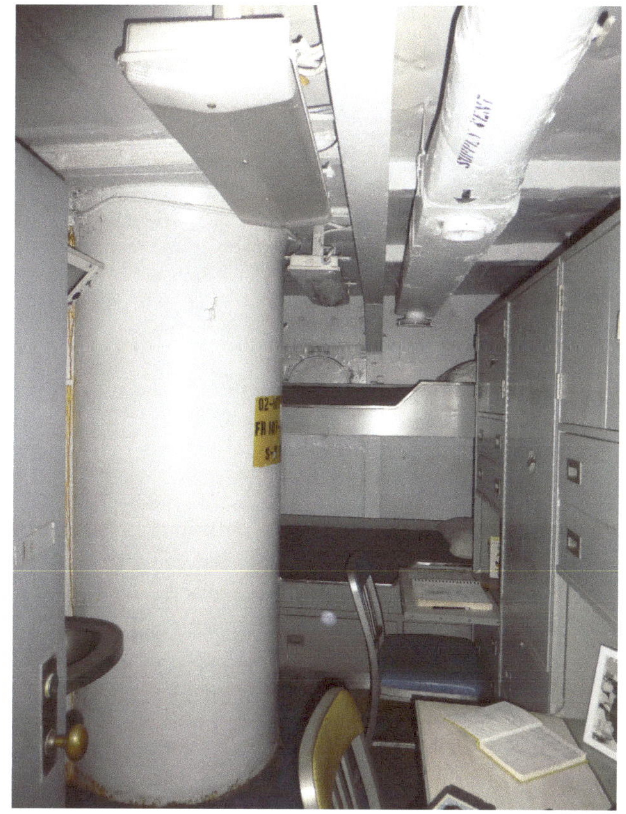

(16) The quality the staterooms vary throughout the ship. This stateroom is the same size as the one next door shown in photograph (15) but it has an armored trunk for director cabling running through it that greatly reduces the useable space.

(17) A 5-inch mount viewed from above. The shield for the gun captain is at the back and the counterweight for the gun captain's hatch is below. When the mount was under local control, the gun captain had to spot the shots and give corrections to the sight setter. The hood provided the gun captain with a modicum of protection. The trainer's sight hood projects from the right side of the mount. There is one hood on the right side and two on the left.

(18) These are views of the former locations of 5-inch mounts 55 (starboard/above) and 56 (port/below). The kingpost supporting the forward davit arm is farther aft on the starboard side than on the port side due to its shorter boat cradle. The large lockers are for life jackets.

(19) This is the area behind the forward end of the starboard boat davit. An expansion joint runs just aft of the planked area. The raised step gives access to the whaleboat platform. The 03 $\frac{1}{2}$ level missile platform extents forward of the deckhouse and the 14-foot whip antenna platform extends even farther.

(20) This is the forward end of the starboard whaleboat platform. The platform wraps around the deckhouse but does not directly connect to 02 level forward.

(21) The starboard whaleboat platform viewed from below.

(22) Port whaleboat platform. One has to cross a considerable gap to climb into the boat from here.

(23) This was originally a covered passageway at the 02 level with the 03 level running over the gap. The view here is to starboard. In the 1980's the 03 level here was raised to match the 03 $\frac{1}{2}$ level. The seams in the bulkhead indicate the original height. The missile platform covered up air intakes so a plenum chamber was created just below the missile platform to feed them. The units on the bulkhead are for air filtration into the plenum chamber. They keep missile exhaust from reaching the engineering spaces.

(24) This area of the 02 level deckhouse was expanded in the 1980's to support the missile platform above. The view here is forward. The bulkhead to the left was the original side of the deckhouse. The space created at the right was used for equipment rooms for the Tomahawk missiles. The extremely high overhead in this passage reflects elevated location of the missile platform above.

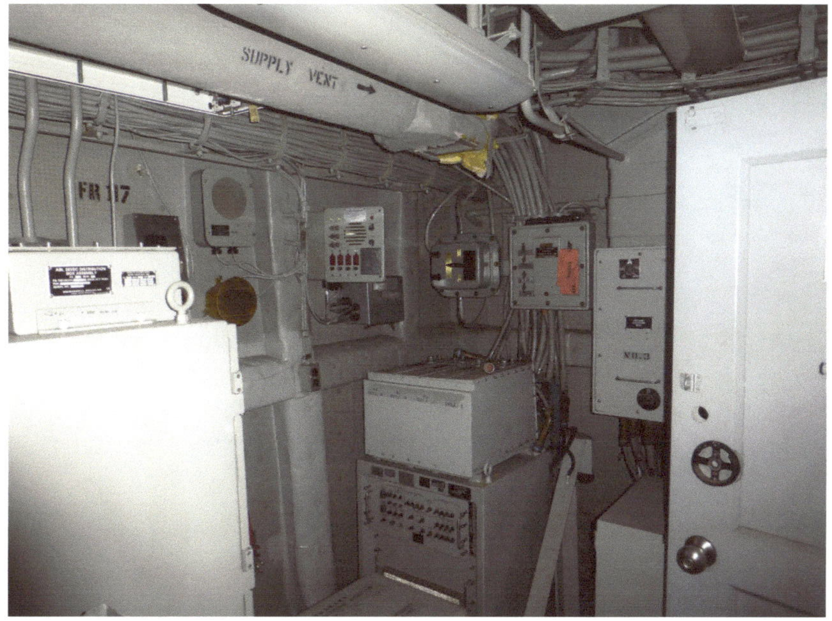

(25) Tomahawk equipment room no. 1 housed electronics equipment.

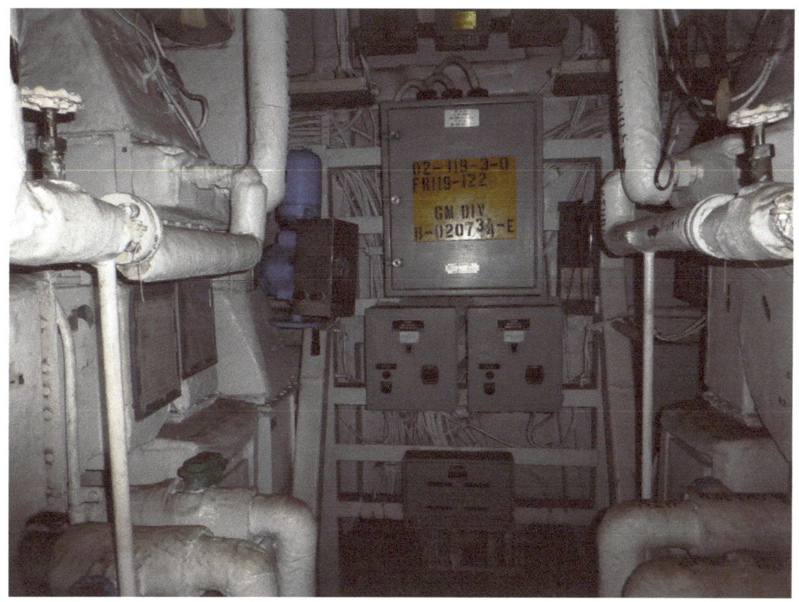

(26) Tomahawk equipment room no. 3 houses heating and cooling equipment.

(27) These are two views of the television studio showing the cameras (forward) and news desk (aft). This was part of a Shipboard Information Training and Entertainment (SITE) system. The TV studio was used to produce content for the crew. The system could also broadcast content supplied to the ship on tape. This compartment was originally a bunkroom for junior officers.

(28) This is the control panel in the adjoining TV control room. The compartment was stripped when the ship was decommissioned. Volunteers recreated the room using similar equipment. This room also was originally a bunk room for junior officers.

(29) The ship had its own FM radio station. This small booth within the TV control room is the broadcast studio for the radio station.

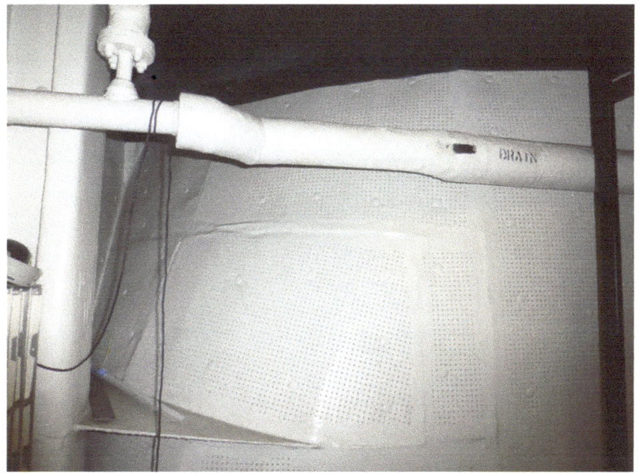

(30) The outboard corner of the aft bulkhead in the TV control room transitions from an angle at the deck to a circular arc at the overhead. This creates a nook that is actually over the adjoining fire control workshop. The circular shape of the upper edge of the bulkhead matched that of a 40mm tub that had originally existed there.

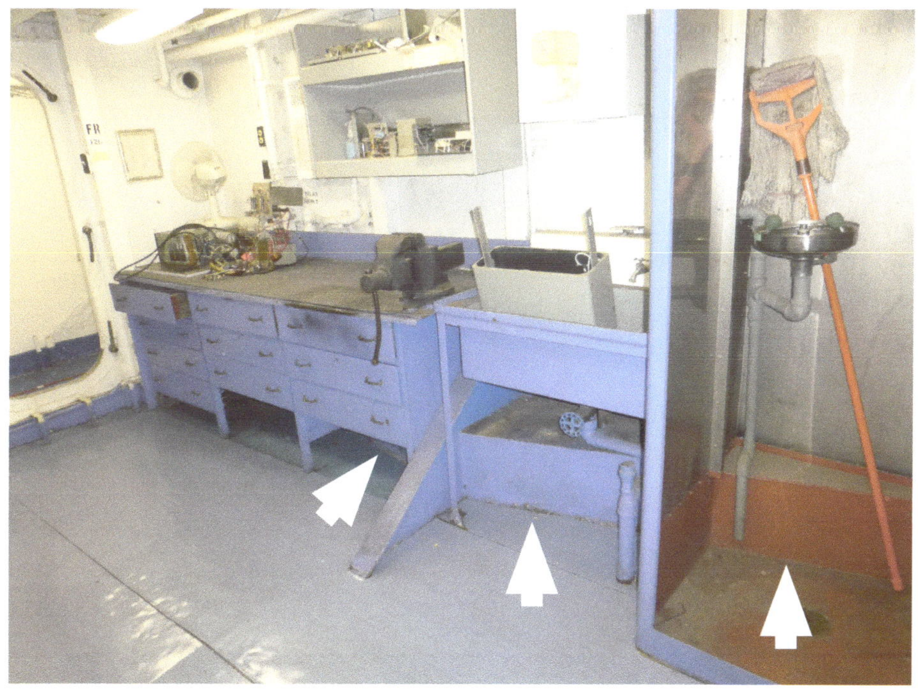

(31) The battery charging and electrical workshop looking aft. The interesting structural feature of this room is at the deck. The triangular area of raised deck provides clearance for the curved path of the uptakes behind the bulkhead and below the deck.

(32) The fire control workshops on USS *New Jersey* have elevated decks. This view is looking outboard to starboard. Originally, there was a 40mm mount directly above and the bulkhead at the right transitions from being rectangular at the deck to circular at the overhead to match the shape of the gun tub. There is a fire control trunk for Harpoon missiles at the top of the photograph. Two Harpoon launchers are directly above. The trunks to the launchers merge then pass through the outer bulkhead in the background.

(33) This is a view of the exterior of the starboard fire control workshop showing the path of the trunks shown in photograph (32). After going down to the 02 level, the trunk turns to enter the former 5-inch gun handling room. From there the cabling passes into the citadel through an armored trunk formerly used as a 5-inch ammunition hoist.

(34) This area or to the weather stored oxygen and acetylene gas bottles used for welding. The ladder leads to the aft missile platform.

(35) The odd shape of the bulkheads in the cleaning gear locker provide clearance for the uptakes forward and below. Originally, there was a practice loading machine for 5-inch guns above and aft of this compartment. Gun crews would use the machine to train for a high rate of fire without expending ammunition. There is a practice decanning scuttle in the overhead of this compartment that was used with training.

(36) This is one of the Tomahawk hydraulic power rooms. The unit to the left is a hydraulic power unit. To the right is a hydraulic reservoir.

(38) This platform was used to control the boat boom used to transfer boats between dollies at the main deck and the water. The boom is secured to the edge of the platform. A winch on this platform and another forward at FR133 were used to traverse the boom. The winch foundation remains but the winch was scavenged when USS *New Jersey* was decomissioned.

(39) This is the 02 boat platform viewed from below showing the boom and its mounting. The upper connection point for the boom is visible just below 03 level.

(39) The glide slope and helicopter control platform overhang turret no. 3. This platform is raised 18 inches above 02 level so that the turret can pass under it. This view is looking aft at the steps up and the back of the glideslope.

(40) The visual glide slope indicator provided vertical guidance to an approaching helicopter. A pilot would see a green (high), amber (on glide path), or red (low) light on the indicator depending up approach angle. The indicator is stabilized so rolling or pitching motion of the ship would not provide confusing guidance to the helicopter. The red lights at the side are wave off lights. They would flash when the controller wanted the helicopter to abort the approach.

(41) A panorama view of helicopter control. From the seat shown the controller could see the helipad and operate the lighting and communications equipment. There is a wind speed and direction indicator to the right of the battle lantern and a compass indicator directly below. Window cleaning controls are between the left and center window. The crash alarm is below the lower left corner of the window to the right. Visitors to the ship discovered that this alarm still works. Rotary dial phones were still in use on USS *New Jersey* in the 1990's. An escape hatch to the 03 level provided an additional exit to the one shown at the left.

(42) The helicopter control windows are heated and have wipers and cleaning fluids so they can be kept clear from the inside. Covers fold up over the windows to protect them when the 16-inch guns were fired.

(43) Just as the helicopter control platform had to be raised to make room for the turret, the aft end of 03 level had to be raised to allow the controller to stand within helicopter control. This is the port side of helicopter control looking forward.

(44) Like turret no. 2, turret no. 3 has a RAS tripod that can be mounted on either side of the turret. The object in the foreground is the turret captain's periscope. There is another periscope out of view to the right for the turret officer. These would have been used to spot the fall of shot in desperate situations—that never occurred—where the turrets had to fight on their own. The turret roof is made of five plate 7.25" plates. The plate joints have scarf joints machined into them. The rows of bolts shown pass through the scarf joints to hold the plates together.

03 Level

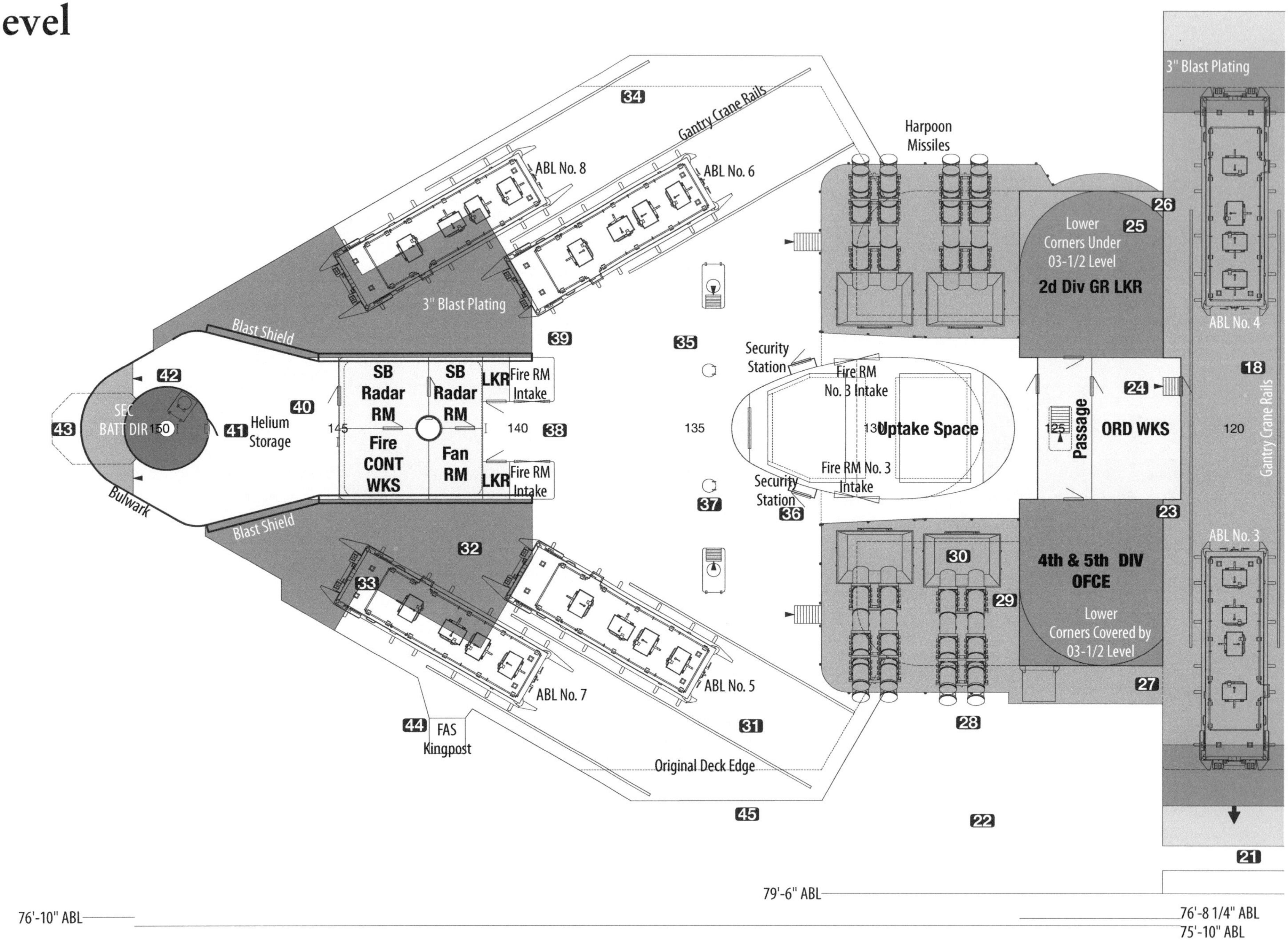

03 Level
1"=15'

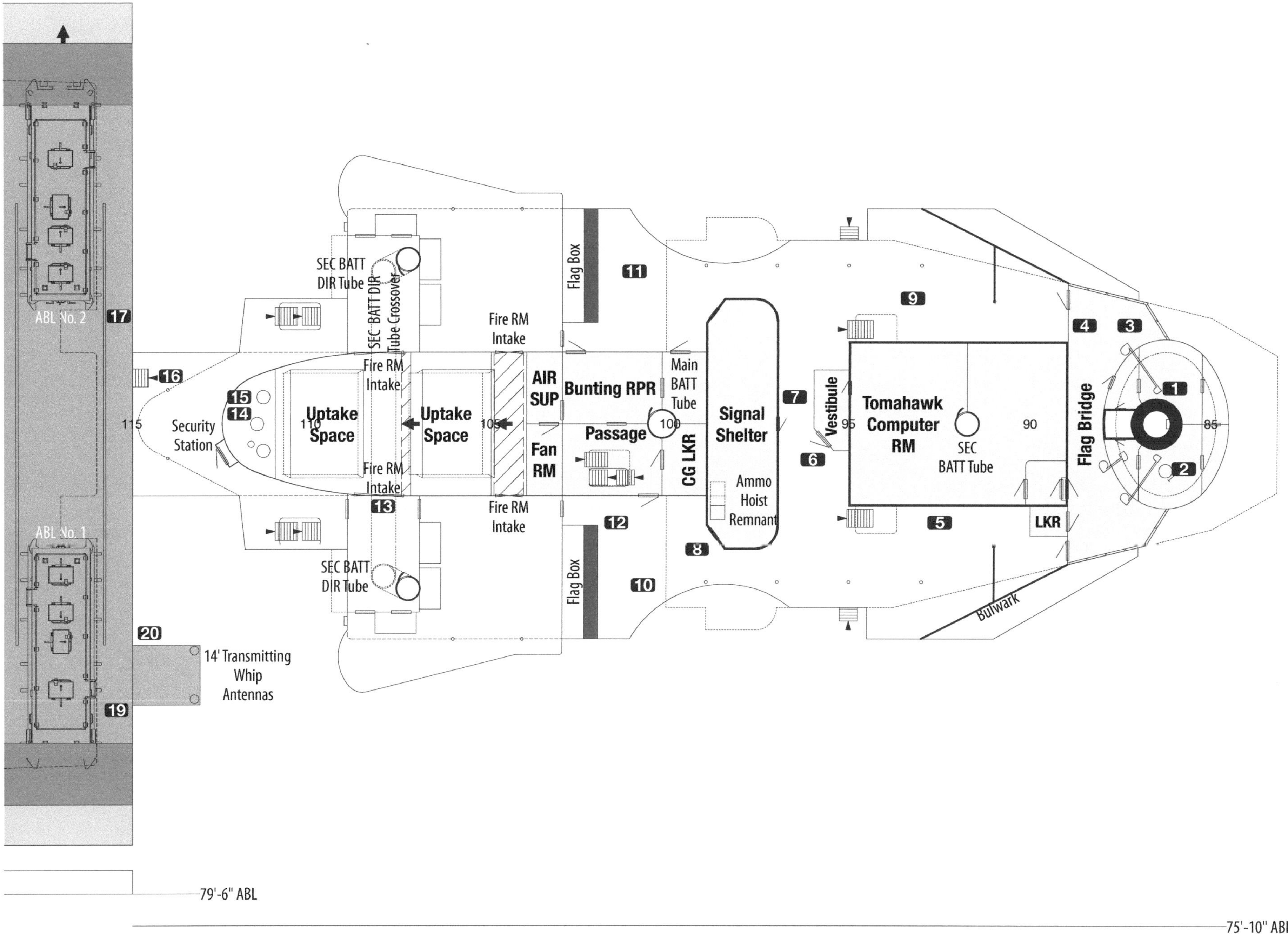
ABL No. 2
ABL No. 1
17
16
19
20
14' Transmitting Whip Antennas
115
Security Station
15
14
110
Uptake Space
Uptake Space
105
SEC BATT DIR Tube
SEC BATT DIR Tube Crossover
Fire RM Intake
Fire RM Intake
Fire RM Intake
Fire RM Intake
13
SEC BATT DIR Tube
AIR SUP
Fan RM
Flag Box
Flag Box
11
12
10
Bunting RPR
Passage
100
Main BATT Tube
CG LKR
Signal Shelter
Ammo Hoist Remnant
8
7
6
Vestibule
95
9
Tomahawk Computer RM
SEC BATT Tube
90
5
LKR
Flag Bridge
4
3
1
2
85
Bulwark
79'-6" ABL
75'-10" ABL

(1) The 03 level of the conning tower is largely empty on USS *New Jersey*. All that can be found is the conning tower door opening machinery and the armored trunks that protect the wiring running to the conning tower. This view is aft at the port side. The drive shafts at the top of the photograph are part of the door opening machinery for the port armored conning tower door at the 04 level. The door visible through the arch provides access from the flag bridge.

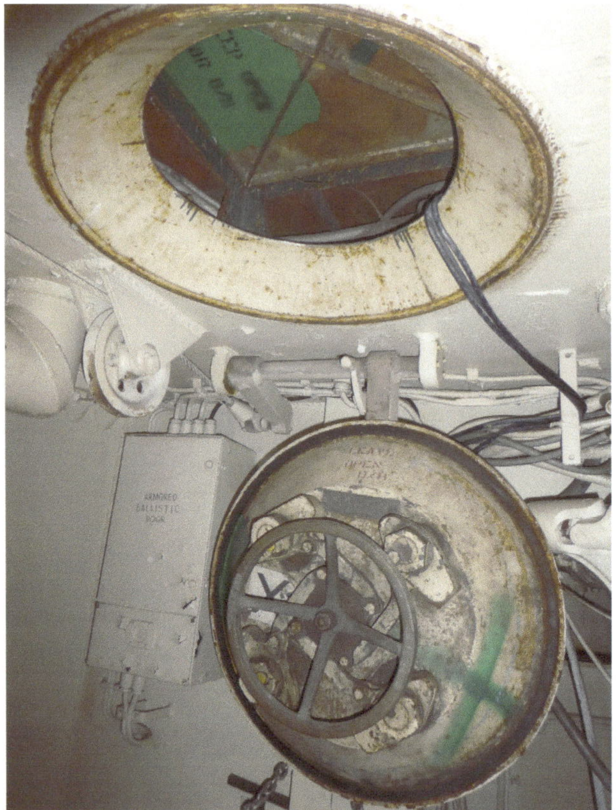

(2) This hatch is in the armored floor of the conning tower. The raised deck of the conning tower is visible through the hatch. A removable plate in the 04 level raised deck must be removed to gain access to this hatch. Electrical components for the armored doors in the conning tower are on the bulkhead. Part of the operating mechanism for the starboard armored door at the 04 level is visible to the right of the hatch.

(3) The port side of the admiral's bridge. A gearbox for the armored door mechanism is at the center of the photograph. The door to the right is the one shown in photograph (1).

(4) The admiral's bridge viewed to starboard. The conning tower cuts the bridge in half. The flag plotting room was originally behind the bulkhead to the right. That area was converted into the Tomahawk computer room in the 1980's.

The connection between the armored trunk at the aft end of the conning tower and the director trunk is at the center of the photograph. The upward slope at the conning tower angle reduces cable chaffing. There is a manhole for accessing the armored trunk directly above in the conning tower casing.

The five stars in the tile at the background were installed by the crew in the 1960's to honor Admiral William Halsey.

(5) The starboard entrance to the admiral's bridge. The structure to the left of the door was originally a stairwell that provided direct access between the admiral's cabin at the 02 level and the 03 admiral's bridge. When the admiral's cabin was eliminated in the 1980's, the ladder was removed and the deck opening was closed off to create a storage locker.

(6) This light lock accessed the flag plotting room when the ship was built. The inner door is armored, as is the rest of the compartment. There is another light lock that gives access to the admiral's bridge at the starboard side. During the 1980's, the flag plotting room became the Tomahawk computer room.

(7) This door leads to the Signal Shelter. The door and the shelter are armored. An armored trunk that formerly housed a hoist for 40mm ammunition is visible in the background. A passing scuttle is to the left of the hoist.

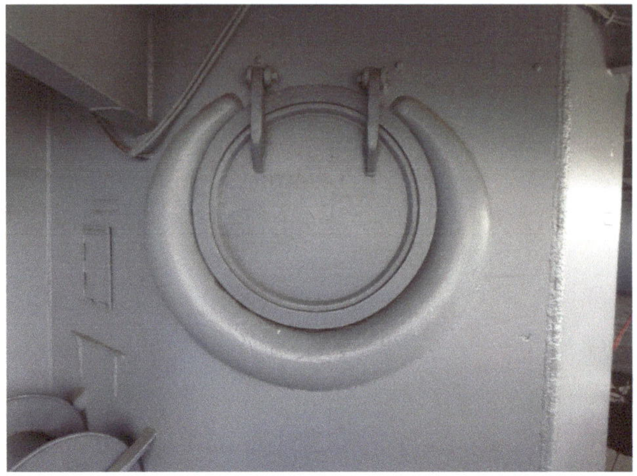

(8) The passing scuttle (visible through the door in photograph (7)). This allows passing messages and signal flags to the flag and signal lamp operators at the 03 level. A crank on the inside opens the scuttle. This scuttle is mirrored on the port side.

(9) The starboard side entrance to the flag bridge. A chair for the admiral is visible poking up above the bulwark. There is another chair on the port side. The locker at the center was designed for storing hand grenades. It was probably used for other purposes.

(10) The port signal flag board. The lines are signal halyards that run up to the signal yardarm. Normally there is a step at the bottom of the flag board but it was removed to replace the decking and had not yet been re-installed.

(11) The starboard flag board viewed from above with the cover removed. The flag board keeps the signal flags organized by having slots for each type of flag. The sailor is standing on the step that had not been replaced in photograph (10). (Nat'l Archives)

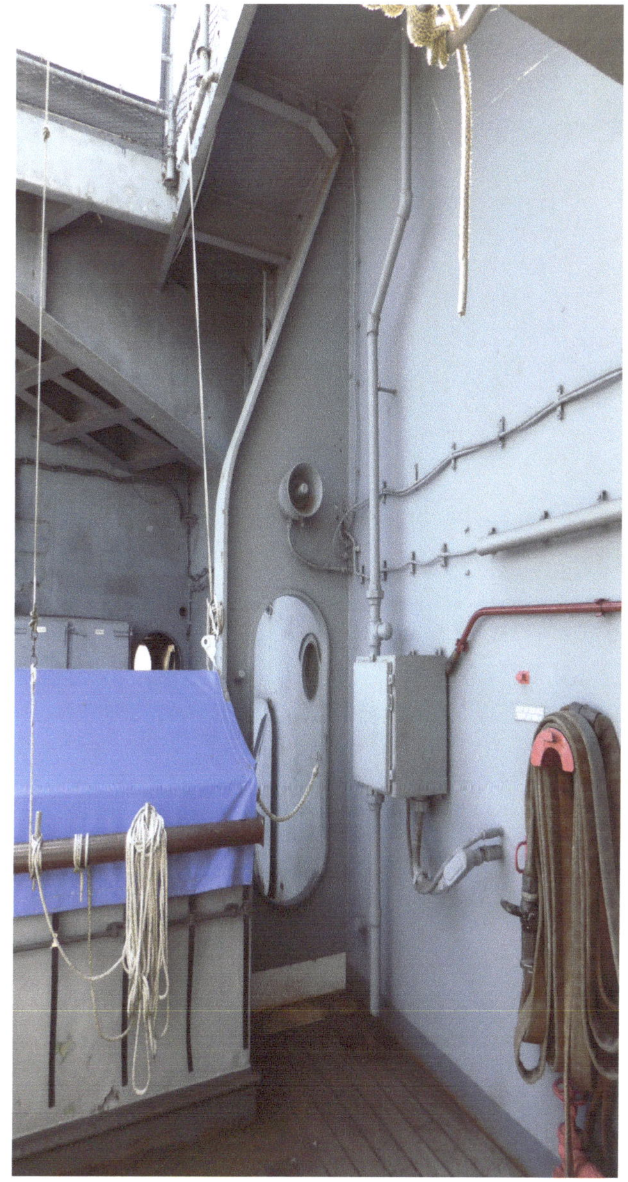

(12) This support is part of the aft foundation of the forward fire control tower. The support has to widen at the 03 level so that it can span the uptake space below. The extensions beyond the side of the deckhouse transfers the weight of the fire control tower to the bulkheads below at either side of the uptake opening.

(13) Between the main deck and 05 level the uptake spaces are open areas with no decks. This is the view down the forward uptake space on the USS *Iowa*. The uptakes for fire room no. 2 are visible to the left. Those for fire room no. 1 are behind the bulkhead to the right. The uptakes casings for the two fire rooms come together at the 04 level. The screened opening in the background is the boiler air intake for fire room no. 1. The intake air flows unconstrained through the uptake space. The armored trunk for the port director Mk.37 is visible beyond the screen. (Michael Getscher)

(14) This view is downward from the 03 level through the aft section of the forward uptake housing. The armored main deck air intake gratings for boiler room no. 2 are visible in the main deck. The four pipes are exhausts that emerge through tubes at the top of the uptake housing. The two outer pipes are atmospheric exhausts for venting steam in the boilers. The inner right pipe is an auxiliary exhaust and the inner left pipe is the forward emergency diesel generator exhaust.

(16) The navy added two platforms for Tomahawk missiles to *Iowa*-class battleships in the 1980's. The forward platform mounted four quadruple Tomahawk Armored Box Launchers (ABLs) and the four quadruple Harpoon launchers. This area originally had 40mm machine gun mounts at the sides that were elevated above 03 level to improve their field of fire. The forward missile platform was created in alignment with this elevated 03$\frac{1}{2}$ level. This modification created a step up from the forward part of the 03 level to the missile platform.

(15) This is the view upwards from the same location as photograph (14). The 05 level is visible at the top. A muffler for the diesel exhaust is located above 05 level. This area serves as the air intake for boiler room no. 2. The port air intake at 04 level is visible the left. Much of the structure shown is made out of brass because it is located near the master compass.

(17) The museum has opened one of the ABLs for display showing a dummy Tomahawk missile breaking through its sealed canister. The forward launchers were arranged so that the missiles fired over the opposite side. This allowed the blast from the launch to be directed over the side and eliminated the need for elaborate heat protection at the center of the deck. *New Jersey* was the first ship equipped with ABLs so it got a mix of early models as they became available. The two forwardmost ABLs on USS *New Jersey* have rungs at their left front and right back ends while the others have a single ladder at the back end on the side closest to the center of the deck.

(18) The starboard ABLs on the forward missile platform. The tracks in the deck were added in 1983. A gantry crane could be erected over the rails that was used to reload the ABLs.

(19) A turbofan engine similar to those frequently used on light business jets powers the Tomahawk missile. However, the missile is launched using solid fueled rocket. The rocket exhaust gas is about 5,000°F and contains hydrochloric acid. To protect the ship's structure, there is three inches of blast shielding over the deck at the rear of the ABLs. The 30-degree upward slope at the edge was intended to divert rocket blast away from the boats below.

(21) The underside of the forward missile platform. The view here is upwards and forward. This area originally contained a platform for two 20mm mounts. In the 1980's the deckhouse was extended upwards from the 01 level to meet the missile platform and create space for equipment rooms for the Tomahawk missiles.

(20) A platform for two 14-foot transmitting antennas was added to the forward port side of the missile platform. The red base indicates a transmitting antenna.

(22) This photograph shows the forward missile platform in September 1982, three months before USS *New Jersey* was recomissioned for the final time. Several features had not yet been installed, including the Harpoon missiles, Tomahawk ABLs, the bulwark around the boat cradle, and the skip box and tracks for 20mm CIWS ammunition—so the door to the 20mm magazine at the 04 level appears to go nowhere.

There is a distinct step down in the deck near the empty Harpoon launchers from the 03 $\frac{1}{2}$ level to the 03 level.

The large triangular structure on the forward boat davit arm is due to its short stop on the deck. The port davit arm stops are tall, like the aft stop here. Also note the lack of diagonal stanchions from the 02 level to the aft 03 missile platform above the large air intake. Those stanchions were added when the missile platform was extended. (Nat'l Archives)

(23) A 40mm machine gun mount originally occupied the location of the ABL at the left. The scuttle in the center originally supplied 40mm to that mount from a handling room. In the 1980's, a platform for CIWS mounts was erected above the former handling room.

(24) In the 1980's the area between the 40mm handling rooms was enclosed to form a workshop. There is a step between the 03 level of the workshop to the 03 $\frac{1}{2}$ level of the missile platform forward. The center section of the missile platform between the former 40mm mounts forms a plenum chamber that feeds the ventilation air for engine room no. 2 and fire room no. 3. The opening below the door covered by expanded metal leads to this plenum chamber.

(25) Conforming to the original deck elevations in this area created construction oddities after the 1980's modifications. This is the forward port corner of the gear locker. Originally this area was open and held a 40mm mount with a circular shield. In the 1980's the straight deckhouse sides were extended vertically to the $03\frac{1}{2}$ level and the shield was extended to the 04 level. This modification created a recess at the corner between the 03 level and $03\frac{1}{2}$ levels,

(29) This is the starboard side of the deckhouse at the $03\frac{1}{2}$ level. The starboard side originally had a curved walkway as well. During the 1980's it was removed so that a lift for 20mm ammunition could be added. The skip box for carrying ammunition is shown locked in place at the 04 level CIWS magazine. A plate in the $03\frac{1}{2}$ level must be removed to allow the car to be lowered. The car is not powered. Instead, it was raised and lowered using a portable davit that could be erected at the 05 level.

(26) This curved walkway originally connected the two port 40mm mounts at the $03\frac{1}{2}$ level. It was left in place during the 1980's modifications and it now wraps around the CIWS platform to connect the forward and aft sections of the missile platform. The area shown is directly above the recess in photograph (25).

(28) This is the underside of the 03 $\frac{1}{2}$ level at the port side of the CIWS platform. The track for ammunition loading runs along the side of the deckhouse down to the 01 level.

(29) This is the starboard aft Harpoon launcher. The launcher can hold four missiles in sealed tubes.

(30) The starboard Harpoon launchers viewed from above. Each launcher has a blast shield to protect the deck and deckhouse. The 04 level wraps around the aft uptake housing. The ladder between the blast shields accesses the 04 level. The 03 level in this area is covered but not enclosed.

(31) The aft missile platform mounts four Tomahawk ABLs. This is the front end of ABL no. 5 on the aft missile platform. The legs on the ABL foundations have different lengths. The end of each leg is aligned to a beam supporting the deck. Because the four aft ABLs are at angles, the legs are different lengths to align with a beam. The doors are for maintenance access and remain closed during firing. The firing end of the ABL elevates over these doors. The air conditioner to the right is a museum addition.

(32) The rear end of ABL no. 5 showing the breech doors. The missile doors have to be opened for firing. The deck has three inches of blast shielding behind the ABLs. The ABL elevation split line is visible on ABL no. 7 to the right.

(33) The roof of ABL no. 7 showing the access doors.

(34) The port side of the aft missile platform looking forward. This missile platform was extended outward in 1983 to accommodate the gantry crane tracks. The Harpoon missile platform overhangs the 03 level Tomahawk platform.

(35) This is the port side of the aft uptake housing at the 03 level. The 04 level extends from the uptake housing to the bulkhead on the Harpoon platform to create a u-shaped tunnel underneath. The door at the center is to a security station added in the 1980's. This houses communications equipment. There is another security station mirrored on the starboard side.

(36) The boiler intakes for fire room no. 3 are at the sides of the uptake housing. They were sealed up when the ship was retired.

(37) The 03 level aft of the uptake housing originally was very narrow and held a 5-inch gun practice loading machine. This is one of two decanning scuttles formerly used for gun loading practice at this level.

(39) During the 1980's blast shielding was added to protect the director towers. The forward part of the shield was mounted directly on the existing fire control tower. A support for the shielding was added to the aft edge of the tower. The 3-inch deck shielding is visible as well.

(38) These are the boiler intakes for fire room no. 4. These intakes flow straight down to main deck. There is a storage locker behind each intake. The aft director tower is at the center.

(40) This is the area aft of the fire control tower showing the supports added for blast shielding. This area was used to store helium for weather balloons in the 1980's. Originally, it held two directors for 40mm mounts.

(41) This is the same area as in photograph (40) looking aft at the base of the aft director Mk.37 (Sky 4) and its armored door.

(42) The aft end of the 03 level had to be raised 12 inches to accommodate the helicopter control station at the 02 level that had to be raised above 03 level to clear turret no. 3. The escape hatch from the helicopter control station is visible in the deck.

(43) Here is how the aft end of the 03 level appeared in 1984. The row of floodlights could illuminate the helipad. The aft director Mk.37 is at the top of the photograph. (Nat'l Archives)

(44) Access to the FAS kingpost is at 03 level. The kingpost was added in the 1980's to facilitate refueling escorts.

(45) The underside of the aft 03 missile platform. Note the angled stanchions that were added in 1983 to support the extended platform. Compare with photograph (22).

(46) This is the forward end of the aft 03 missile platform. The angle of the forward edge shows where the platform was extended.

04 Level

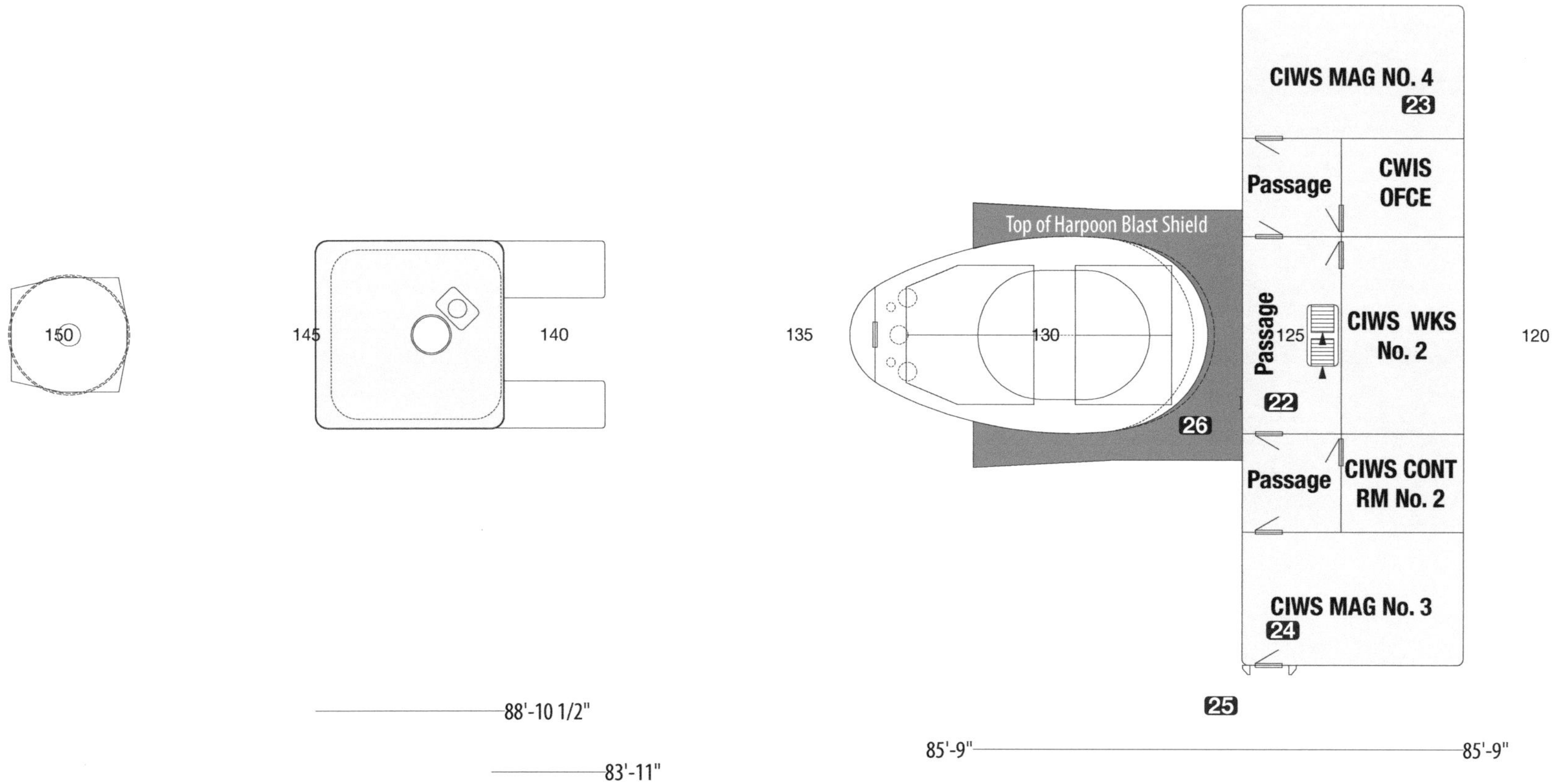

04 Level
1"=15'

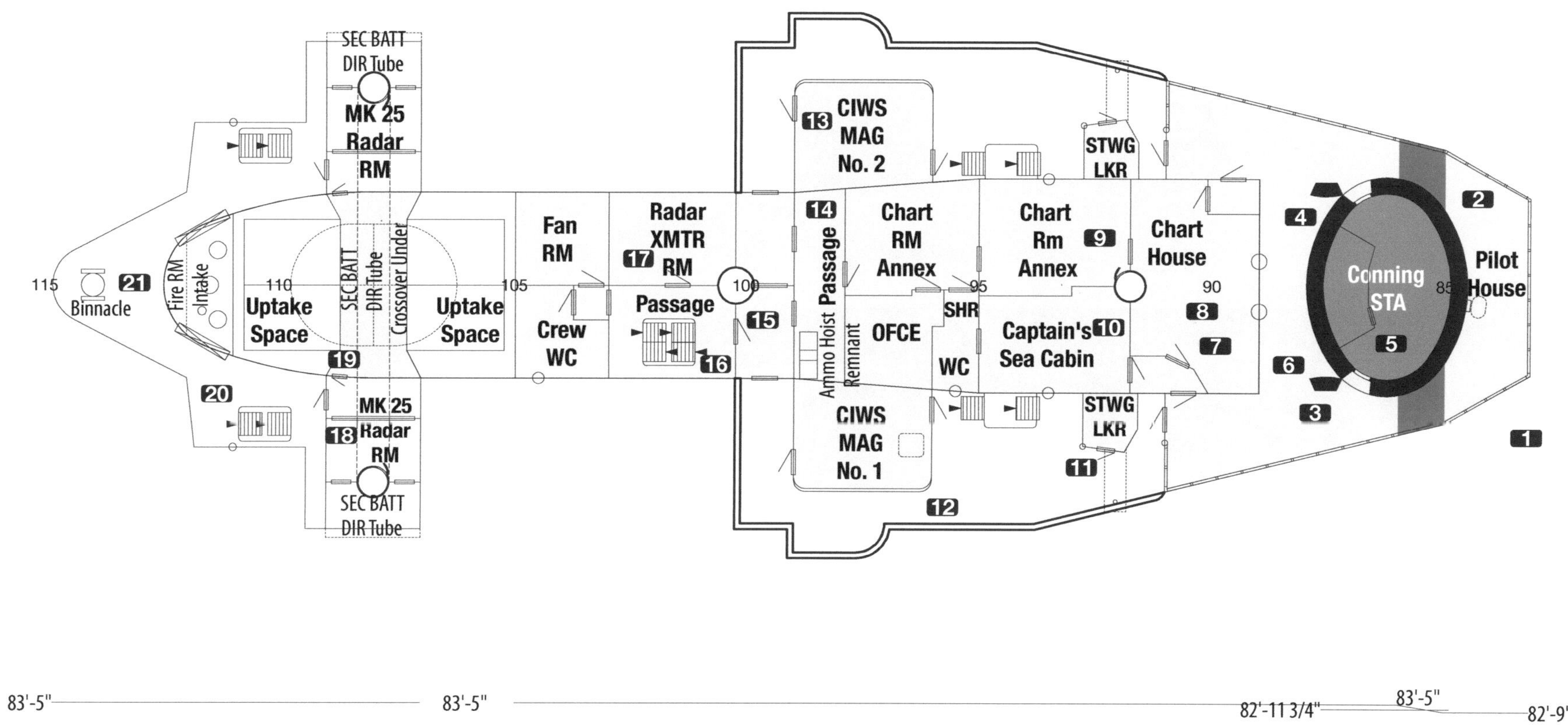
SEC BATT
DIR Tube
MK 25
Radar
RM
CIWS
MAG
No. 2
STWG
LKR
Chart
RM
Annex
Chart
Rm
Annex
Chart
House
Pilot
House
Conning
STA
Binnacle
Fire RM
Intake
Uptake
Space
SEC BATT
DIR Tube
Crossover Under
Uptake
Space
Fan
RM
Crew
WC
Radar
XMTR
RM
Passage
Passage
Ammo Hoist
Remnant
OFCE
SHR
WC
Captain's
Sea Cabin
CIWS
MAG
No. 1
STWG
LKR
MK 25
Radar
RM
SEC BATT
DIR Tube
115
110
105
100
95
90
85
1
2
3
4
5
6
7
8
9
10
11
12
13
14
15
16
17
18
19
20
21
83'-5"
83'-5"
82'-11 3/4"
83'-5"
82'-9"

(1) The flat front bridge was created in June 1945. The front face is not centered on the centerline with the starboard side being one inch wider than the port side. The two rows of railing around the face provide a place to stand and to hook up to while cleaning the windows. The 35-foot whip antennas at the front face were added in 1968.

(2) The flat-faced bridge allowed one to see from one side to the other. The captain's chair is at the starboard side of the bridge. The armored conning tower is to the right. The front of the bridge is six inches below the 03 level. The bridge slopes down at the sides of the conning tower.

(3) At 17.3" the conning tower has the thickest armor on the ship. Five plates make up the conning tower sides. The plan of the conning tower forms four tangent arcs that approximate an ellipse. The side plates are joined using hourglass-shaped keys and are welded at the edges.

One of the plate joints passes through the center of this door opening. The weld seams are visible above and below the opening. At the top of the opening, the outline of the key faintly visible. The wedge shape at the side of the door allows the door to stay in place if the hinge is shot away. The door operating machinery is located in the conning tower support at 03 level. The removable crank shown extends and retracts the bolts. Each of the three bolts has to be operated separately.

(4) The conning tower door had a mill weight of 5,354 lbs. A two and half ton door flapping around in a seaway would create a life-threatening hazard. When the door was open it was secured to this fixture.

(5) The main feature of the 04 level conning tower is the steering stand with its engine order telegraph. There is a radar repeater behind the telegraph. There is a limited direct view through vision slits. Two periscopes provide a magnified view outside. The removable crank in the foreground can open the door manually.

(6) There is a narrow passage between the aft end of the conning tower at the right and the charthouse at the left. This area was originally open to the weather but was enclosed shortly after the ship was commissioned. The handle at the upper left, just below the girder, blows the ship's horn.

(7) This is the charthouse looking to port. The door in the corner is to the port side light lock. There is another light lock to starboard. There is plenty of room to spread out charts and to store them.

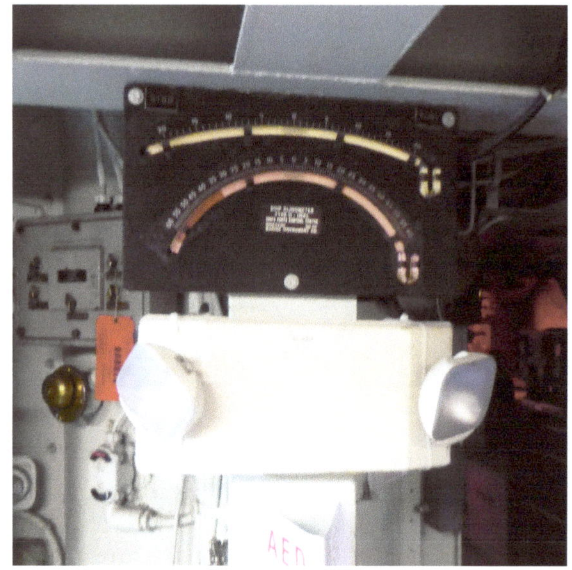

(8) This inclinometer in the chart house indicated the amount the ship was rolling.

(9) The chart house annex was originally a radar room and the navigator's sea cabin. The use of radio navigation systems, such as LORAN, diminished the need for having the navigator immediately on the bridge. Eliminating these spaces allowed the chart house to be expanded.

(10) The captain's primary job is to operate the ship. When the ship was at sea the captain slept in this sea cabin so that he could quickly reach the bridge if summoned by the officers on watch. The archway in the center leads to a bathroom.

(12) This is the starboard bridge wing. There is a mounting point for an alidade and a captain's chair. In the 1980's the bridge wings were extended aft from Frame 99. The wings on the other *Iowa*-class battleships were extended aft and outward during the 1980's.

(11) Storage lockers at either side of the bridge are a unique and original feature of USS *New Jersey*. They were used to support 35-foot whip antennas installed in 1968.

(13) The port CIWS magazine. This area was originally open with a 40mm mount cantilevered over. The tapered beam at the top hints at the original structure. This area was enclosed during the 1980's to create this magazine. There is a hatch in the overhead to haul ammunition from the magazine to the CIWS mount directly above

(14) This area was originally a breezeway that connected the two bridge wings. The remnant of the 40mm ammo hoist that runs from second deck passes behind the bridge. There was originally an archway to the left (forward) of the hoist trunk that leads to the bridge wing. CIWS magazines were added to both sides in the 1980s' at the other side of the bulkhead so the archway was closed off.

(15) In the 1980's this vestibule was converted into a breezeway to link the two bridge wings after the archway shown in photograph (14) was blocked off. The view here is aft. Archways were cut into the deckhouse aft of the new CIWS magazines. When the ship was mothballed these archways were sealed.

(16) Dual ladders leading up the forward director tower allow traffic to flow in both directions.

(17) The radio transmitting room. The equipment spaces tend to be very small within the forward director tower.

(18) The radar equipment below the starboard Mk.37 director (Sky 2) was upgraded over time. However, this has always been a radar equipment room. The director tube runs through the room and is open here showing its $1\frac{1}{2}$" thickness.

(19) The uptake casings for boiler rooms nos. 1 & 2 meet here at 04 level. The crossover trunk, where the duplicate cables for the two side Mk.37 directors (Sky 2 and Sky 3) pass to the opposite side, is visible below.

(20) This is the starboard air intake for fire room no. 2. There is another intake mirrored on the port side. The intake air for fire rooms nos. 1 and 3 flows through their uptake spaces. The intakes for fire rooms nos. 2 and 4 flow outside their uptake spaces.

(21) The ship's master binnacle compass. It supplies heading information to compass indicators and fire control systems throughout the ship. There is no steel within twelve feet of the compass. Brass was used instead for the superstructure in this area.

(22) This is the transverse passage at the 04 level below the aft CIWS platform.

(23) There are CIWS magazines at both sides of the 04 level below the CIWS platform. This is the port magazine. The racks held 20mm ammunition.

(24) 20mm CIWS ammunition was loaded through this door in the starboard CIWS magazine.

(25) Ammunition was hauled up to the CIWS magazines using this skip box that runs from 01 level to 04 level. The skip box is parked at the door shown in photograph (24). The skip box is not powered. It had to be hoisted up using a portable davit erected at 05 level.

(26) These are three views from the top of the Harpoon missile platform shields. (Left) The 04 level here creates a breezeway over 03 level around the aft smoke pipe casing. This area is accessed from ladders behind the missile launchers. (Below) Directly above is the mainmast. The appellation *mainmast* refers to its location. The mainmast is modest in size and is much smaller and much less important than the foremast. The platform at the center of the mainmast was a late addition for a Pioneer RPV communications antenna. Fan antennas run from both ends of the mainmast to the foremast. (Right) The right ladder leads from the 04 platform leads to the 05 CIWS mounts. The left ladder leads up the smoke pipe housing.

05 Level

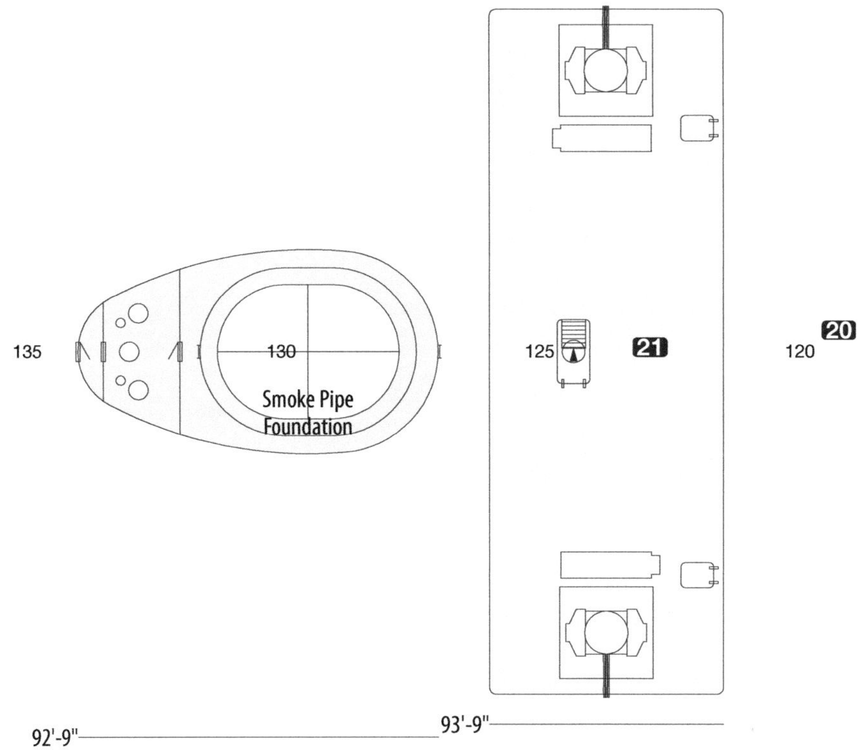

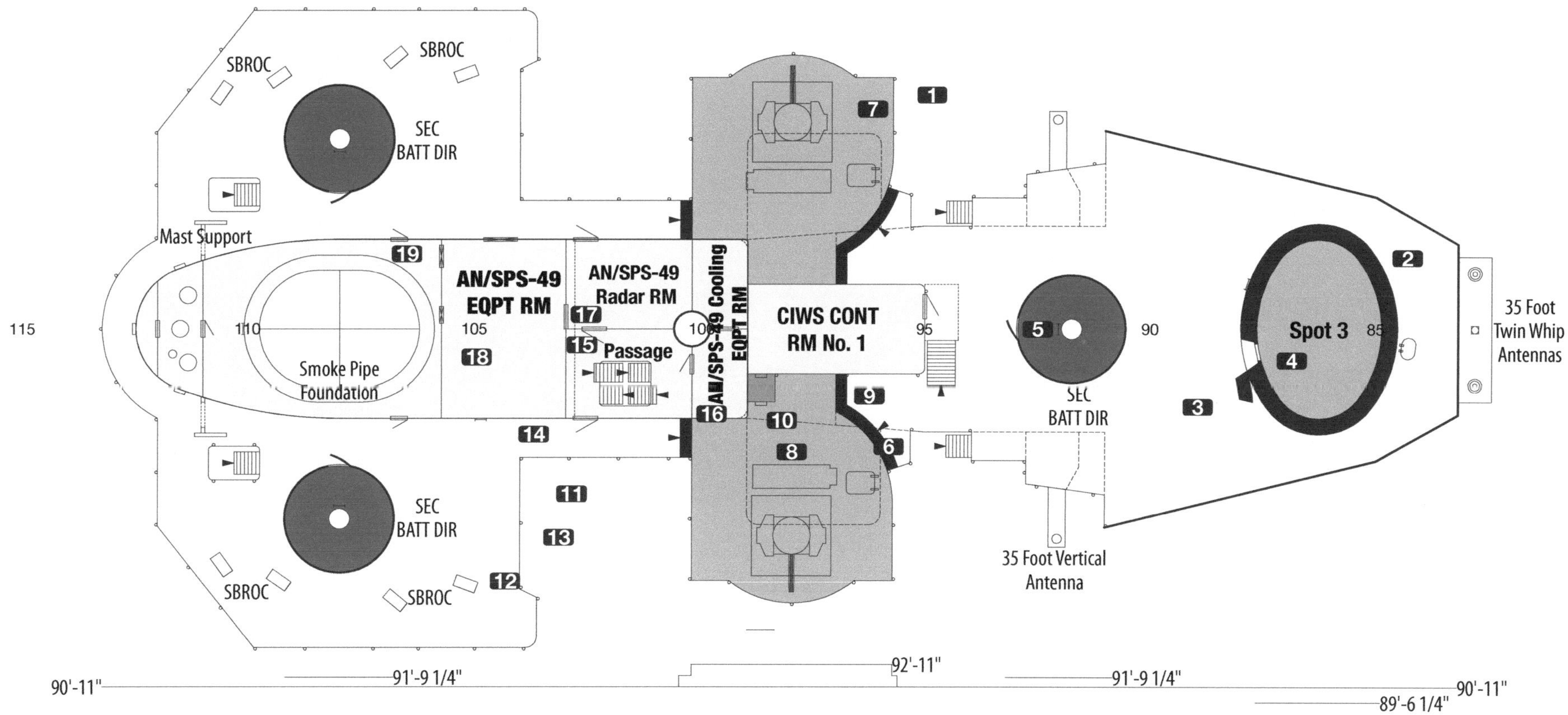

SBROC
SBROC
SEC BATT DIR
Mast Support
AN/SPS-49 EQPT RM
AN/SPS-49 Radar RM
AN/SPS-49 Cooling EQPT RM
CIWS CONT RM No. 1
Passage
Smoke Pipe Foundation
SEC BATT DIR
Spot 3
35 Foot Twin Whip Antennas
35 Foot Vertical Antenna
SEC BATT DIR
SBROC
SBROC
115
110
105
100
95
90
85
1
2
3
4
5
6
7
8
9
10
11
12
13
14
15
16
17
18
19
90'-11"
91'-9 1/4"
92'-11"
91'-9 1/4"
90'-11"
89'-6 1/4"

05 Level
1"=12'

(1) The forward section of the 05 level dominates this 1984 photograph. USS *New Jersey* was passing through the Panama Canal at the time. (Nat'l Archives)

(2) There is an exposed conning station at the forward bulwark. The tallest device at the center is a pelorus. The instruments to the left are indicators for the rudder angle, engine orders, and propeller revolutions. The conning station lacks a steering stand and an engine order telegraph so orders had to be transmitted from here by telephone.

(3) The top level of the conning tower is the fire control level. It houses the director Mk.40. There is a single 5,084 lb. mill weight door. There is also an escape hatch that leads to the lower level of the armored conning tower. Other than being slightly smaller and lacking a vision slit, this door is similar to and operates in the same manner as the armored doors at 04 level. Prior to recommissioning in 1982, a commercial marine radar was added to the roof of the conning tower. (Ryan Szimanski)

(4) The director Mk.40 inside the conning tower was known as *Spot 3*. This director could spot fall of shot but lacked a rangefinder. While this director does not have the view and range of the Mk.38 directors in the fire control towers, it is located in the most heavily armored location in the ship. If an *Iowa*-class battleship had been in an intense ship-to-ship battle that resulted in both Mk.38 directors being disabled, it is likely the target distance would have closed to the point where the Mk.40 director could be very effective. During the destruction of *Bismarck* the opposing battleships came within 3,000 yards. (Ryan Szimanski)

(5) The forward Director Mk.37 (Sky 1) support tube looking upwards into the director. The Rangefinder Mk.42 is visible through the hatch. The hatch has to be closed to operate the rangefinder. The circular rack gear used to traverse the rangefinder is visible at the top of the photograph. Two holding down clips that keep the rangefinder from pulling out also are visible, wrapped around the rack gear. The ladder slopes backwards, making it difficult to climb.

(6) This photograph shows how the 35-foot whip antennas at the bridge sides are mounted on an extension to the 04 level deck locker. The blue base indicates this is a receiving antenna. The antenna at the opposite side is a transmitting antenna. The two lockers shown at the 05 level stored illumination pyrotechnics.

(7) The forward starboard Phalanx Close in Weapon System (CIWS) mount. The Phalanx is a last line of defense weapon against incoming missiles, with an effective range of about 1,500 yards. The Phalanx operated independently from the ship's fire control systems. It just had to be turned on. The white dome at the top contains two radars. One is a search radar that scans for targets. When the CIWS identified a target, a targeting radar took over for aiming and firing the weapon.

The Phalanx has a six-barreled hydraulically powered Gatling gun that fires 20mm projectiles at 3,000 rounds per minute. The six bolts follow a cam path that opens the breech for extraction of the spent cartridge and closes the bolt for loading the next cartridge as the barrels rotate. The cartridges are electrically fired.

The mount holds 989 round so there is only about 20 seconds of firing time before reloading. Ready service lockers for reloading are located behind the mount and there is a magazine directly below. There is a hatch to the magazine in front of the mount with a fitting for a portable davit to raise ammunition to the gun next to it. The box at the rear of the mount is the shock mounted electronics unit.

(8) The ammunition for the Phalanx is stored in a drum under the gun. The Phalanx uses belt fed ammunition, unlike similar aircraft guns that use a linkless feed. There are two cartridge paths because spent cartridges are returned to the drum. The gun is externally powered so any cartridges that misfire are simply extracted and returned to the drum.

(9) The forward Phalanx guns are on elevated platforms that originally held 40mm mounts. One has to go up two steps to reach the guns. These are the steps to the starboard mount. The lockers held ready use ammunition.

(10) This box is the terminus of the 40mm hoist trunk that ran to second deck. The ladder provides access to the ship's bell at the 07 level.

(11) The starboard Super Rapid Bloom Offboard Countermeasures (SRBOC) platform. Each platform has four, six barreled SBROC mortars. These mortars can fire projectiles that release chaff to confuse radar guided weapons or flares to confuse heat-seeking weapons. The lockers on the platform contained reloads. The starboard Director Mk.37 (Sky 2) sits in the center of the platform.

(12) A close up view of a SRBOC mortar. The barrels have been closed for museum display. When loaded, the projectiles extended past the muzzle.

This photograph shows crewmembers of USS *Wisconsin* reloading SBROC mortars in August 1991. (Nat'l Archives)

(13) The underside of the SRBOC platform hints at its original configuration. There were curved platforms fore and aft of the Mk.37 director. The forward platform had a 20mm mount and the aft platform had a searchlight.

(14) A walkway connects the SRBOC platform to the forward part of the 05 level. There are stairs on the aft side of the CIWS platform as well as the forward side. The door to the left accesses the stairway passage of the forward fire control tower.

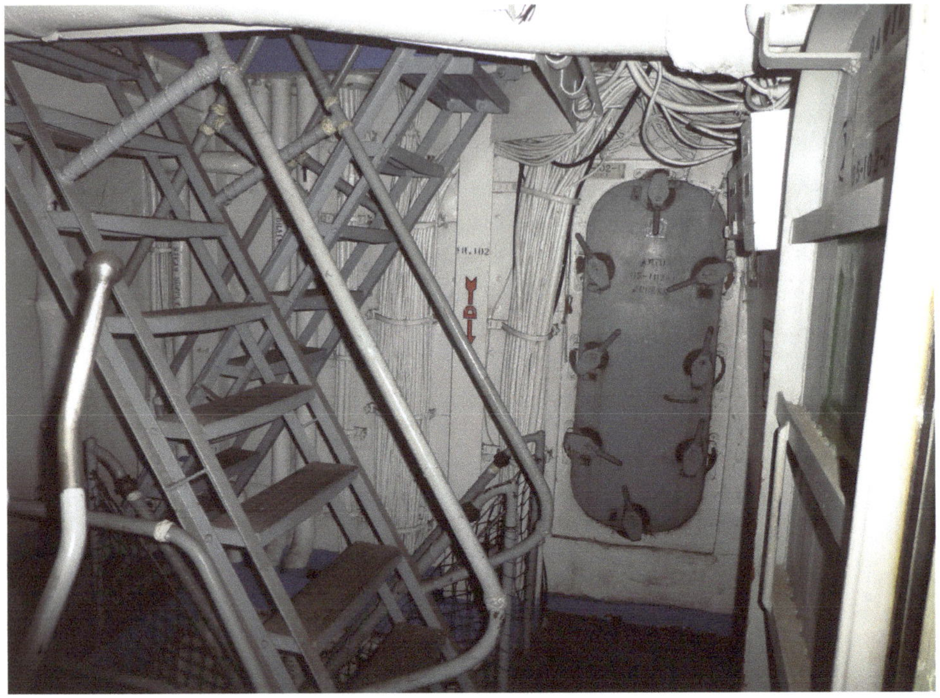

(15) This is the door shown in photograph (14). The ladders shown here are the route up the fire control tower.

(17) The Navy scavenged the equipment from the AN/SPS-49 radar room when the ship retired. This view is to port with the door to weather deck in the background.

(16) The 05 level of the forward fire control tower was devoted to equipment for the AN/SPS-49, the large air search radar at the top of the foremast. This is the cooling equipment.

(18) The AN/SPS-49 equipment room. The arch to the right leads to the radar room. There is a removable plate to weather partially visible behind the open equipment door. The Navy cleared out nearly every piece of equipment when the ship was mothballed. The only exception is the electrical equipment at the upper left with the dangling cables whose removal was obstructed by the light fixtures. Providently, this equipment controls the rotation of the radar antenna. The museum was able to reconnect it to get the radar dish rotating again in May 2021.

(19) The four uptakes for boiler rooms nos. 1–2 join together at the deck level to form the smoke pipe. However, each uptake has a separate path within the smoke pipe. Cooling air is drawn in from the top of the smoke pipe casing. At the 05 level the cooling air enters the rectangular holes at the base of the smoke pipe where it flows through the outer casing of the uptake to main deck.

(20) The aft CIWS platform was constructed in the 1980's above the former location of 40mm mounts. As with the forward mounts, the aft mounts sit directly over their magazines. The four lockers between the mounts held ready use ammunition. The large hatch with a scuttle provides personnel access from below. There are racks aft of the personnel hatch for storing two J-davits but the davits were in storage.

(21) These are views of the aft Phalanx mounts starboard (left) and port (below). Each mount has a deck hatch next to it for ammunition loading.

06 Level

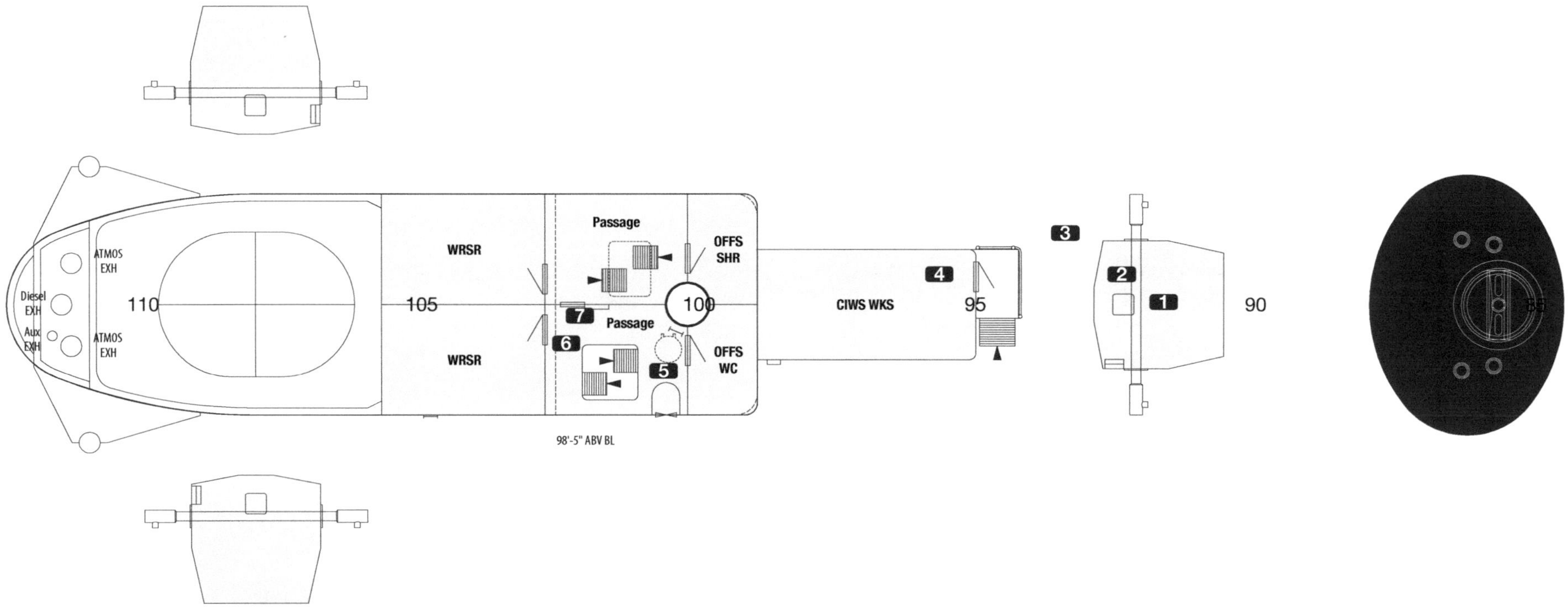

06 Level
1"=12'

(1) A panorama view of the front of the forward Director Mk.37 (Sky 1). The Mk.37 director can operate in four modes. First, in automatic mode the fire control computer aims the director. If the target becomes obscured, the computer points the director where it thinks the target should be and the director crew can manually correct the aim. Second, in local mode the pointer and trainer move the director with power drive. Third, in manual mode the point and trainer move the director through direct gearing. Finally, in slewing mode the director follows the director officer's slewing sight.

The director officer's position is at the left. The officer controls the slewing sight, the large box at the center of the photograph. When the slewing sight is active, the director follows where the sight is pointed. This sight allows the director to move quickly to a target or to change targets. Once the director is on target, the officer transfers control to the pointers (controlling elevation at the center) and the trainer (controlling traverse at the trainer) for precise aim.

(2) The Rangefinder Mk.42 divides the director in two. The Director Mk.37 was designed in the 1930's before there were practicable radars. The optical rangefinder was expected to be the primary means of determining range to target. Once radar became available it was quickly incorporated not the directors. The directors onboard *Iowa*-class battleships had radar from the start. The Radar Mk.25 is located to the right of the rangefinder. Sky 1 upgraded to this radar during the Korean war.

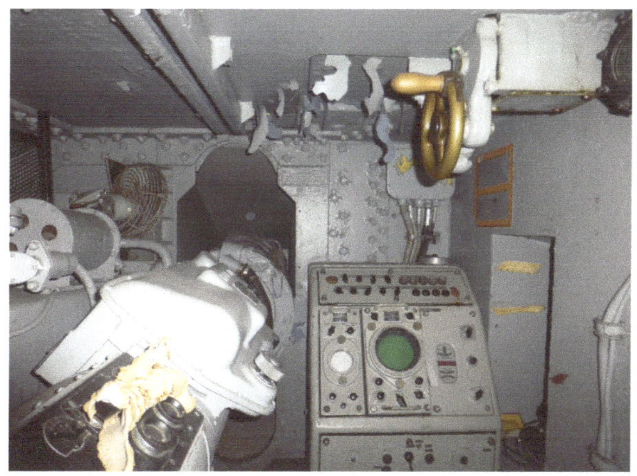

(3) This structure at the face of the forward fire control tower has compartments at the 05 and 06 levels. The 05 level compartment was a 40mm magazine during World War II and became a CIWS control room in the 1980's. The navy completely stripped it when the ship was decommissioned. The forward end at 06 level originally housed a radar transmitting room while the rest of the 06 level was open. The entire structure was enclosed at 06 level during the 1980's to become a CIWS workshop.

(4) This the CIWS workshop was largely cleared out when the ship was mothballed. A workbench remains to suggest its final purpose. This area was enclosed in the 1980's.

(5) A circular opening is visible at the starboard side of the forward director tower. That opening leads to this ventilation trunk that provides cooling air to the electronics equipment at 07 level.

(6) The 06 level has two officer's cabins that appear to have never been upgraded along with a shower and water closets that fell into disuse. The cabins are aft of the structural bulkhead for the fire control tower. This bulkhead is sloped so this recess had to be added to install a vertical door.

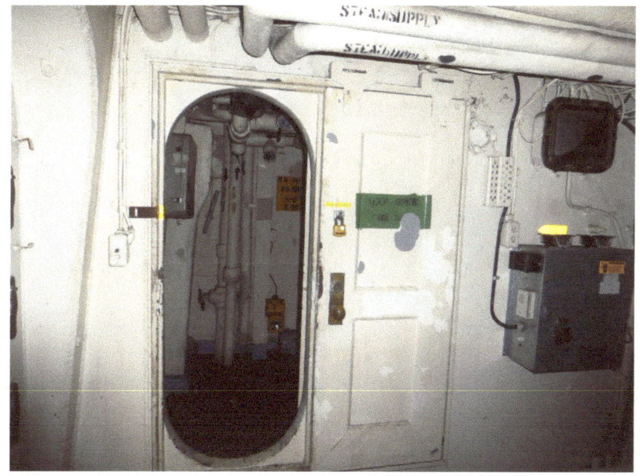

(7) A bulkhead divides the fire control tower into port and starboard halves. The ladders in the fire control tower are on the starboard side to 06 level and on the port side above that. This unusual sliding door provides access between the two sides.

07 Level

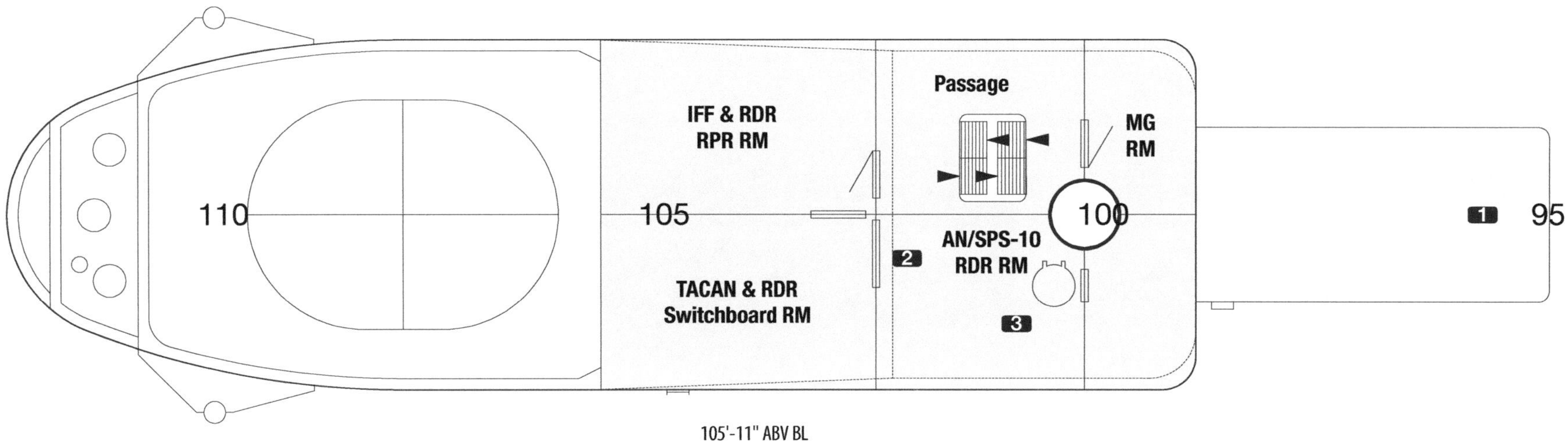

07 Level
1"=8'

(1) The ship's main bell. There is another, smaller bell, at the forward end of the discone antenna at main deck.

(2) 07 level was used to house various electronic systems over the years. In World War II it housed a radar room. During the 1960's the port side of this area held the equipment for controlling QH-50 drone anti-submarine Helicopters (DASH). DASH was originally designed to carry anti-submarine torpedoes. Some were modified to carry a television camera so that they could be used for target spotting. This variant was known as *DASH Snoopy*. USS *New Jersey* never carried DASH drones. Instead, the drones were based on destroyers and USS *New Jersey* took over control after they were launched.

In the 1990s this area held IFF (Identification Friend or Foe), TACAN, and AN/SPS-10 radar equipment. The Motor Generator Room forward supplied 400 Hz power for radars. The Navy emptied out the 07 level when the USS *New Jersey* was retired. This is the view aft from the AN/SPS-10 radar room.

(3) 07 level was a secure area. During the 1980's modernization, one of the doors from electronics areas to the passage was blocked off. This escape scuttle in the deck was added to create a second exit route.

08 Level

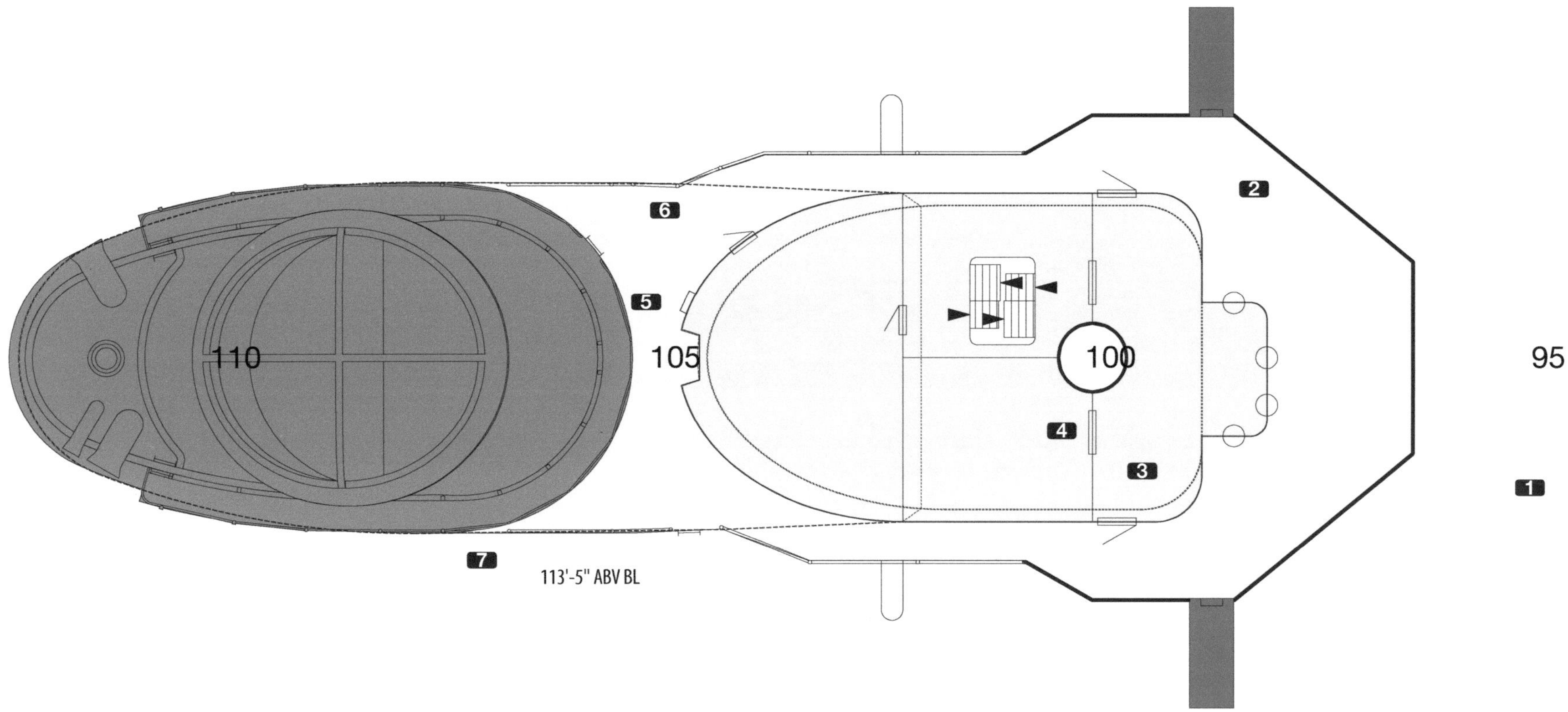

08 Level
1"=8'

(1) The 08 level housed the primary conn for the ship. This functional description is misleading because the primary conn was rarely used except when navigating rivers and inlets. The ship was normally steered from the secondary conn at the 04 level.

Adding to the confusion, some sources call this the *secondary conn* with the primary conn at the 04 level. Nonetheless, the switches for routing controls designate this as the *primary conn.*

(2) The primary conn has both enclosed and open areas. This is the open-air area of the primary conn.

(3) The steering stand (center) and engine telegraph (left) in the enclosed area of the primary conn. USS *Iowa* is similar. USS *Missouri* and USS *Wisconsin* have a semicircular enclosure housing the viewing scuttles.

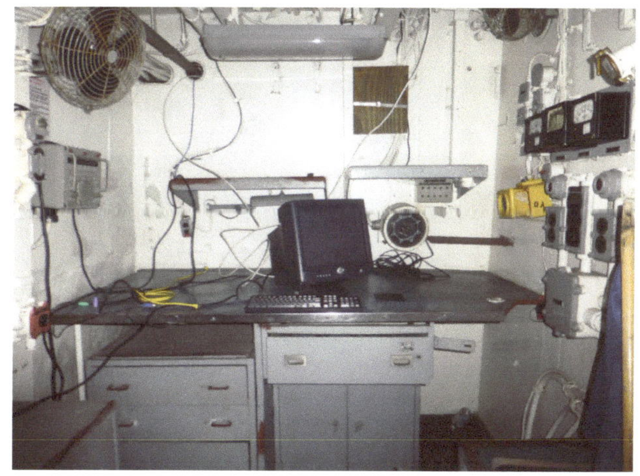

(4) This small charthouse is located directly aft of the conn. The meteorological facility is in the aft part of the level 08 enclosure. This aft area was originally a second sea cabin and water closet for the captain.

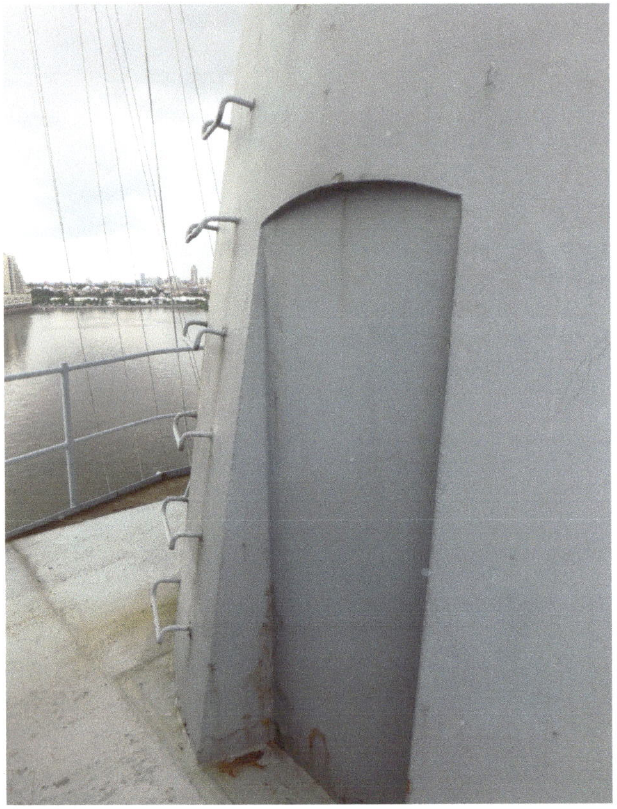

(5) *Iowa*-class battleships were built with this recess at the aft end of the director tower at the 08 level to accommodate foremast wiring. Unlike those of her sisters, the foremast of USS *New Jersey* never extended down to 08 level to make use of this recess.

(6) This ladder from the 08 level gives access to the walkway around the forward smoke pipe cowl.

(7) This is a downward view of the walkways around the forward smoke pipe. A cover with exhaust piping running from each of the four smoke pipe divisions was welded over the smoke pipe opening when the ship was retired. The railing around the top of the smoke pipe was used to secure a canvas cover over the opening. Below that are shorter foot rails at the side that provided a place to stand while tying down the canvas cover. The ladder at the aft end of the walkway gives access to the interior of the smoke pipe cowl.

09 Level

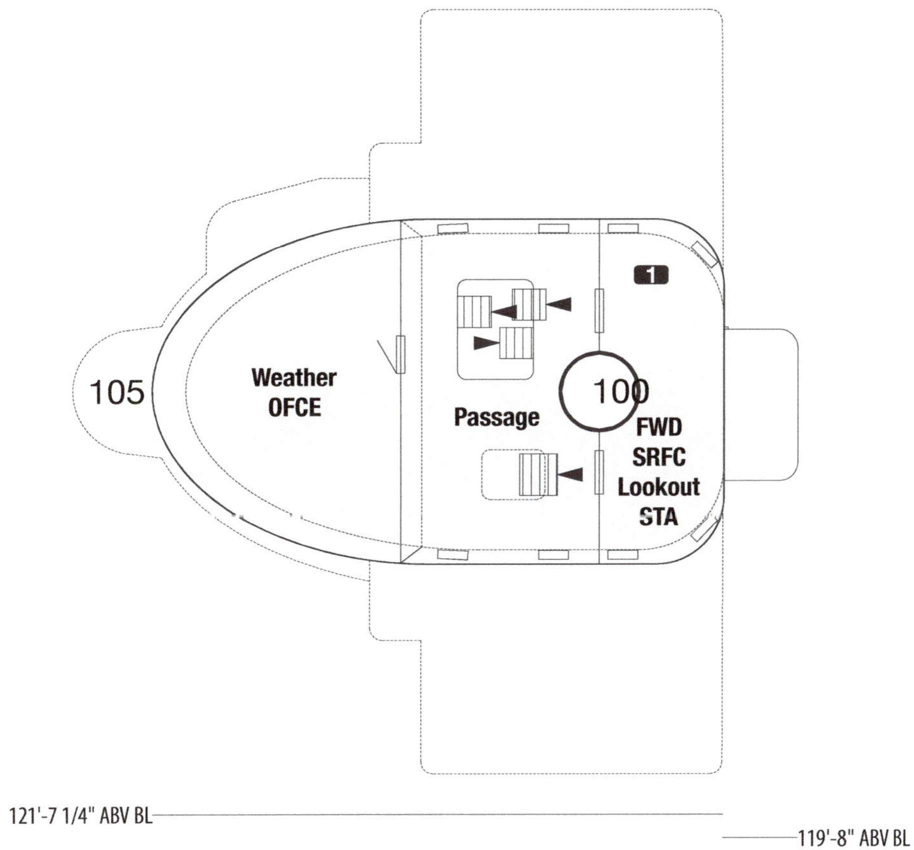

09 Level
1"=8'

(1) The forward part of 09 level was originally a lookout station. It is ringed with viewports and stools. After radar eliminated the need for these positions, the viewports were sealed and this space became unused. The aft area of 09 level was originally the VHF radio room. During the 1980's that space served as the weather office. The navy clear out everything from this area when the ship was decommissioned.

010 Level

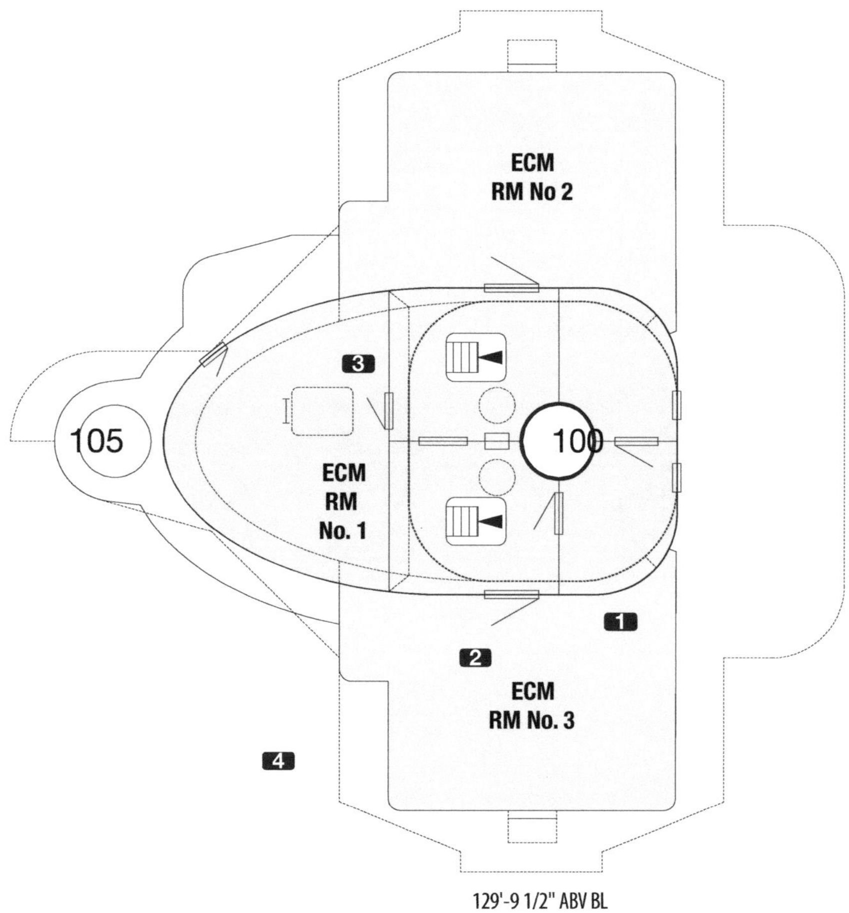

010 Level
1" = 8'

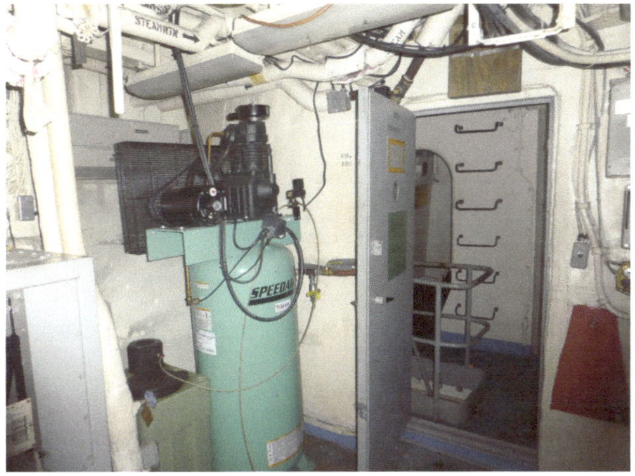

(2) All of the ECM equipment was removed from the 010 level when the ship retired. The Speedaire air compressor in ECM room no. 3 is a museum addition to blow the ship's horn. The door leads to the central passageway. Inclined ladders end at the 010 level so the rest of the way up is though vertical ladders. The rungs in the centerline bulkhead visible through the door access the 011 level. There is a second rung ladder mirrored on the opposite side of the bulkhead.

(1) The enclosed area of the 010 level was originally an unassigned space containing four lookout viewports. The whistle platform wrapped around the outside of the forward director tower at 010 level. During the 1960's two ears were added to the sides of the 010 level to house electronic countermeasures (ECM) equipment. This is ECM room no. 3 in the starboard ear looking inboard. The expansion caused the two viewports at the side to be enclosed.

(3) In ECM Room no. 1 there is a vertical ladder to the 011 level. There is a small door leading to the remnant of the whistle platform that now supports the forward mast.

(4) This view is of the underside of the 010 level. The platform at the aft end of the tower was originally the whistle platform and it extended all the way to the forward face of the fire control tower.

011 and 012 Levels

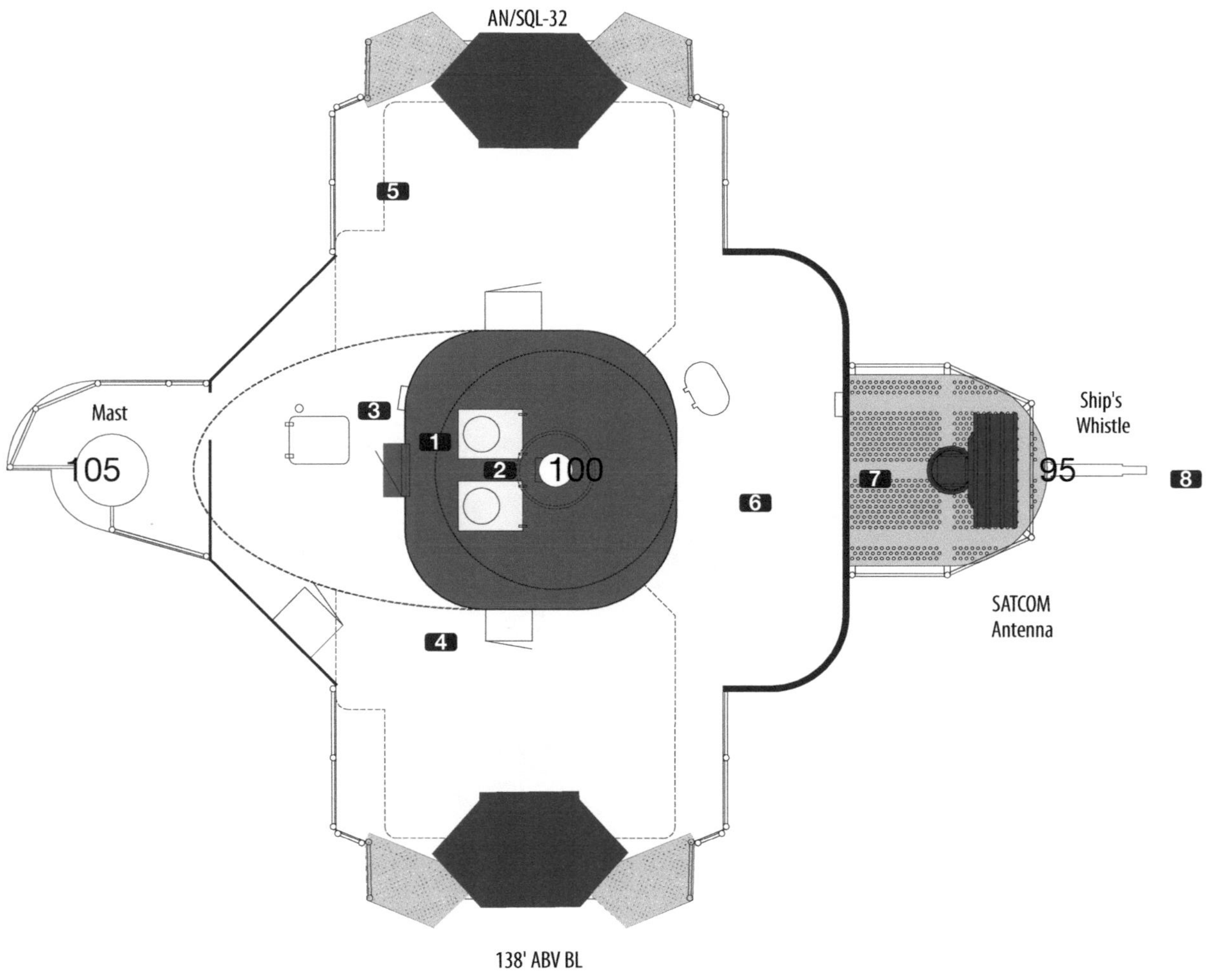

011 Level
1" = 8'

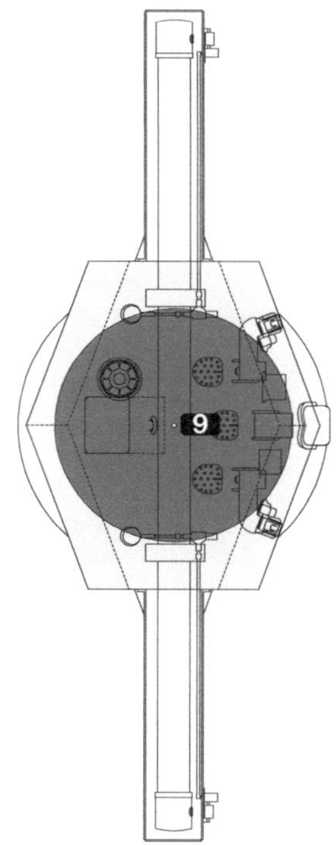

012 Level
1" = 8'

(1) There is a raised platform over the enclosed area of the 011 level. This photograph shows one of two interior deck hatches used to access the 011 level and the elevation of the platform above the deck. This is an unassigned area of the ship.

(2) This is the access ladder to the Director Mk.38 at the 012 level. The open deck hatch provides a view into the director. This structure rotates within the 011 level.

(3) An armored door gives access from the enclosed direct tower to the weather at the 011 level. The curved structure above the door allows the door to stay open in any position. The lever at the center of the door retracts a spring-loaded prong that fits into holes in the curved structure.

(4) A view aft at the 011 level. The door shown in photograph (3) is at the right. The prong that holds the door open is on the fact of the door. The hatch to the left leads to ECM room no. 1 at the 010 level.

(5) The AN/SLQ-32 electronic warfare system was installed on USS *New Jersey* during the preparations for reactivation in the 1980's. Antennas for this system are located the port and starboard sides of the 011 level. There are retractable maintenance platforms below the antenna that give access to its front faces. The platforms are shown extended.

(8) This photograph shows the underside of the 011 level. The ship's whistle is at the forward edge with the SATCOM platform immediately behind. A foremast yardarm is visible above the deck.

(6) The 011 level was originally the air defense station. During World War II this area was manned by aircraft spotters and was protected by a shield. Only this forward segment of the original shield remains. The ladder gives access to the SATCOM and ships' whistle platform.

(7) The forward AS-3018/WSC-1 satellite communications antenna. The ship's whistle is forward of the antenna. An identical antenna is located on a platform at the aft smoke pipe housing.

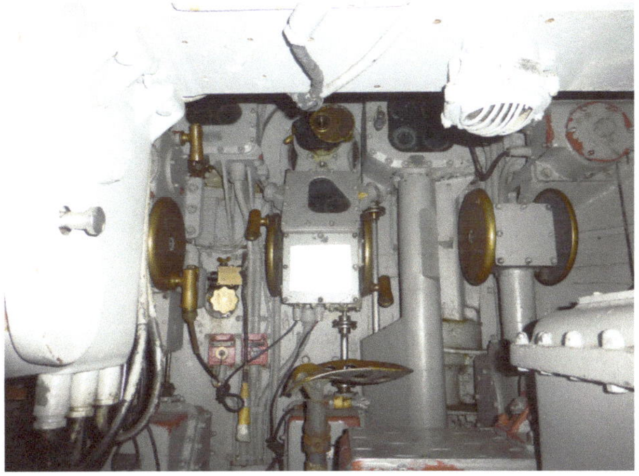

(9) The rotating director Mk.38 makes up the 012 level. This view is forward within the director. This director was known as *Spot 1*. The director can rotate up to 198 degrees to either side so that it faces the target. Rotation could be controlled either by the Rangekeeper in the plot room or by the trainer whose position and control wheels are at the right. The stereoscopic Rangefinder Mk.48 is directly behind these positions and is not visible. The rangefinder was not gyroscopically stabilized. The cross leveler (center) and pointer (right) kept the rangefinder at the correct elevation and rotation by correcting for the horizon. If the horizon was not visible, they could match pitch and roll angles transmitted from the stable vertical in the plot room.

There is a spotter's periscope at the rear of the director. This was used to spot the fall of the shot so that aiming corrections could be sent to the plot room. The rangekeeper in the plot room sounded a buzzer in the director just before it estimated a salvo would arrive on target so the spotter would know which shots came from his own ship. The main battery directors have a Radar Mk.13 (originally Mk.8) radar mounted on their roofs that largely replaced the optical rangefinder.

The main battery directors on USS *Missouri* and USS *Wisconsin* have vertical rear plates that give them more internal room than the sloped sided directors have on USS *Iowa* and USS *New Jersey*.

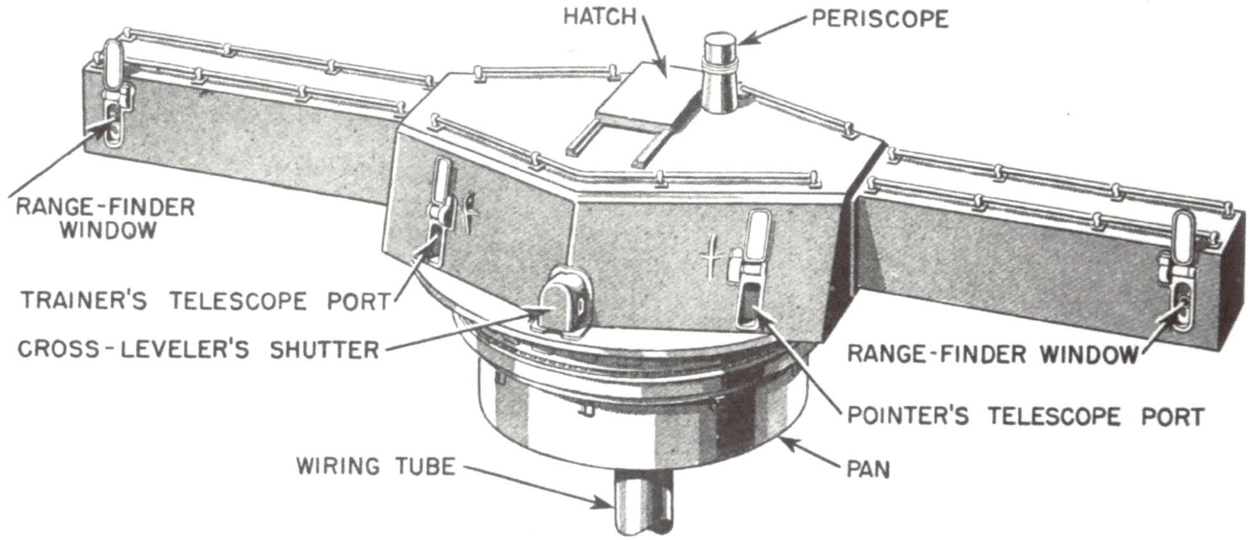

This illustration shows the major exterior features of the Director MK.38. (U.S. Navy)

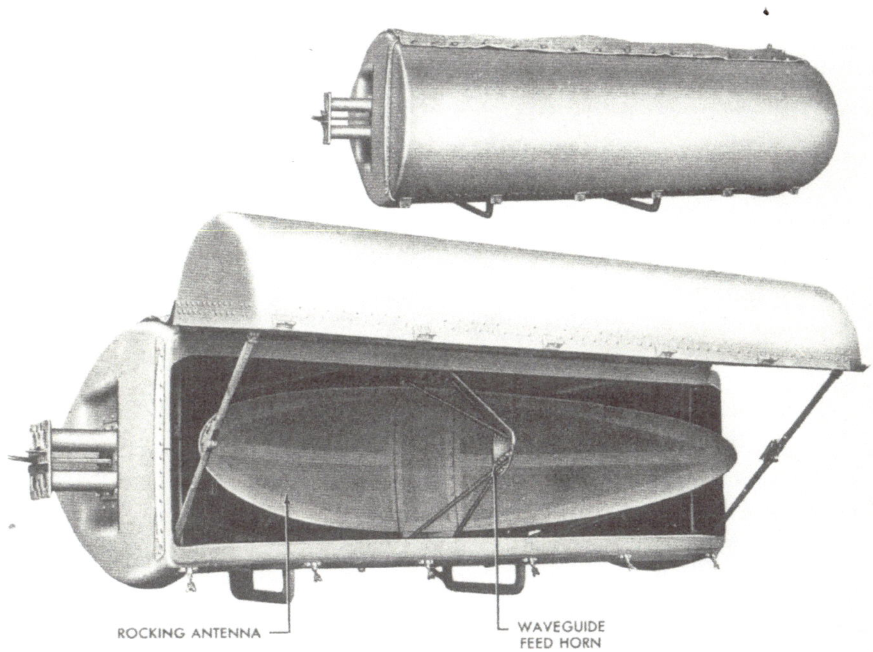

A Radar Mk.13 is mounted on top of the Director Mk.38. This illustration shows how the radar appears under its protective cover. (U.S. Navy)

Sixteen-Inch Gun Turrets

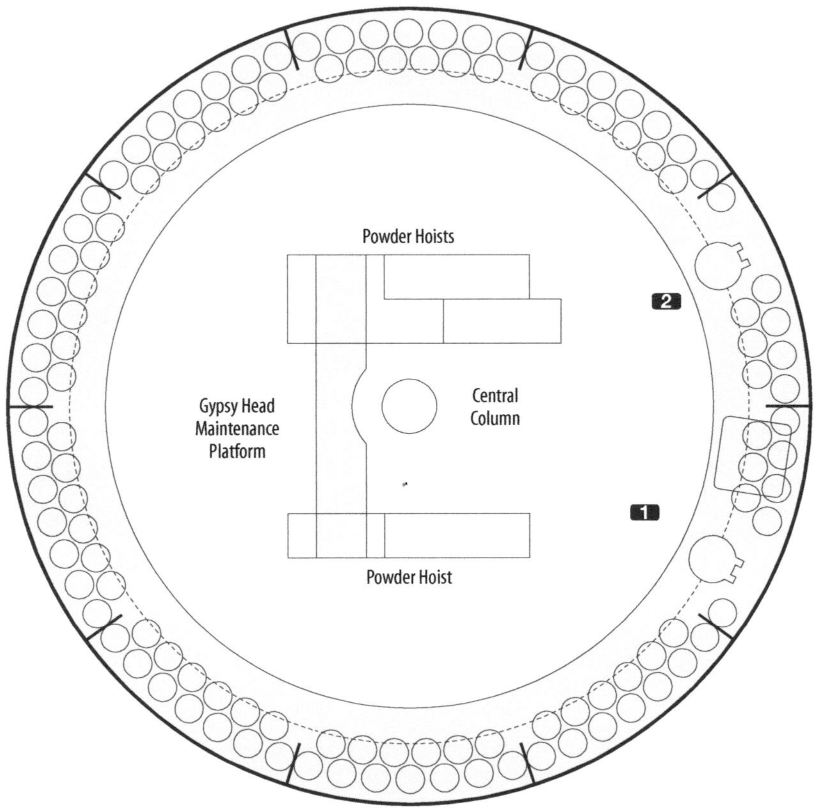

Shell Stowage Platform
(Turret No. 2 Only)
1" = 8'

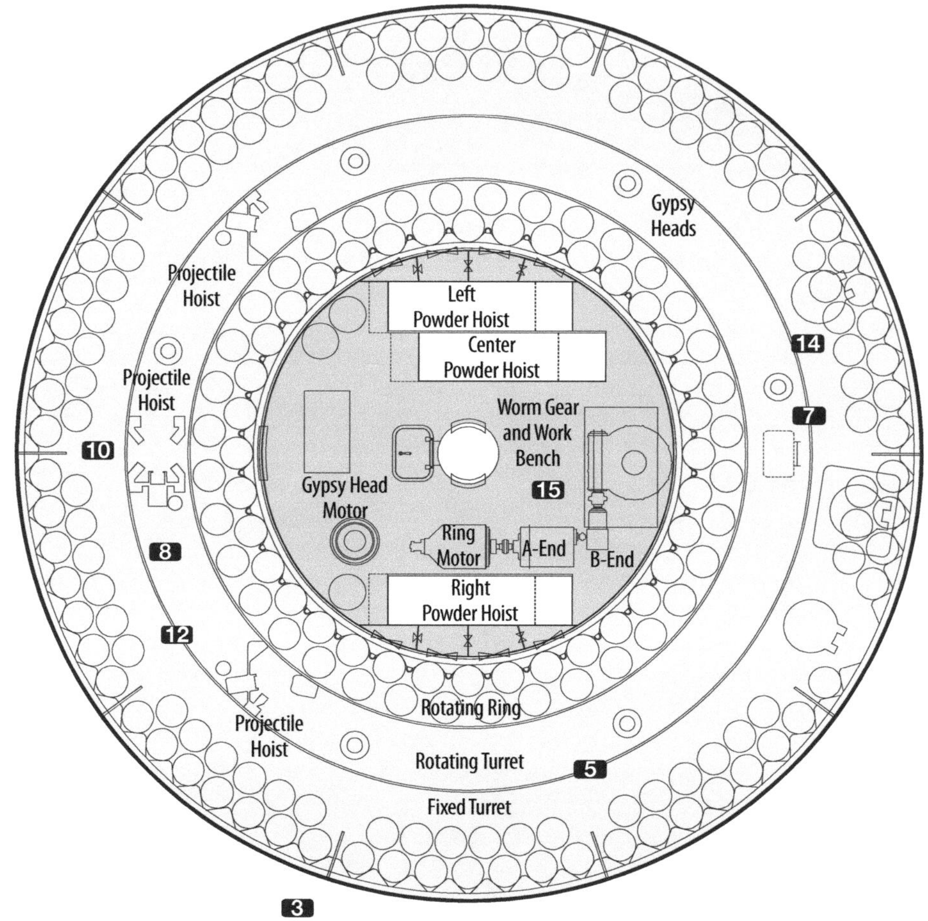

Lower Shell Flat
Turret No. 2 Shown. Others are similar.
1" = 8'

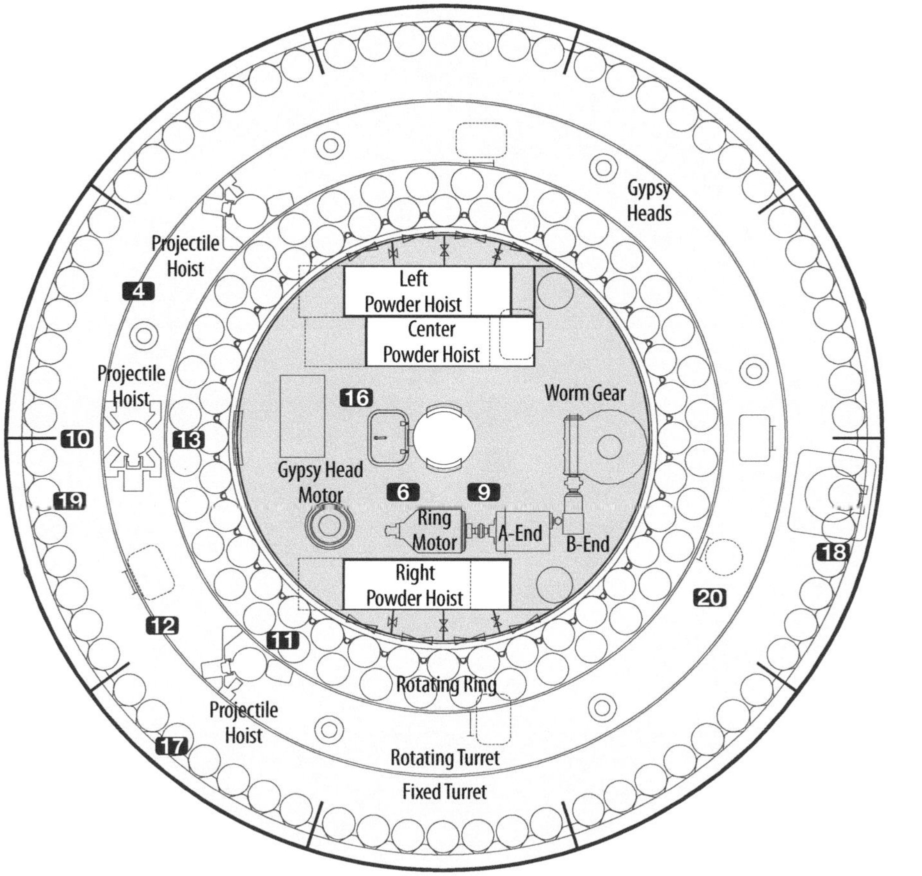

Upper Shell Flat
Turret No. 2 Shown. Others are similar.
1" = 8'

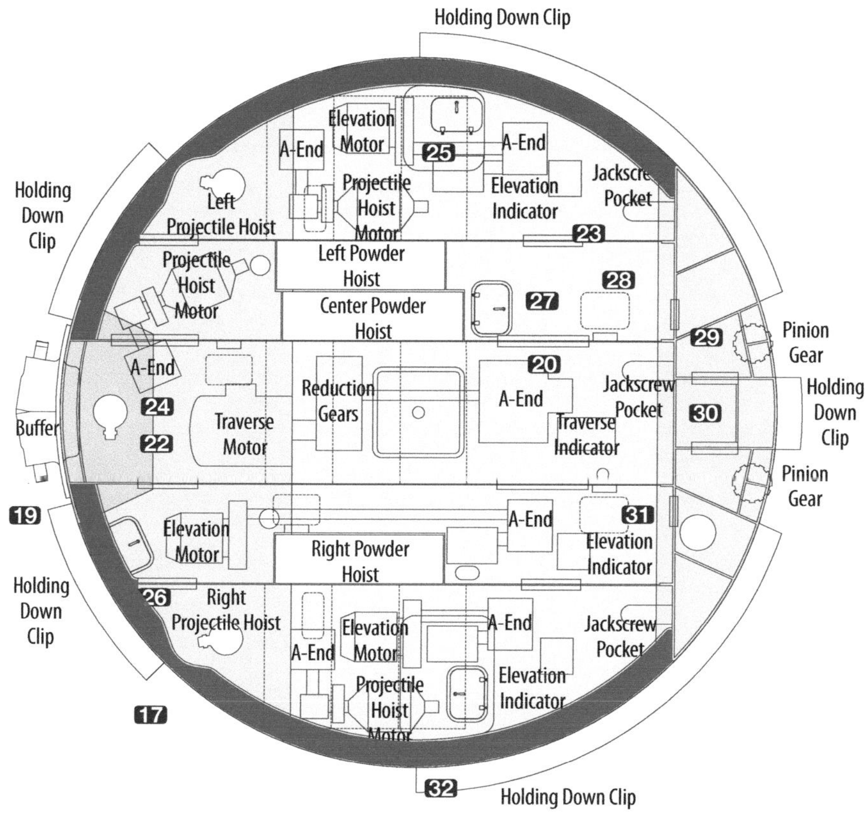

Electric Deck
1" = 8'

Powder
Hoist
A-End
Left
Powder Hoist
Center
Powder Hoist
Powder
Hoist
A-End
Powder
Hoist
A-End
42
Right
Powder Hoist
10'-6" Above Pan Plate
Left
Gun Pit
Gun
Pocket
Gun
Elevation
B-End
Left
Protectile
Hoist
Jackscrew
Left
Powder Hoist
Center
Powder Hoist
Traverse
B-End Motor
Jackscrew
Center
Protectile
Hoist
37
Center
Gun Pit
Gun
Pocket
38
39
36
34
Worm
Wheel
Housings
Gun
Elevation
B-End
Right
Powder Hoist
35
Traverse
B-End Motor
33
Right
Protectile
Hoist
Right
Gun Pit
Gun
Pocket
Jackscrew
Gun
Elevation
B-End
40
Left
Powder Hoist
Center
Powder Hoist
Right
Powder Hoist
41
6'-6" Above Pan Plate

Pan Plate
1" = 9'

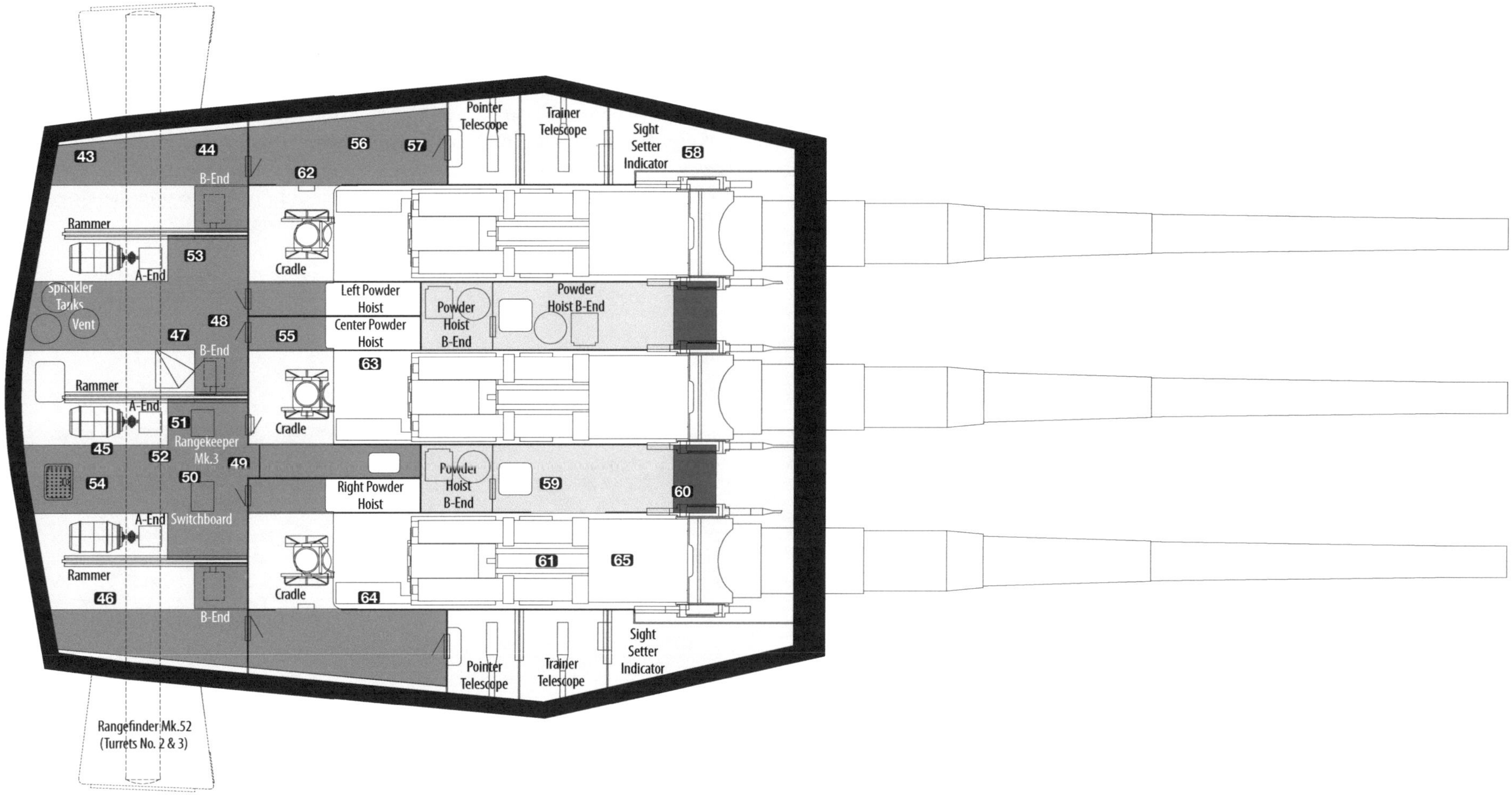

Shelf Plate
1" = 10'

Lower Shell Storage Platform

(1) The gunhouse of turret no. 2 sits much higher than those of the other turrets so that it can fire over turret no. 1. The extra height creates room for a shell storage platform above the powder-handling flat that the other turrets lack. This is forward half of the shell storage platform in turret no. 2 looking to port. The shell storage platform is part of the nonrotating turret that forms a balcony over the powder-handling flat. This creates an open vertical space within the powder flat that is only found in turret no. 2. One hundred and twenty projectiles were stored in two staggered rings against the turret support. The powder hoists and the turret's central column are at the left of the photograph.

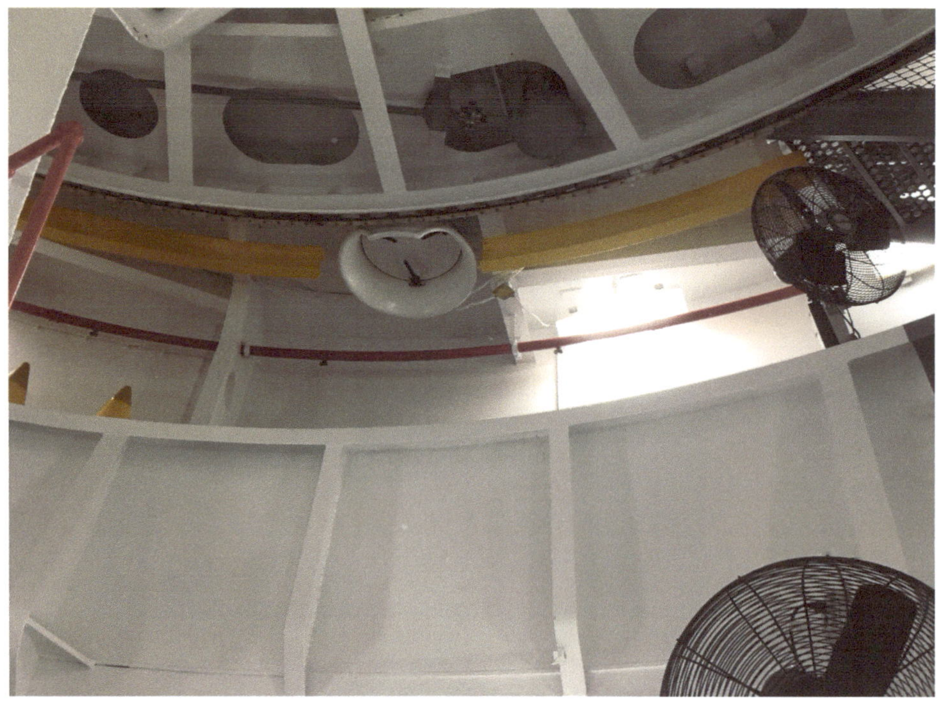

(2) This is the shell storage platform viewed from the powder flat below. A trolley track runs around the platform to move projectiles. Projectiles were hauled up to the shell flat above through the bell mouthed hatch between the track segments. The circular beam inside the trolley track marks the end of the rotating turret. A drive shaft and gearing for a gypsy head at the shell flat above are visible in this area.

(3) A panorama view of the lower shell flat of turret no. 2 looking to the left side of the turret. The upper and lower shell flats are nearly identical in their layout. American battleships were unusual in that projectiles were stored vertically within the turrets. The projectiles shown are fiberglass replicas.

The projectile flats are divided into three concentric rings and a circular central core. The divisions between the rings are clearly visible in the photograph. The outer ring is part of nonrotating turret and is fixed to the surrounding turret support bulkhead. The outer ring was used to store projectiles. The amount of storage in the outer ring varies among the upper flats due to their position relative to the upper taper of the fixed turret. The upper flat of turret no. 2 is the smallest because it is highest within the conical part of the turret support. It only holds a single row of projectiles while the other flats have a staggered double row.

The middle ring rotates with the turret and houses the three projectile hoists. The right hoist is at the center of the photograph, the middle hoist is to the left, and the left hoist is partially visible in the background.

The inner ring is within the rotating turret but it has a motor that allows it to be rotated independently from the turret. The central ammunition racks are part of and rotate with the inner ring. The hoists and ammunition racks are positioned so that, by rotating the inner ring, a projectile could be moved within inches of the opening of all three hoists at once.

The small opening (circled) near the outer edge of the inner ring to the left of the projectile with a rope around it is the centering pin. The centering pin keeps the ring from rotating when the turret is not in use. Ship movement in a seaway could induce rotational torque. While rotation machinery would stop the ring from rotating, the pin prevents the force from being transmitted to the machinery where it could cause damage.

The inner bulkhead marks the start of the central core. The central core houses the machinery that rotates the inner ring and the motor that drives the gypsy heads used to move projectiles. There is a step down from the inner ring to the central core due to the height of the rotation mechanism for the inner ring. The stairs shown in the right background are a museum addition.

(5) This photograph shows how the gypsy heads were used to parbuckle projectiles on the shell flat. It was taken on USS *Iowa* in 1988. The gypsy head spins continuously. Pulling the end of the rope wrapped around the gypsy head creates enough friction for the gypsy head to pull the rope with sufficient force to move a projectile. In this case the sailor is using the gypsy head to parbuckle a 2,700 lb. practice projectile from the outer storage ring to the inner storage ring. Ammunition was generally loaded from the inner ring. As ammunition was consumed, projectiles were moved from the outer ring to replenish the inner ring. The decks of the shell flat projectile rings are made of brass because projectiles moving across a steel deck could cause sparks. The decks were lightly oiled to reduce friction when moving projectiles. (Nat'l Archives)

(4) This is the upper shell flat of turret no. 2 of USS *Iowa* in 1986. Projectiles were stowed vertically and were chained to bulkheads to hold them in place. The inner ring could hold 72 ready-use projectiles in double, staggered rows. Because turret no. 2 sits higher within the tapered portion of the fixed turret, it only has room for a single row in its upper flat holding 70 projectiles. The outer rings of the other five shell flats store projectiles in double rows.

Most of the projectiles shown are 1,900 lb. high capacity rounds. The cap over the nose protects the fuze and would be removed before loading into the gun. Two 1,900 lb. and one taller 2,700 lb. practice rounds are in the background center.

At the center are the operator's controls for the projectile hoist. The controls on the two flats are mechanically linked by the rod extending through the deck. The three-position control handle controls a hydraulic piston that drives the hoist. The control is shown in the center neutral position. If moved to the right, it raises the hoist's piston. When moved to the left, it lowers the piston. At the bottom center there is a gypsy head used to parbuckle projectiles on the flat. (Nat'l Archives)

(6) The gypsy heads run at a constant speed so there is no need for electric-hydraulic drive as with most of the turrets' mechanical systems. Each shell flat has an electric motor in the central core oriented vertically that drives its gypsy heads. This is the gypsy head motor for the upper flat of turret no. 2.

(7) The gypsy heads on a shell flat are linked together and to their motor using shafts and gears. This is one of the gears below a gypsy head of the upper shell flat of turret no. 2. The input drive shaft is at the left. The output shaft driving the next gypsy head is at the right.

(8) This is the ring operator's station for the lower shell flat of turret no. 3. The ring operator's station is at the right side of the center hoist and the hoist operator's position is at the left side. It would have been unsafe to rotate the ring while a projectile was being parbuckled into a hoist. The three lights at the top indicate when each hoist is ready for the ring to be rotated. The lever in the foreground controls the direction to rotate ring. The ring motor automatically stops when the next projectile storage locations are aligned with the three hoist apertures. The buttons below start and stop the ring and gypsy head motors.

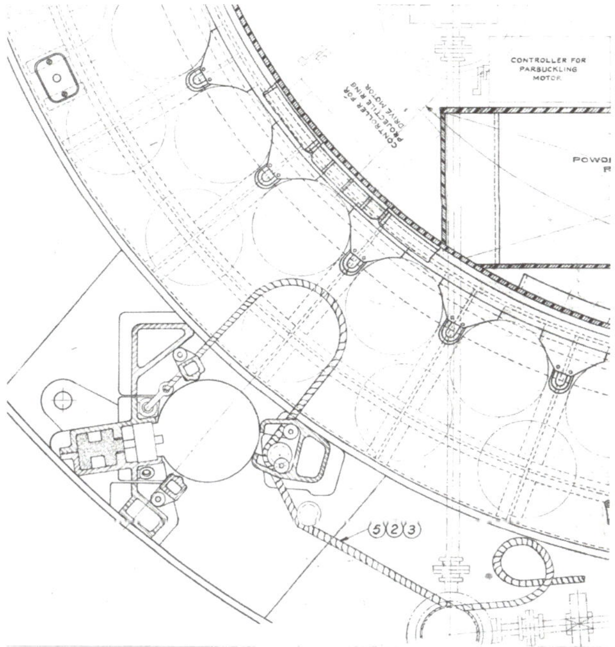

Moving projectiles from the inner ring to the hoist was surprisingly simple. There was a rope that had one end tied to the hoist and the other end wrapped around a gypsy head. A crewmember threw the center of the rope behind the projectile. Another crewmember tightened the rope around the gypsy head. This caused the projectile to shoot into the hoist. It took seconds to move a projectile from the inner ring to the hoist. (Plan 7212ZF)

(9) The motors that drive the inner ring are located within the central core. This is the upper shell flat of turret no. 2. At the center is the electric motor that powers the system. Its output shaft is connected through gears to the A-end pump directly behind. The B-end hydraulic motor is in the background and is oriented at a right angle to the A-end. The B-end drive shaft is covered in protective material. It drives a worm gear that turns a worm wheel in the gearbox at the background left. The gearbox drives a pinion gear against circular rack gear attached to the inner ring.

The central column is at the center left. It ends at the box structure at the top of the photograph.

(10) One difference between the upper and lower shell flats is that the upper flats have a heavy metal flag that can be placed over the hoist tube. Without the flag, a projectile would tip over as it moved into the hoist. When the turret had a full crew complement both the upper and lower shell flats were manned. However, projectiles could only be loaded from one shell flat at a time. The flag is one of the reasons for this. When the flag is down, it blocks projectiles from coming up from the lower shell flat. When a shell flat was on standby, its crew could use the time to refill the inner storage ring or, in the case of turret no. 2, haul projectiles up from the lower projectile storage platform.

Normally, the hoists were loaded from the inner ring because a projectile could be easily positioned directly in front of the hoist doors by rotating the ring. However, the projectile hoists have openings that allow them to be loaded from either inner or outer ring. The openings have shutters that close the openings when the hoist is operating.

(11) A pawl on a moveable pawl carrier in a projectile hoist. The hoist uses a ratcheting action to move projectiles through the hoist tube to the guns. Each hoist can hold five projectiles. One side of the hoist tube has fixed pawls at the projectile stowage positions. The other side has a moveable pawl carrier that the hoist piston drives up and down. When the hoist operator moved the piston up, the pawl carrier moved all the projectiles in the hoist up to the next stowage position. The projectiles pushed aside the fixed pawls as they went by. After a projectile passed, the fixed pawl snapped back out and prevented the projectile from sliding backwards. When the operator moved the piston down, the projectiles pushed aside the movable pawls as they passed downward. Once the moveable pawl went beneath a projectile, the pawl snapped out. Interlocks prevent raising the projectiles when the hoist is full. The hoist also had the ability to withdraw pawls so the hoist could be operated in reverse.

The hoists on the *Iowa*-class battleships are very similar to those used in *North Carolina*-class and *South Dakota*-class battleships. In fact, they have many common parts. The major difference is the hoists in the *Iowa*-class have a longer path to the guns so they hold five projectiles rather than four in the earlier battleships. Otherwise, the hoist capabilities among those ships were functionally the same.

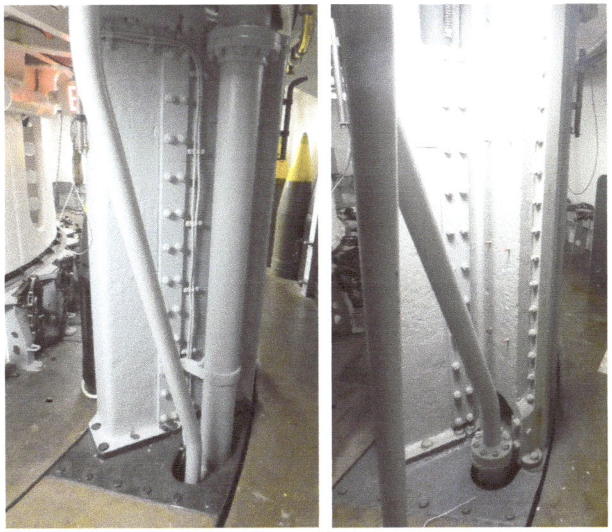

(12) A visible difference between the upper and lower shell flats is that the lower flats (left) have a piston next to each hoist while the upper flats (right) have a guide for a piston rod.

(13) These photographs show a practice projectile moving through a projectile hoist from the lower shell flat to the upper flat on USS *Iowa* in 1988. In the first photograph the pawl carrier is at the left in the down position and a fixed pawl is at the right. In the second photograph the pawl carrier has moved up, taking the projectile with it and the projectile is pushing the fixed pawl aside. (Nat'l Archives)

(14) To reach their stowage locations, projectiles had to be lowered from the main deck down to the second platform. From there, they had to be raised back up to the shell flats. The shell flats have electric winches to move the projectiles up to the flats where they can be parbuckled into their final position. This winch is at the lower shell flat of turret no. 2. There is a projectile loading hatch in the deck near the winch.

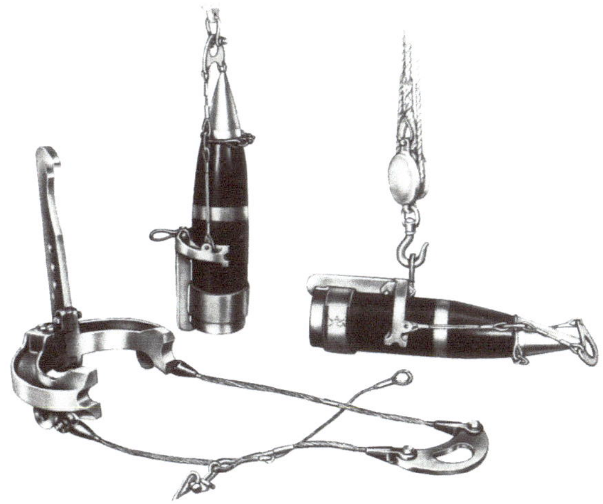

Because the projectiles are smooth for aerodynamics, they have no location for attaching a hook. While the projectiles were moved, they were held in a shell carrier that wrapped around them and provided lifting points. The shell carrier was designed for projectiles to be moved while oriented either vertically or horizontally and to easily transition between the two. Projectiles were usually oriented in the direction they were moved so they could pass through hatches and doors. (U.S. Navy)

(15) Each turret has a workbench in the central core of the lower flat. The workbench is mounted above the worm gear box for rotating the inner ring. The three powder hoist motor controllers are above. The ring motor B-end is at the lower right.

(17) One of the false myths about the *Iowa*-class is that the turrets are held in place just by gravity. The turrets actually have holding down clips like this one that keep the turret in place. There are seven holding down clips arranged around the turret circumference. The holding down clips are attached to the electric deck bulkhead and extend outwards beneath the roller track support. There is a $\frac{1}{32}$" gap between the clip and the track support.

To the left is the conical top of the fixed turret. The brackets at the top left support the roller track upon which the turret rotates. The rack gear used to rotate the turret is visible inside the track support. The circular bulkhead of the electric deck is at the right.

(16) Access to the shell flats from below was through a rung ladder on the central column. This is the ladder at the upper shell flat of turret no. 2.

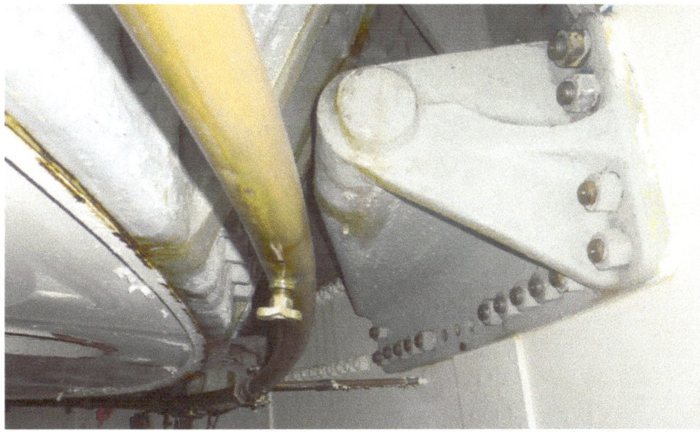

(18) Each turret has two stops to prevent it from rotating too far and causing damage to the ship. The stops are attached to the fixed turret near the normal position of the front face. This is one of the stops in turret no. 3 viewed from the upper shell flat. When USS *New Jersey* was built, each turret could rotate 150 degrees to either side. In 1982 this was reduced to 125 degrees for turret no. 3 so that the guns would not strike the new boat winch and glide slope platforms at the 03 level. This change was made by moving the stops outwards 25 degrees. The inside face of the stop had to be grinded so that it would fit flush over the butt plate on the fixed turret. The two rows of bolts behind indicate the old position of the stop.

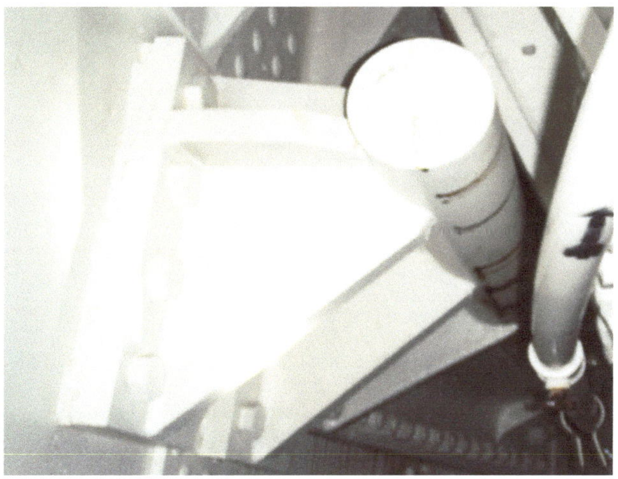

When USS *Iowa* was refitted in the 1980's, the yard did not move the stops on turret no. 3. Instead, it created and installed new stops having a completely different shape than the originals. (Michael Getscher)

(20) Moveable ladders at the upper shell flat provide access to the electric deck above. This is turret no. 2. There are five access points. The relatively large number of ladders is due to the difficulty in moving around the electric deck.

(19) Each turret has a buffer attached to the electric deck at the back centerline. This is the buffer on turret no. 2. The buffers have plungers at both sides. When a buffer strikes a turret stop, the plunger depresses and cuts off the turret rotation. This view is from the upper shell flat looking upward

This photograph shows the upper shell flat of turret no. 1 while the turret was under construction. The turrets were too heavy to be lifted by yard cranes in one piece. The turret was divided into upper and lower sections at the electric deck for installation. The lower turret was craned in first then the upper turret was in lowered on to the roller track. Finally, the lower turret was jacked up so that the two sections could be joined. Here only the lower section of the turret is in place.

A. Turret traverse rack gear
B. Roller carriage and conical bearings (covered with tarps)
C. Turret stops
D. Holding down clips (to be installed)
E. Powder hoists
F. Central column
G. Transfer Panel
H. Ring motor A-end
I. Worm gear (not in position)
J. Ring motor B-end (not in position)
K. Powder hoist motor controllers
L. Gas ejection system relay tank
M. Projectile Hoist Locations

(Nat'l Archives, Phila.)

Electric Deck

(21) The turret traverse machinery is located on the centerline of the electric deck. The front of the turret is towards the left of this photograph. At the center is the A-end hydraulic pump for turret traverse. This pump drives two B-end hydraulic traverse motors on the pan plate level above. The two vertical pipes emerging from the pump are the pressure–suction pipes whose pressure differential drives the B-end motors. At the top of the photograph are the tee-joints that split the pressure differential pipes to connect with both of the two motors.

The turret trainer's position is to the left of the vertical pipes. Normally the turret traversed automatically in response to orders received from the fire control system. Alternatively, an operator sitting at this position could traverse the turret manually. In this mode, traverse orders from the fire control system were displayed on an indicator. The operator would traverse the turret by rotating the brass wheels until the turret was positioned as ordered. The control wheel has very fine resolution. One complete turn of the control wheel turned the turret just two degrees.

(22) This view of the turret traverse machinery is along the centerline of turret no. 2 looking forward. The 300 H.P. electric motor that powers the turret traverse is in the foreground, just forward and offset to the left of the motor are the reduction gears. The A-end pump shown in photograph (21) is in the background. The center gun pocket is above the pump. The gun pockets are depressions in the pan plate that give the guns space to recoil while at their full 45° elevation.

(23) There are three gun layer's positions in the turret. The gun layers could control the elevation of the guns. Like the turret traverse, the guns were normally elevated automatically in response to orders from the fire control system but they could be operated manually as well. This is the left gun layer's position. The front of the turret is to the right. The jackscrew that elevates the guns runs through the pipe that is partially visible above the operator's seat.

(24) The A-end pump of the center projectile hoist extends through the arch at the center. The view is towards the left side of the turret. The A-end creates the pressure required to raise and lower the piston at the lower shell flat that ratchets projectiles through the hoist tube. The electric drive motor for the A-end is visible through the right side of the arch. The traverse drive motor shown in photograph (22) is at the right. The ladder on the bulkhead leads to hatch in the center gun pit visible in photograph (36).

(25) This view is to the rear along the left side of turret no. 2. The 50 H.P. electric motor that powers the elevation of the left gun is at the center. The 75 H.P. electric drive motor for the left projectile hoist is at the left and the A-end for the hoist is in the background, directly behind the electric motor. The seven-gallon drum on top of the A-end is an expansion tank holding hydraulic fluid. The left projectile hoist is visible in the background.

(26) This is a forward view of the right side of the electric deck. The right projectile hoist is in the foreground with the A-end of the hoist motor is immediately behind that. The right gun pocket is visible above. Access to the right gun pit is through the door in the gun pocket.

(28) This is the same passage looking forward. The arch in the foreground leads to the pinion mountings for the traverse gears. The mounting for the turret traverse gear is visible through the arch at the right side of the opening.

(27) The only open space of any kind in the electric deck is this narrow passage at the left side of the turret. The view here is to the rear of the turret. The deck hatch leads to the central core of the upper projectile flat. The center powder hoist is behind the hatch. The arch to the left is the same one shown in photograph (21). To move between two points within the electric deck usually requires exiting to the upper shell flat at the closest hatch and reentering the electric deck at the hatch closest to the destination.

(29) A view through the arch shown in photograph (28). The left pinion gear mounting is at the center. The pinion gear that drives the turret traverse is partially visible below the curved protective sheet.

(30) Here are the pinion supports looking to the right side along the forward end of the electric deck. The pinion is supported above and below the gear at the electric deck. The curved protective sheet at the center obscures the right pinion gear but a small bit pokes out from under the sheet.

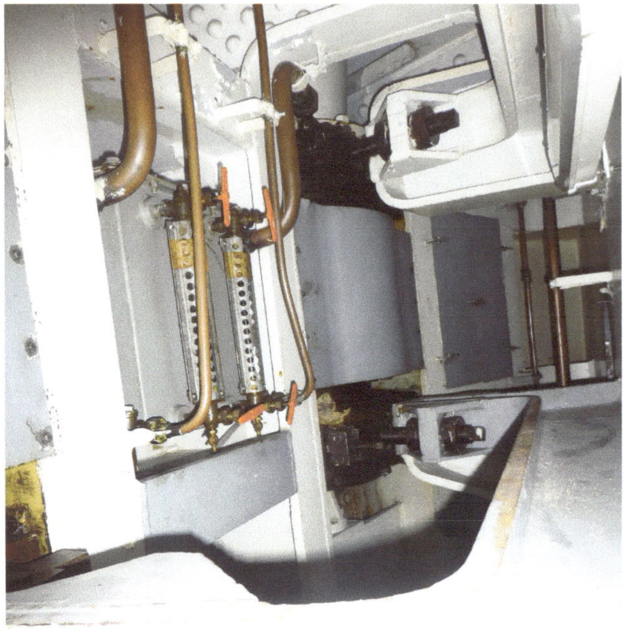

(31) The center gun layer's position showing the controls. When the guns are not in automatic mode, the gun layer rotated the brass wheel to elevate the guns. The dials above indicated how the gun was to be positioned. The firing key attached to the wheel could fire the gun. Normally, this firing key was disabled. However, when the turret was under local control, the turret officer would enable the firing key when he wanted the gun to be fired.

(32) This photograph shows how the rotating turret meets the nonrotating turret. Note the liberal use of grease. The view is looking outwards from a maintenance opening at the electric deck.

The inclined lower roller bearing track is at the bottom of the photograph. The lower track is the extreme top of the nonrotating turret. Is slopes slightly downwards toward the outer edge at 1°37'47" so that it has a conical upper surface.

Seventy-two conical bearings roll on top of the track of which two are visible here. Each bearing weighs 546 lbs. The bearings have a smaller radius at the inside of the track than at the outside so that they will naturally follow a curved path along the roller track.

The horizontal band at the center of the bearings is part of the roller carriage. The roller carriage retains the bearings at their proper spacing. There is a similar band at the outside of the bearings. The bearings are not uniformly spaced in order to reduce the likelihood of firing or seas causing the track to deform.

The upper bearing track is at the top of the photograph. This is part of the rotating turret and is attached to the bottom of the pan plate. The upper track slopes slightly upwards towards the outer edge. The entire rotating turret below this point is suspended beneath the pan plate.

There are six short removable segments in the lower bearing track to allow damaged bearings to be replaced through this opening. There is no record of this ever having been done in the *Iowa*-class battleships.

Pan Plate

(33) The A-end pump at the electric deck drives two B-end hydraulic motors on the pan plate. This is the drive shaft for the right motor on turret no. 2 looking down and forward. The drive shaft runs through the protective sheath at the center of the photograph and connects the right B-end motor to the right worm wheel that turns the traverse pinion gear. At the top of the photograph (forward end) is the turret brake. The visible open hatch leads down to the electric deck.

(34) The left worm wheel housing for turret no. 2 viewed forward. This unit is mirrored on the right side. The drive shaft (photograph (33)) enters the housing near the far bulkhead and rotates a worm gear oriented horizontally. The circular bolted plate at the top of the housing covers the end of the pinion shaft driven by the worm gear. The pinion shaft is oriented vertically and has a pinion gear extending from the bottom (photograph (29)). The pinion gear meshes with a rack gear fixed to the non-rotating part of the turret.

This arrangement serves two major functions. First, it is a reduction gearing system that decreases the torque required to turn the turret. The drive shaft from the B-end motor has a 175 R.P.M. maximum speed. The gearing from the drive shaft produces a 4° per second ($\frac{2}{3}$ R.P.M.) rotation in the turret. Second, it creates a one-way system. The drive shaft can move the turret but the worm gear prevents any external force on the turret from acting on the drive shaft.

(35) This is the right B-end motor that drives the turret traverse, looking down and forward. The drive shaft emerges at the other end of the motor (photograph (33)). The two pipes at the center transmit the pressure differential from the A-end that drives the motor (photograph (21)). The large pipe at the right is for ventilation. It is the same pipe shown in photograph (41).

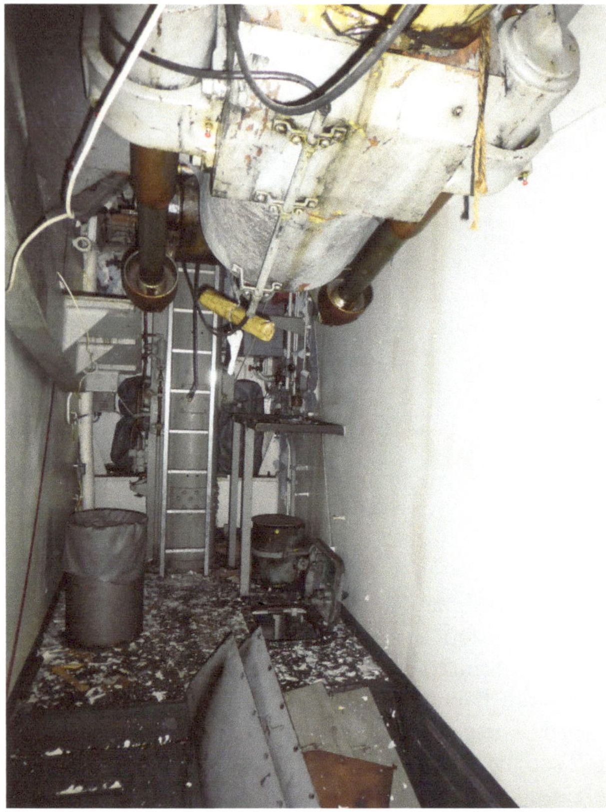

(36) The center gun pit of turret no. 2 looking towards the rear. The underside of the gun is at the top of the photograph. The gun pocket is visible at the bottom. Gun pockets are recesses in the pan plate that allow the guns to recoil at their full 45° elevation. Some loose metal plates are sitting in the pocket at the right. The bulkheads at either side are the gun girders. The gun girders are part of a large weldment that had to withstand the force of the gun recoil. They are $1\frac{5}{8}$" thick forward and $1\frac{1}{4}$" aft. The diagonal line in the bulkhead at the right is where the thickness changes. The curved line along the bulkhead at the left indicates the gun recoil limit. The recoil system has 48 inches of travel but the actual recoil was a few inches shorter than that. If a gun were to move 48 inches it would strike metal on metal and cause damage. The beam running along the left bulkhead is the loader's platform. It is mounted to the underside of the slide so it rotates with the gun. When the gun at the 5 degree loading position, the gun captain stands on this platform to open the and close the breech. The ladder at the center leads to the shelf plate where the gun captain, rammer operator, and cradle operator are positioned. The projectile hoist is directly behind the ladder.

The primer man's platform is to the right of the ladder. When the breech opens it rotates downwards and automatically ejects the spent primer. The primer man reaches up and inserts a new primer the primer lock mounted at the breech.

The 16-inch gun is above. The two rods at the sides are the counter-balances for the breech.

(37) Here a sailor is inserting a primer into the primer lock on USS *Iowa*. The primer's walls expand to seal its small opening through breech when the gun is fired. The cocking lever for mechanical firing is to the left of the sailor's hand and the loading platform is above his head. (Nat'l Archives)

A 7.62mm bullet casing (left) and a primer (right). Igniting the primer is the first in a series of rapid steps that fires the projectile. When fired, the primer sends a jet of hot gas through a tube starting at the breech that ignites a pad of black powder on end of the first powder bag. The black powder ignites the main charge within the powder bags.

The primer appears similar to a blank rifle cartridge. However, its structure is entirely different. The primer can be fired either electrically or mechanically. The primer contains insulation so that it can form part of an electrical circuit. Passing a current from the case to the center of the base ignites guncotton wrapped around a filament that triggers a series of charges within the primer. Alternatively, a hammer on the primer lock can push in a plunger at the base that crushes an internal percussion cap that triggers the primer. The primer has to withstand the force generated by six 110 lb. powder bags so its walls are much thicker than that of a rifle cartridge and is noticeably heavier.

(38) The center gun pit of turret no. 2 looking forward. The jackscrew that elevates the gun is at the right. The B-end hydraulic motor that drives the jackscrew is immediately to the right. The unit mounted to the right bulkhead with the large button in the center is the slide stop. This cuts the motor off when the gun reaches full elevation. The door in the background leads to the training gear housings (photograph (34)).

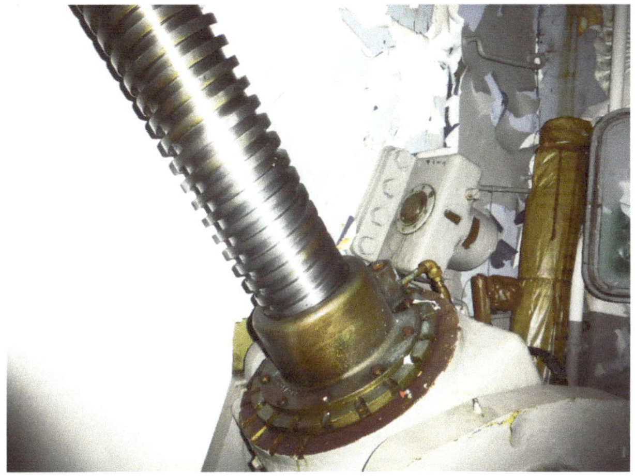

(39) A close up view of the elevation jackscrew and one of the slide stops. There is a slide stop on each side of the gun pit. As the gun elevates, the bottom end of the jackscrew descends through the electric deck (photograph (23)). All guns had a maximum elevation of 45 degrees. The guns in turrets nos. 1 and 3 had a maximum depression of -2 degrees while those of turret no. 2 could depress to 0 degrees. The guns could be elevated at 12 degrees per second.

(40) The areas at the turret sides at the pan plate hold ventilation and linkages from the sights

(41) There are intermediate levels above the pan plate between the guns. This is the right intermediate level 6'-6" above the pan plate looking aft. The right powder hoist is to the left. The rung ladder in the background on the bulkhead to the right lead to the center gun pit. The B-end traverse motor is located directly below this location. Forward of the hoists this level is only four feet high.

(42) This another intermediate level 10'-6" above the pan plate looking aft. This level is only about four feet high. At the center is the A-end pump that drives the right powder hoist. The electric drive motor is located at the level above. Protective material covers the drive shaft at the top of the pump. A bracket to the right of the circular opening is for mounting a bunk. During World War II and Korea a few sailors slept in the turrets.

Shelf Plate

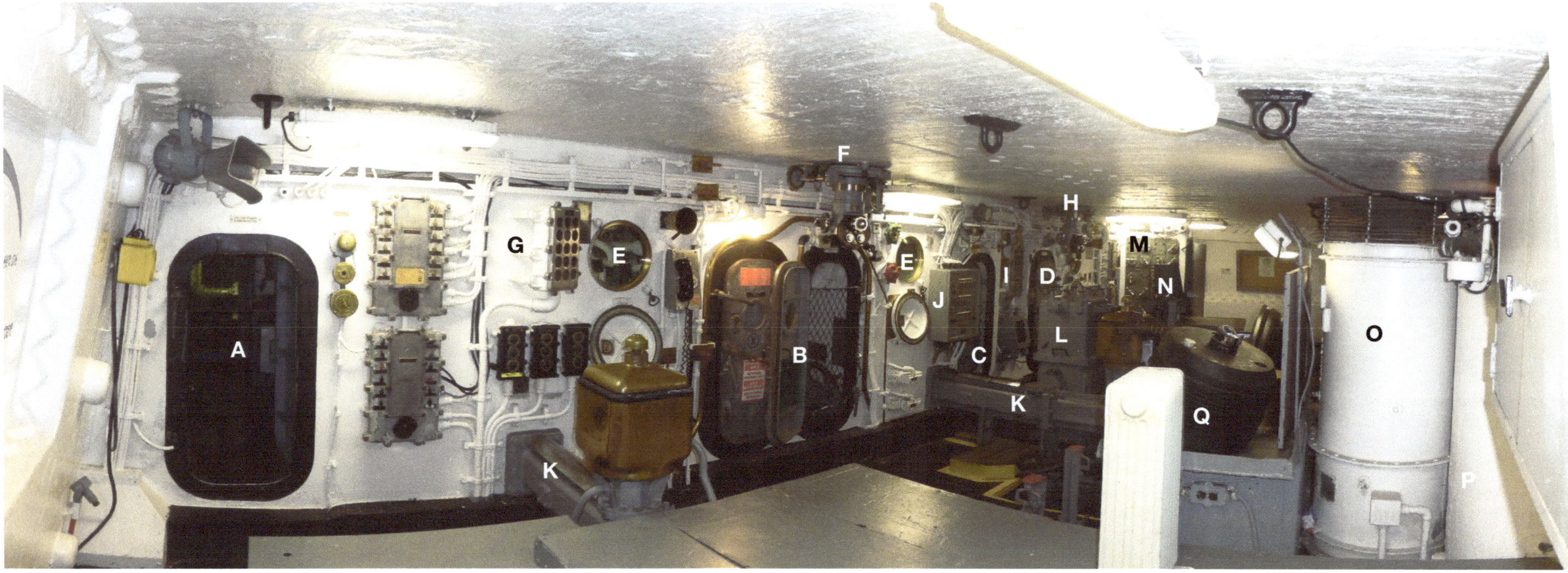

(43) The shelf plate is the foundation for the armored gunhouse. It is the highest level inside the turret and is the only level above the weather deck. Much of the shelf plate has raised platforms for equipment and crew positions. A transverse bulkhead creates the turret officers' booth at the rear of the turret. The booth was the station for the officers in command of the turret and it could serve as the last ditch command center for controlling the main battery. This is a panorama view of the officers' booth in turret no. 1 taken from the left rear corner.

A. Door to left sighting stations and left gun pit.
B. Doors to the left and center powder hoist operator positions
C. Door to center gun pit
D. Door to right powder hoist operator's position.
E. Gun pit viewing scuttles
F. Turret captain's periscope
G. Turret captain's indicators
H. Turret officer's periscope
I. Turret officer's indicators
J. Announcing System
K. Chain rammers
L. Rangekeeper Mk.3
M. Switchboard
N. Radar
O. Vent
P. Sprinkler tanks
Q. Museum displays

(44) The officer's booth of turret no. 1 (top) and turret no. 2 (bottom) looking to the right. Turret no. 1 originally housed a coincidence Rangefinder Mk.53. The rangefinder and its two hoods were removed from all the *Iowa*-class battleships when they prepared for the Korean War because they shipped water, had poor visibility, and added significant weight to the ships. The other two turrets retain their original stereoscopic Rangefinder Mk.52. Consequently, the officers' booth for turret no. 1 is relatively spacious while those for the other two turrets are very cramped.

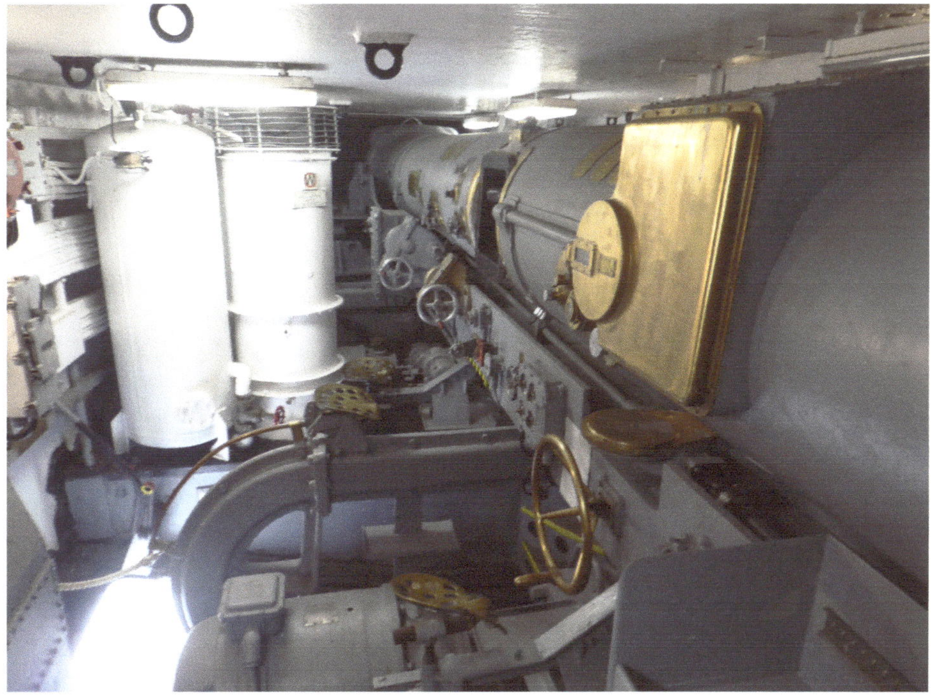

(45) The rangefinder in turret no. 2 looking to the left. The three operators have bicycle seats. The trainer sat in the closest seat. The trainer could rotate the rangefinder to keep it on target. The rangefinder is capable of rotating six degrees to either side of the turret centerline.

The rangefinder operator sat in the center seat mounted on the rammer. The operator measured the range to the target. The rangefinder tube contains two telescopes that look out from shuttered openings at the front the turret hood. Synchronized prisms at each end of the rangefinder control the direction of view from each side. The rangefinder merges the two telescope views at the operator station. When the telescopes are aimed directly forward, the views are misaligned because the prisms are forty-four feet apart. The operator rotates prisms inwards and outward to change the views until they are aligned. When the two views match, the rangefinder mechanically determines the distance by multiplying the 22-foot distance from the centerline to the prism by the tangent of the angle of prism rotation.

The rangefinder has to be able to measure extremely small angles to be effective. Because range to the target would be much greater than the length of the rangefinder, the angles of prism rotation were very small. Over the usable range of the rangefinder, the prism rotation is less than half a degree. Consequently the machinery in the rangefinder required a precision in angle measurement a better than a ten-thousandth of a degree.

The pointer sat at the farthest seat. The rangefinder could elevate or depress 15 degrees from horizontal. Normally, a gyrostabilizer maintained the rangefinder elevation automatically. If the automatic system failed, the pointer could take over the task manually.

The difference between a coincidence rangefinder and a stereoscopic rangefinder is in how the operator merges the images. A coincidence ranger finder splits the image into upper and lower halves. The operator aligns the two halves to get the range. A stereoscope rangefinder does not divide the image at all. Instead, the operator superimposes the views from the left and right sides of the rangefinder using a binocular sight.

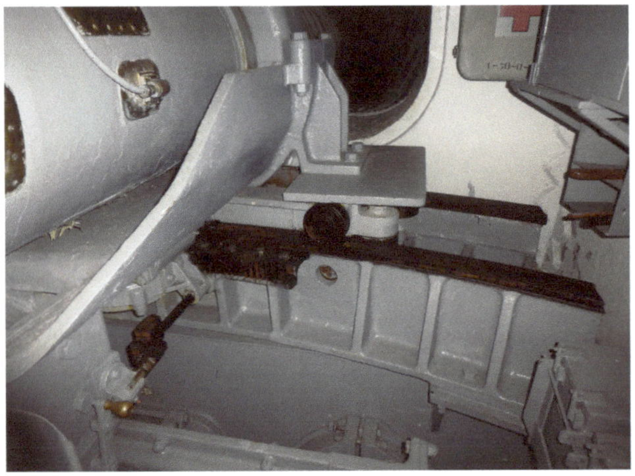

(46) The rangefinders rotate on tracks at either side of the turret. These are the tracks at the right side of turret no. 2. The opening in the turret side where the rangefinder enters its hood is visible above. The flexible gas seal extends from the rangefinder tube to the inside of the hood.

(47) This is the left chain rammer in turret no. 1. The rammer pushed projectiles and powder bags into the gun breech. The structure on top of the rammer is the hydraulic fluid supply tank. The large curved structure is the rammer chain path. When retracted, the chain runs from the upper left (where the chain path is at the bulkhead) to the right where it makes a 180-degree turn downward and under the raised platform to an opening in the bulkhead. The chain is similar to a bicycle chain except that the links are much larger, it does not form a closed loop and the links are designed so that the chain can only bend in one direction. Because the chain path can curve, a chain rammer requires less than half the length of an equivalent piston-driven rammer.

The rammer uses electric-hydraulic drive. The electric motor is at the curved end of the rammer. The A-end hydraulic pump is immediately to the left with the drive shaft covered in protective material. The B-end hydraulic motor is under the raised platform below the upper step tread. The motor drives a sprocket gear that pushes the chain into the gun pit through an opening in the bulkhead below the platform. The chain can push a 2,700 lb. projectile because it cannot bend upwards and the spanning tray holding the projectile prevents it from bending downwards.

A raised platform covers most of the officer's booth. The steps next to the rammer illustrate the difference in height between the shelf plate and the platform. The main hatch is partially visible to the right. The two shelf plate hatches are the only entrances to the turret above the second platform.

(48) The turret captain's Periscope Mk.28. The turret captain is second is command of the turret. The periscope allows the turret captain to identify targets and to spot the fall of shots when the turret is under local control. The turret officer in command of the turret has an identical periscope on the right side.

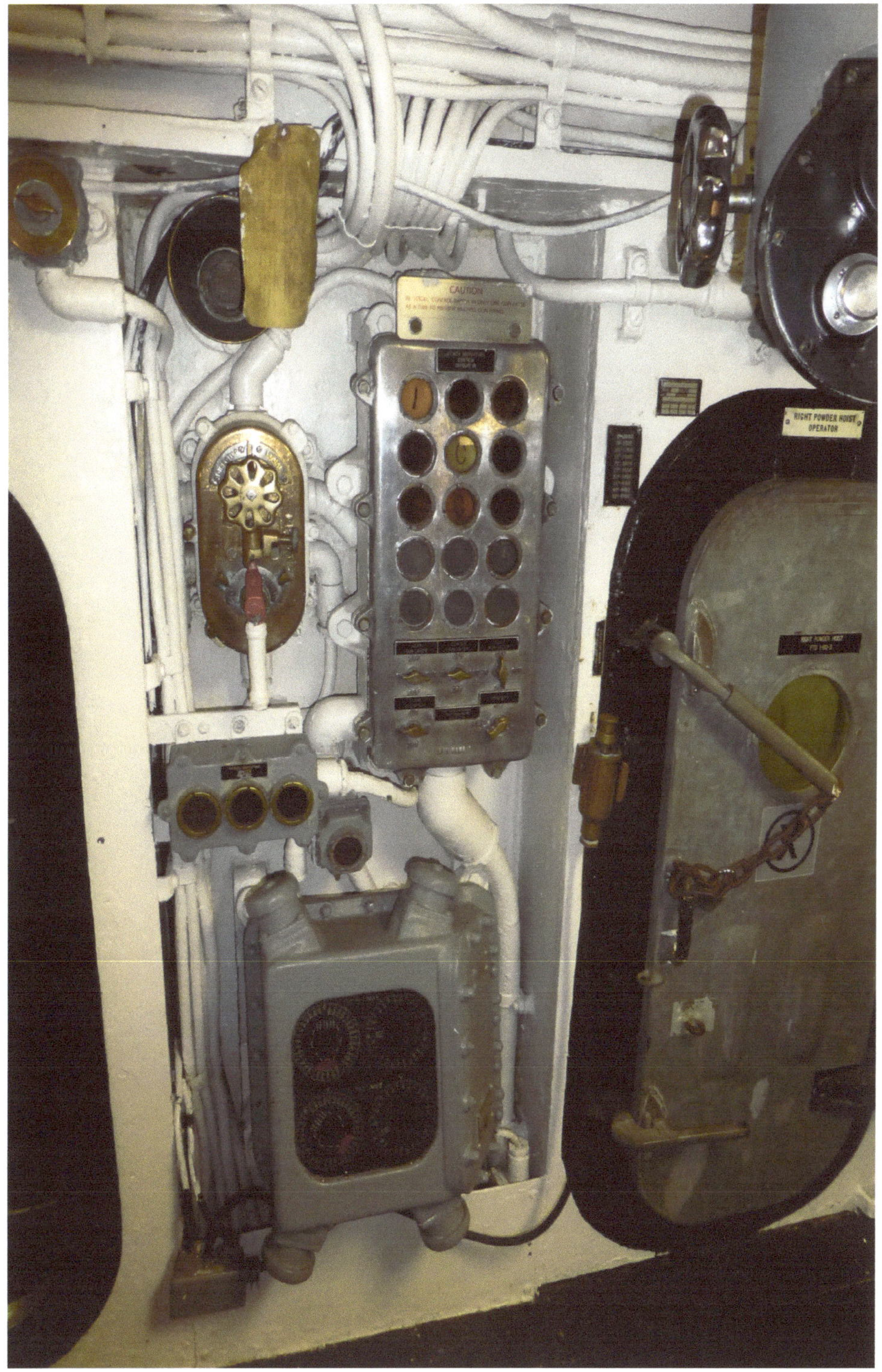

(49) Controls and indicator panels at the turret officer's stations. The upper left panel is the selective switch. The upper switch can be set to off, local control (where the turret controls itself), or director (where the turret is controlled remotely). Normally, this switch would be set to director when the turret was in action. The lower switch determines whether the guns are fired from the ship's power or from a battery.

The indicator panel is at the upper right. The upper part of the indicator panel has lights that show the status of each gun. The lower part of the panel has more switches that control how the turret operates. When under director control, the turret can operate in indicating or automatic mode. In indicating mode, the elevation and traverse orders are transmitted to an indicator. The trainer and gun layers manually adjust the turret traverse and gun elevation to match the orders. In automatic mode, the traverse and elevation orders are transmitted directly from the sight setter to the turret machinery. Normally, the turret would be under director control of a plot room and in automatic mode. Each turret has duplicate sights and sight setter stations on the left and right sides. The bottom right switch determines whether the left or right side is active. The bottom left switch determines whether the turret traverse is in indicating or automatic mode. The row of switches above determines whether each gun is in indicating or automatic mode and whether the gun can be fired by its gun layer. The gun captain has a similar indicator panel but it does not have the control switches.

The lights at the center left are recoil indicators. They illuminate when the guns are recoiled.

The unit at the bottom left is a multiple turret train indicator. This unit is set up to display the train angle for turrets nos. 1 and 2. It also displays the ordered angle so, when in indicating mode, the turret officer can see if the trainer is responding to train orders.

(50) When the turret was under director control, the switchboard determines who controls the turret. The turret can be controlled remotely by one of the two plot rooms or one of the other turrets. This is a close up view of the switchboard in turret no. 2. Each turret function has a separate switch for routing control. The crew has color coded the switch setting for the turret to be controlled by the forward plot, aft plot, turret no. 2 and turret no. 3.

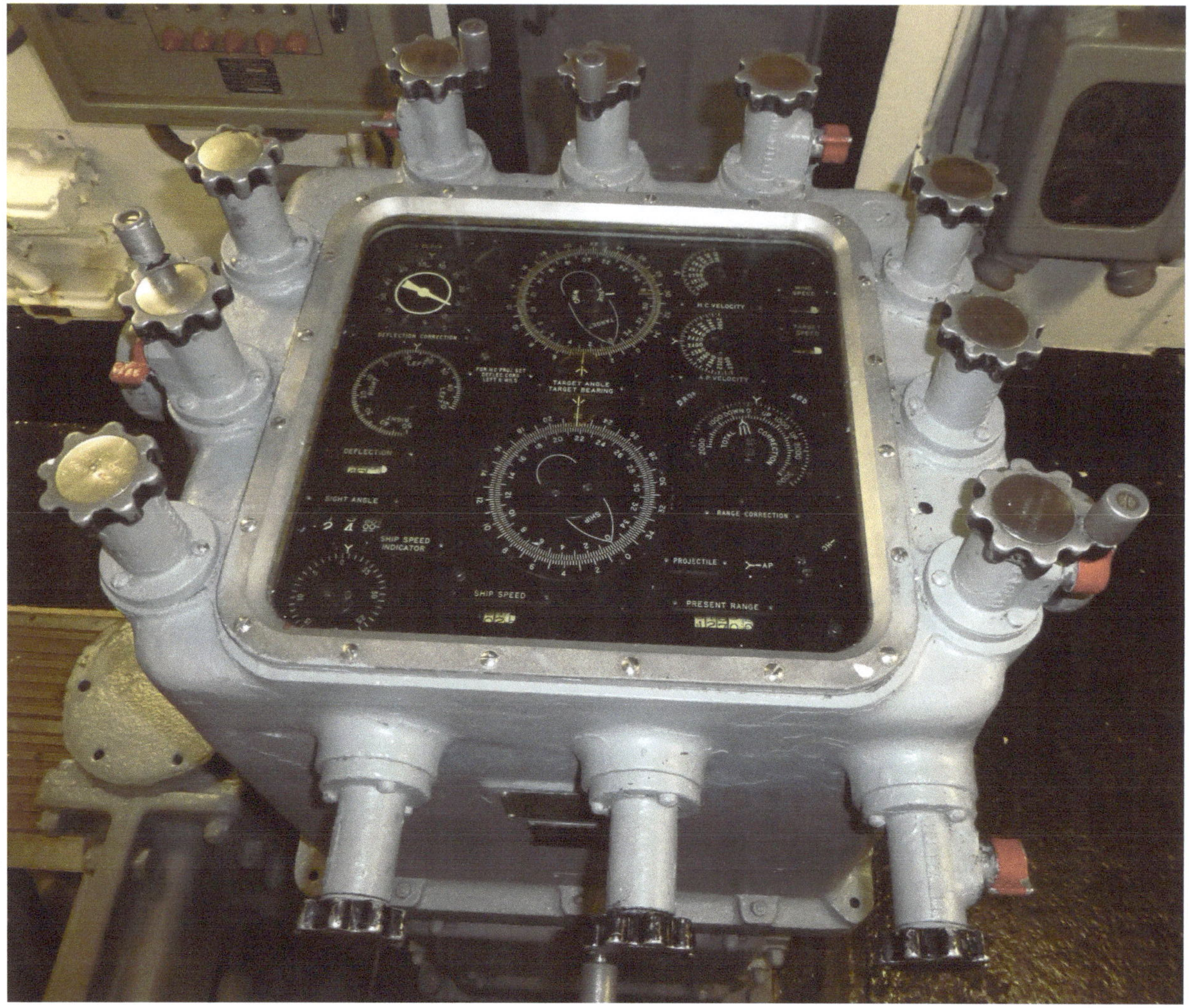

(52) When the *Iowa*-class battleships were reactivated in the 1980's, they used World War II era ammunition. There was concern that such old propellant would not perform consistently. Radars were installed on the roof of each sixteen-inch gun turret to measure the initial velocity of the projectiles when they were fired. This allowed the actual projectile velocity to be entered into the rangekeepers, rather than the estimated velocity.

(51) In a desperate situation, any of the turrets could take over the role of the plotting rooms and control all three turrets. In place of the Rangekeeper Mk.8 in the plotting rooms, the turrets are equipped with the older and much smaller Rangekeeper Mk.3. The Rangekeeper Mk.3 takes into account thirteen manually entered variables and three automatically entered variables (that could also be entered manually). The Rangekeeper could continuously calculate where to aim the guns to hit a moving target.

(53) This view, containing sailor art in turret no. 1, also shows some details of the turret construction looking to the left side. After the rangefinder was removed, the opening was filled in with plating secured by welds along scalloped plates. Around that opening is the $\frac{3}{4}$" backing plate that supports the 9.5" class A side armor and obscures the armor from view. During construction, circular covers were welded over the bolts that secure the armor plates to the backing plate. Three of these covers are visible in this photograph. The gap between the backing plate and the armor was filled with cement.

The roof is 7.25" of Class B armor. A seam between roof plates runs through the upper center of the photograph. The ends of bolts screwed through the scarf join between the roof plates protrude through the armor. The rear plate is at the left. It is 12" of Class A armor. The rear plate was welded to the turret using scalloped plates.

(54) The emergency turret hatch is covered by a raised platform. This cover plate can be pulled open to gain access to the hatch below. The hatches for turrets nos. 1 and 3 swing towards the front of the turret. The hatches for turret no. 2 swing towards the back of the turret.

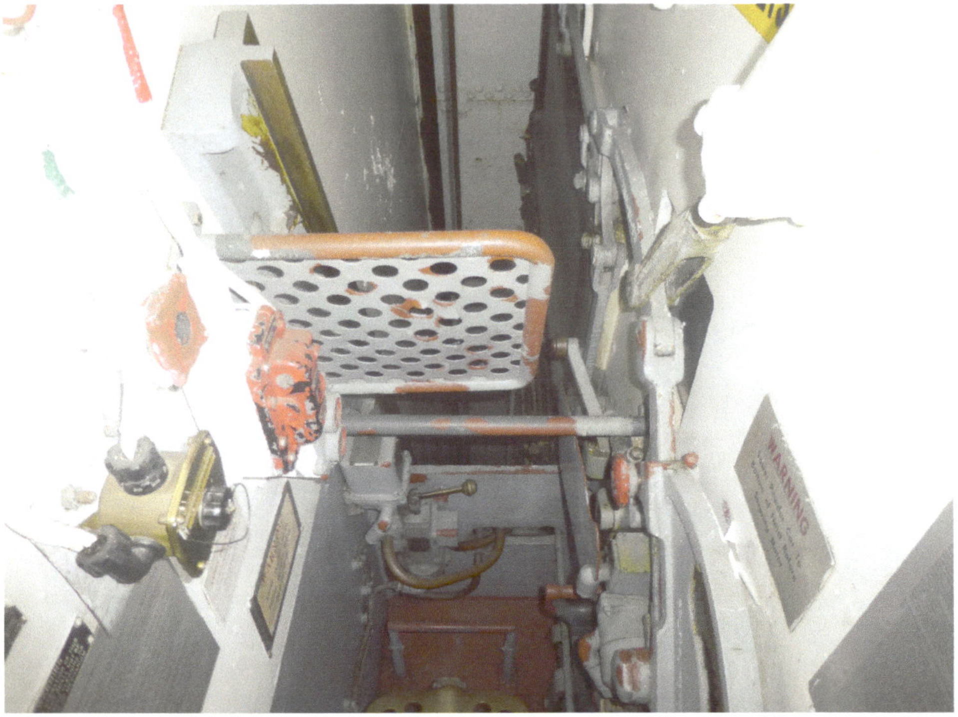

(55) This is the center hoist operator's station. The operator's bicycle seat and footrest are at the bottom of the photograph. The handle at the right that looks like a shovel handle controls the direction of the hoist. The operator only controls the direction. The hoist machinery automatically controls the acceleration and braking. The door for moving powder from the hoist into the gun pit is at the right. The rammer operator operates this door.

(56) This passage connects the left sighting station with the turret officer's booth in turret no. 3. The view here is to the rear. The deck is a raised platform above the shelf plate that is at the same level as the platforms in the officer's booth. The right side of the turret has a passage that is a mirror of this one. At the right of the photograph is the backing bulkhead that supports the side armor. Two rows of covers over the bolts holding the side armor in place are visible. The left gun pit is to the left. The entrance to the gun pit is through a step down that is partially visible just to the left of the door.

The left centering pin for the turret is in front of the ventilator with the tool used to raise and lower it. The tool was normally stowed on the bulkhead. When lowered, the centering pins extend through the bottom of the shelf plate and into the barbette (left), preventing uncommanded turret movement from transferring force to the traverse machinery. An angle bar was welded to the turret pin during decommissioning to prevent the pin from being raised. (Ryan Szimanski)

(57) A panorama view of the left sight stations of turret no. 1. The pointer's station and telescope is in the foreground. The pointer's control wheel has a firing key that the turret officer can enable when the turret was under local control. The trainer's telescope is in the background. The equipment in this area is mounted at the shelf plate so there is a step down from the elevated platform in the access passage. This area is mirrored on the right side of the turret. When the turret was under director control, the crew at these sights were on stand-by duty. When the turret was under local control, the sight crew on one side of the turret was active and the crew on the other side was on stand-by. The turret officer determined which side was active when in local control. The system divided the aiming duties such that the trainer controlled the horizontal aim and the pointer controller the vertical aim.

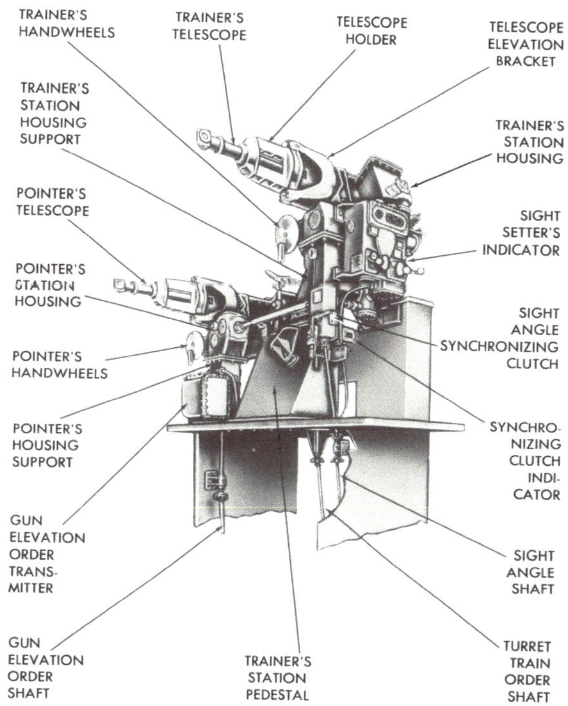

This diagram shows the major functions at the turret sight positions. The right side is depicted. (U.S. Navy)

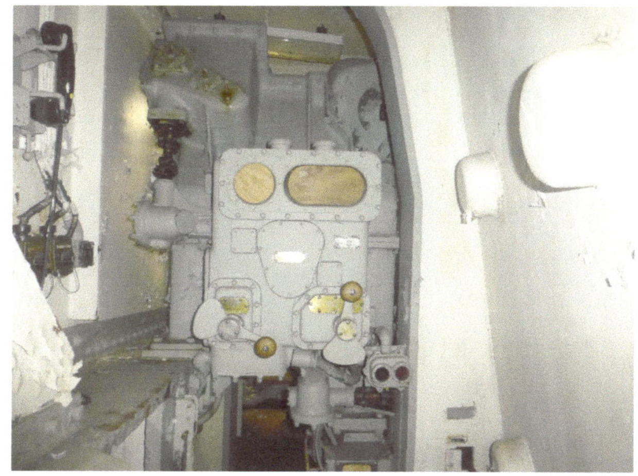

(58) The left sight setter's position of turret no. 3 looking to the rear. The crank at the left controls deflection and the one at the right controls elevation. When the turret was under director control, orders from the fire control system were displayed on dials above (covered in the photograph). The sight setter used the cranks to follow the pointers on the dials. The fire control system sent the same orders to all the turrets. The site setter adjusted their aim to take into account the turret location. When the turret was in local control, the sight setter would receive corrections over the telephone. The sight setter's correction would adjust the aim of the turret's sights.

(61) This is the right gun pit of turret no. 2 viewed towards the rear of the turret. Three men worked in that area to load the gun: the gun captain, cradle operator (at the left), and rammer operator (at the right). At the center is the cradle with its spanning tray and the hoist tube extending below. The cradle is the last stop for a projectile in the hoist. The projectile arrives oriented vertically at the center gun and tilted 16°40' to the rear at the outer guns. The cradle rotates to orient the projectile at a 5-degree angle from horizontal to align it with the breech. As the cradle rotates, the spanning tray extends into the open breech.

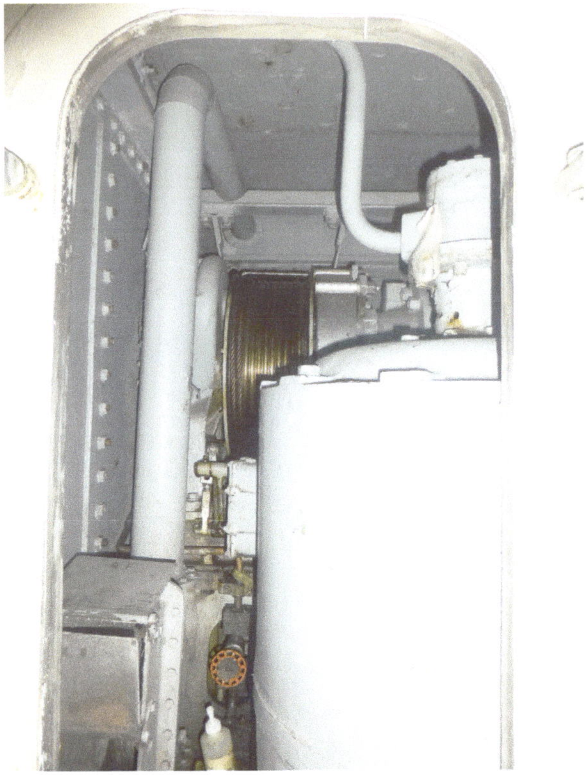

(59) Right powder hoist machinery looking aft. The cylinder at the right of the opening is the electric motor that powers the system. The motor drives the A-end pump that is at the level below (photograph (42)). The B-end motor is directly behind with the hoist cable reel at the left. The hoist works much like an elevator where a cable raises and lowers the car through the shaft. The area to the left of the photograph opens into the right gun pit.

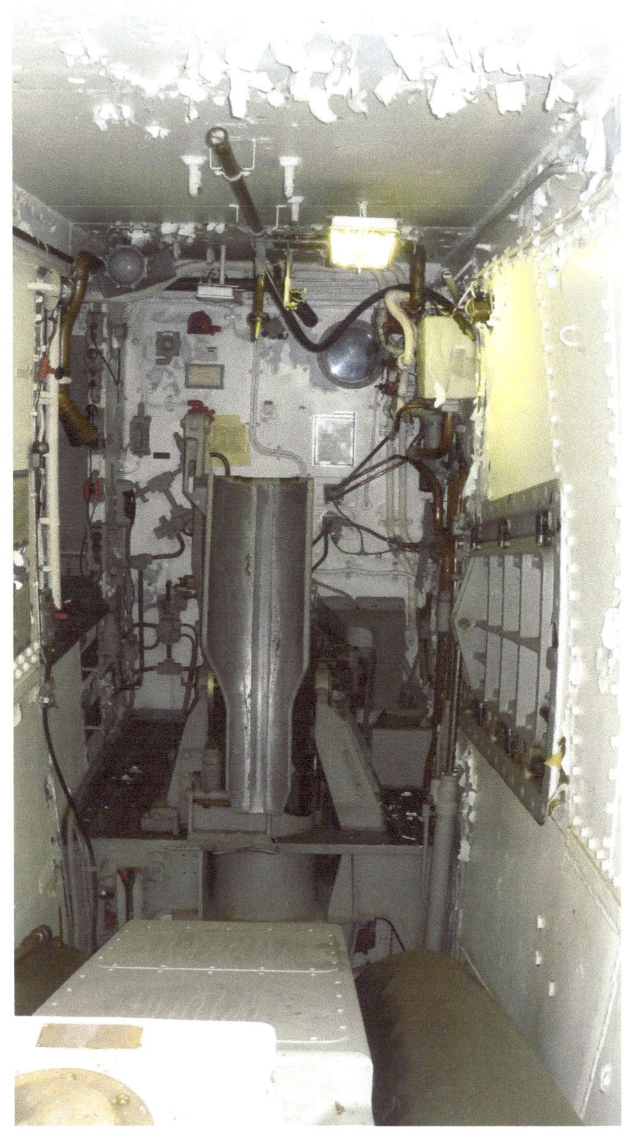

(60) At the center is the $2\frac{1}{2}$" backing plate that supports the 17" Class B front armor plate. The armor plate is bolted to the backing plate with cement filling the gap. Five of the covers welded over the bolts are visible. The large structure to the right is the left trunnion bearing block for the right gun. This is the point where the gun elevates. The circular structure contains a bearing ring. The trunnion on the slide fits snugly within the bearing ring. The heavy bar along the top (with bolts at either end) locks the slide in place. Part of the slide for the right gun is visible behind the bearing block. The recoil system on the slide absorbs most of the energy from firing the any. The bearing block has to absorb any remaining energy so it is five inches thick. The ring of bolts at the left secure a plate that gives access to the right bearing block of the center gun. This is part of a flame-tight bulkhead that separates the guns.

(62) The cradle operator (foreground) and rammer operator (background) positions for the left gun of turret no. 1 looking inboard. The layout of the loading machinery is mirrored from that shown in photograph (60). The projectile cradle is at the center. The pedal at the right corner unlocks the cradle and the lever on the bulkhead rotates the cradle. The spanning tray is attached to the cradle at the left. As the cradle rotates, the arm at the rear of the cradle extends the spanning tray so that fills the gap from the cradle into the gun breech.

The rammer operator's seat is in the rear corner. The rammer control is on the bulkhead next to the seat. This controls the speed and direction of the rammer. The operator controlled the distance of ramming and there is no safety mechanism that prevented over-ramming powder bags and crushing them against the projectile.

The crank above manually operates the door to the powder hoist. The lever above the crank operates the powder door hydraulically. The window in the bulkhead allows the powder hoist operator to see into the gun pit.

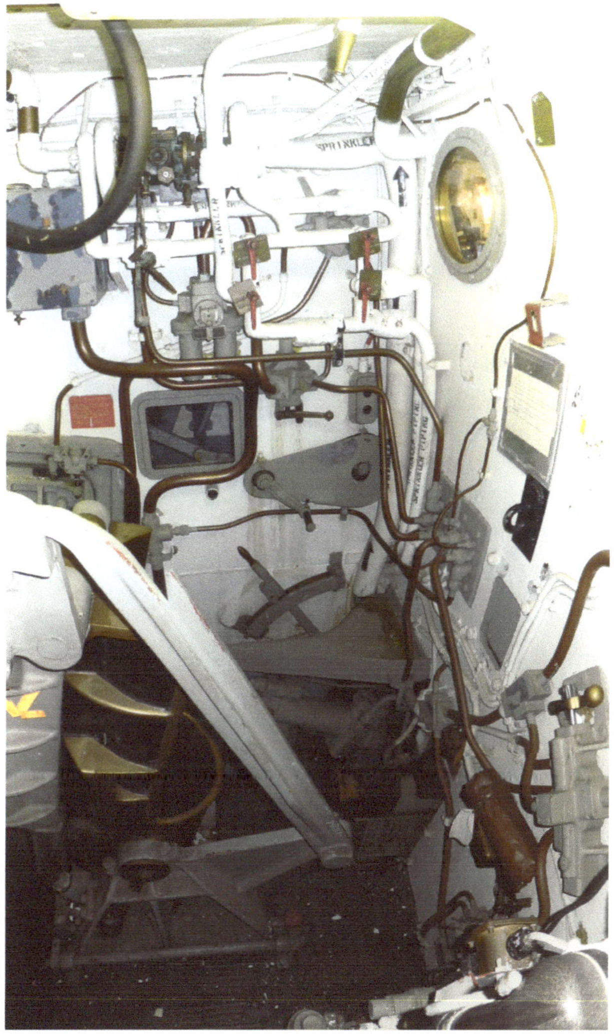

(64) The gun captain used this device to signal the gun was ready to fire to the rest of the fire control system. Functionally, this switch controlled whether the gun elevated to its ordered position (READY) or the 5-degree loading position (SAFE). The switch has an internal lock so that it could only be moved from SAFE to READY when the breech was closed. The switch is spring loaded to move to the SAFE position. When the gun was fired, the switch unlocked, the switch sprung to SAFE, and the gun moved to the loading position. After the gun was loaded and the breech was closed, the gun captain moved the switch to READY and the gun elevated to its ordered position. The knob below manually unlocks the switch in case it failed to do so automatically.

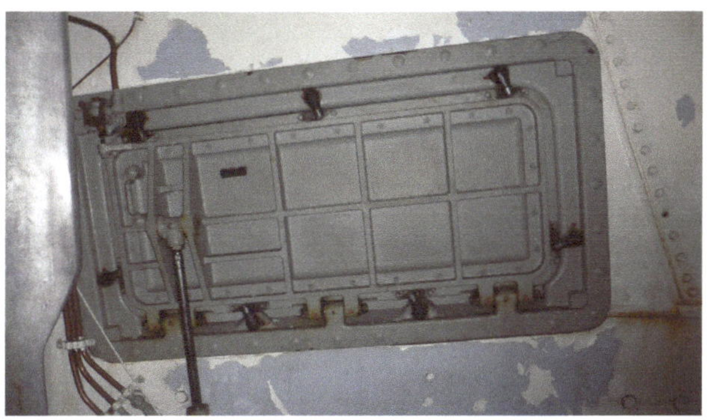

(63) The powder hoist door for the center gun of turret no. 2. The door is hydraulically operated and is controlled by the rammer operator. The door also serves as the path for moving powder bags to the spanning tray.

(65) These hinged plates form the upper gas seal for the gun. The hinge maintains the seal as the gun rotates. The corresponding seal underneath the gun uses a simpler wiper blade that rotates against a curved plate.

This is a diagram of a 16-inch gun assembly and slide. The slide supports the gun assembly and is part of the turret. The guns elevate by rotating the slide along its trunnions within the trunnion bearing block (photograph (59)). The slide houses the recoil and counter-recoil system. The gun assembly moved in and out within the slide (labeled Gun Slide Cylinder on the diagram) as the gun is recoiled and then was moved back into battery. The slide also has small loading platform that extends past the breech end of the gun. When the gun is in loading position, this platform gives the gun captain access to the breech opening mechanism.

The gun assembly has three major components: the gun, yoke, and breech. What the Navy calls the *gun* is just the barrel. The gun rests in the slide and, on its own, can move freely within it. The guns were installed or removed by pulling them through the gun ports at the front of the turret into the slide. Each gun is 816 inches long and weighs about 118 tons. They are 48.2" in diameter at the breech and 23.5" in diameter at the muzzle. It is impracticable to build a gun barrel this size as a single piece. Instead, it is made up of several components. A key component is the removable barrel liner. The liner forms the rifled path that forces the projectiles to spin. Each time the gun was fired that liner eroded. Eventually, the liner had to be replaced. The liner was designed to be removable so that the rest of the gun could be reused.

The yoke is the part of the gun assembly that keeps the gun in place within the slide. The yoke was attached to the breech end of the gun after it had fully passed through the slide and was held in place on the gun using a key and grove. The yoke connects to the recoil and counter recoil cylinders on the slide.

The breech mechanism closes the inside end of the gun during firing. This consists of a 1,400-lb. breech plug and counterbalances for the plug. The breech plug fits inside the end of the barrel. The counterbalances that allow the plug to be easily opened and closed by one person are attached to the yoke. (U.S. Navy)

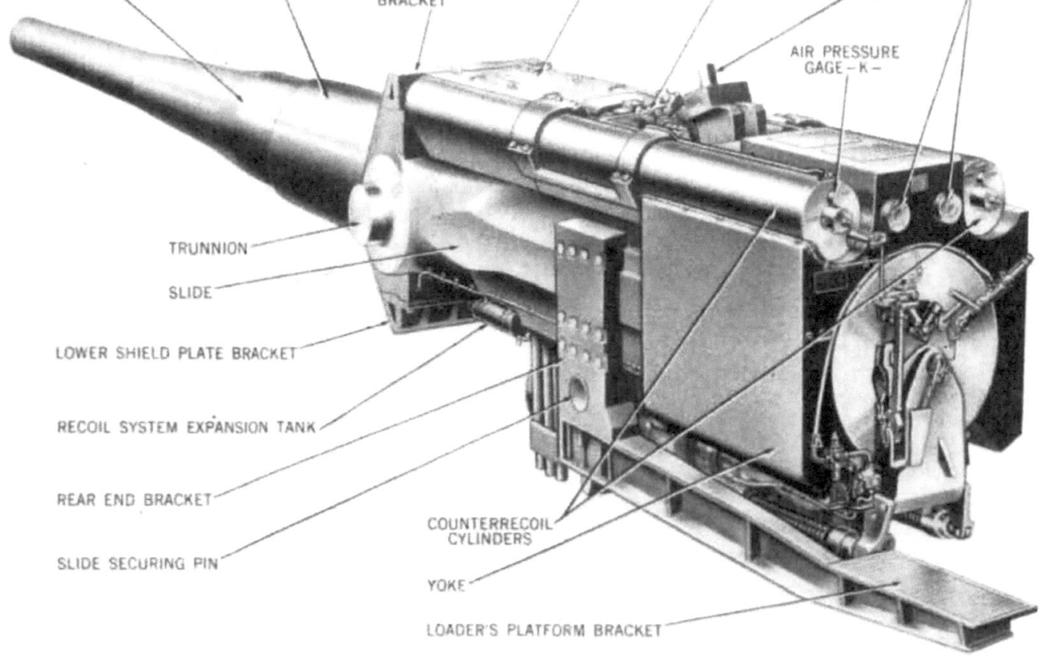

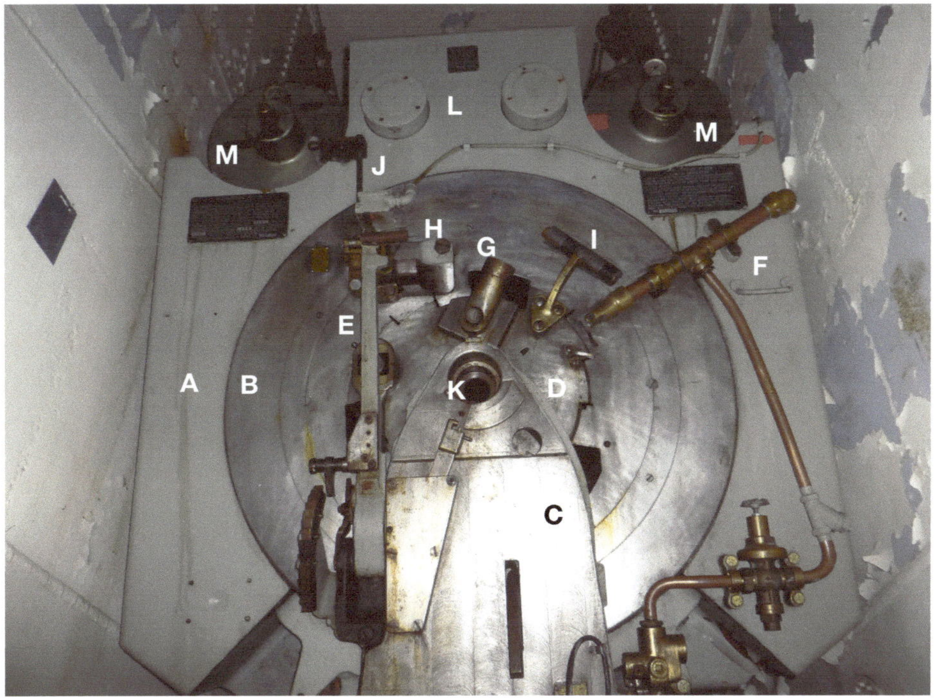

The breech end of the center gun of turret no. 2. This particular MK. 7 16-Inch/50 caliber gun (No. 285) was made at the U.S. Naval Gun Factory at the Washington Navy Yard. The U.S. Army Watervliet arsenal also made sixteen-inch guns for the *Iowa*-class.

A. Yoke
B. Gun
C. Breech Mechanism
D. Breech Plug
E. Breech Operating Lever
F. Compressed Air Line
G. Cam Roller
H. Cam
I. Opening Handle
J. Salvo Lock
K. Opening for Primer Lock
L. Yoke Rods
M. Counter-recoil Cylinders

The operating lever opens the breech by rotating to unlock then moving it outward and downward. This lever also releases a blast of compressed air into the breech to blow any remaining burning material out the muzzle. When opening and closing, the cam and cam roller transition the operating motion between rotation and swinging downward.

The salvo lock is a safety feature that prevents the breach from being accidentally opened after the gun it is loaded. The breech locks when it is closed. Firing the gun unlocks the salvo lock. The salvo lock could be unlocked as part of safety procedures followed if the gun failed to fire.

The yoke rods link the yoke to the counterrecoil cylinders on the slide that return the gun to battery after firing.

The primer lock that goes into the center of the breech has been removed from this gun (see photograph (37)).

An open breech plug on USS *Missouri*. The breech plug locks have a Welin screw. The screw has three, 5-step segments with seventeen threads. The advantage of this system is that it locks nearly all the way around the breech opening but it only has to be rotated 24 degrees to unlock.

The mushroom is at the center breech plug and assists in sealing the breech during firing. It has a stem with a tube that extends to the face of the breech to an opening that is barely visible at the center. The primer fires through this tube to ignore the powder. (Nat'l Archives)

This is a mushroom in a museum display. A flexible ring called the *gas-check pad*, made out of asbestos and tallow is behind the mushroom. The expansion of the propellant pushes the mushroom against the gas-check pad, causing the pad to expand to seal the breech. The tube shown secured in the vise normally would run through the breech plug. This tube allows the primer to ignite the powder bags. This system is known as a *De Bange gas check*.

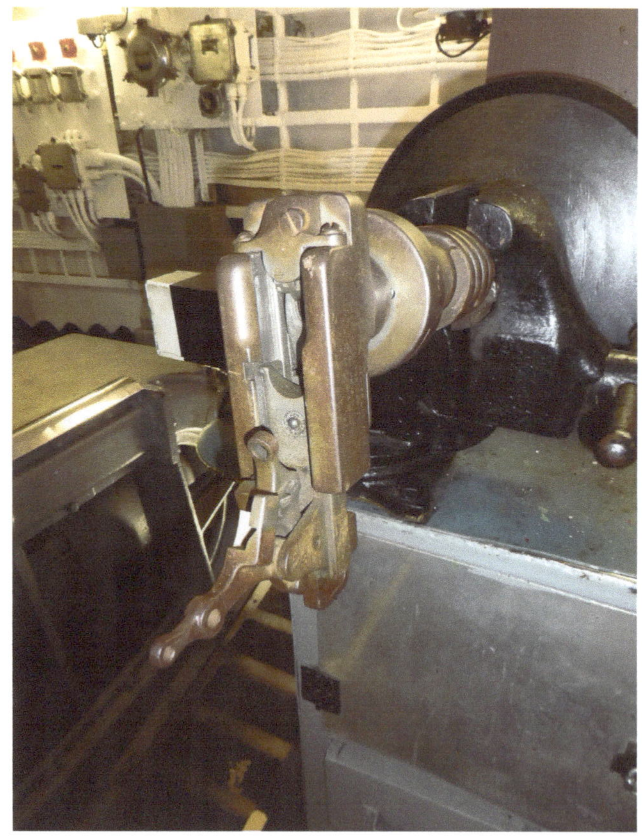

This is the other end of the same mushroom showing the combination primer lock. The primer charge for firing the gun could either be triggered electrically or mechanically. The primer chamber is closed by a wedge that slides up to close and down to open. Here, the lock is shown in the open position. After the gun was fired the lock opened and the spent primer was automatically ejected. The extension at the lower left is the cocking lever. A lanyard could be attached to the opening at the end of the cocking lever. The part with the open loop directly above is the firing hammer. The firing pin is the circular object to the right of the hammer. The part of the hammer that struck the firing pin has been removed. It would have fit within the circular opening of the hammer as shown in photograph (37).

To fire the gun mechanically, a crewmember would pull the lanyard to move the cocking lever backwards. After the lever moved far enough back, the spring-loaded hammer would release and strike the firing pin. This pushed a plunger at the base primer that crushed an internal percussion cap and set off the primer charge.

Normally the gun was fired electrically. The firing pin is insulated from the rest of the primer lock and the plunger at the base of the primer is insulated so that current can pass from the primer case and the plunger. Thus applying current between the firing pin and the body of the primer lock creates an electrical circuit that ignites the primer charge.

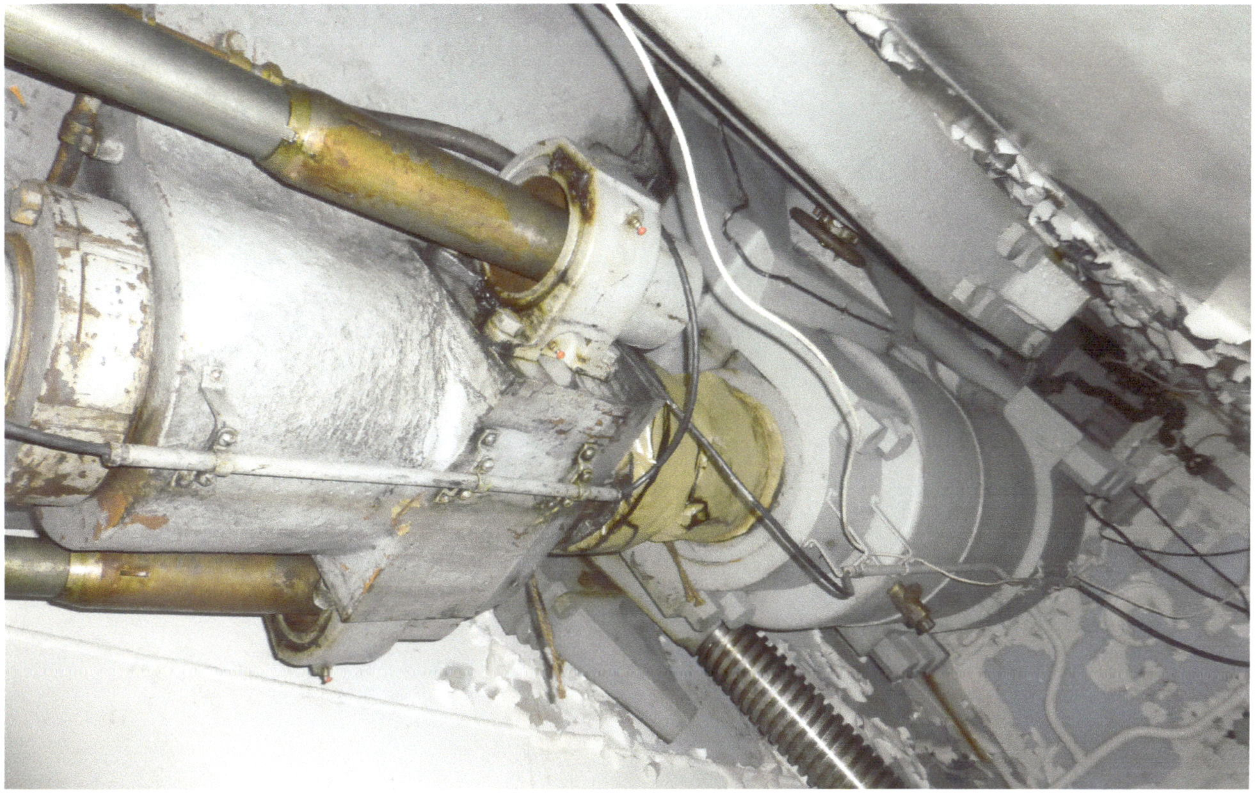

This is the underside of the center gun of turret no. 2. The recoil cylinder on the slide is at the right and the yoke is at the left. The rod linking the two is covered with protective material. The rod is connected to a piston within the recoil cylinder. The recoil cylinder is filled with hydraulic fluid and the piston is perforated with small holes. When the gun recoils, the yoke pulls the piston backwards. In order for the piston to move, the hydraulic fluid has to pass through the holes to flow to the other side of the piston. The flow of fluid from one side of the piston to the other retards the recoil to less than four feet.

At the right of the piston there is a small wheel. This the slide-securing pin release that locks the gun at zero degrees elevation. Rotating the wheel extends a pin from the slide into the gun girder or retracts the pin. When the pin is extended, the gun cannot elevate and the pin prevents the force produced by induced barrel movement from being transmitted to the elevation machinery.

The two counter-balance and closing cylinders for the breech mechanism are on either side of the yoke. These allow one person to easily open and close the breech.

The elevation jackscrew is attached to the left side of the slide and connection between the loading platform and the side is at the right. The jackscrew and loading platform are reversed on the left gun. Part of the trunnion bearing block is visible in the far corner.

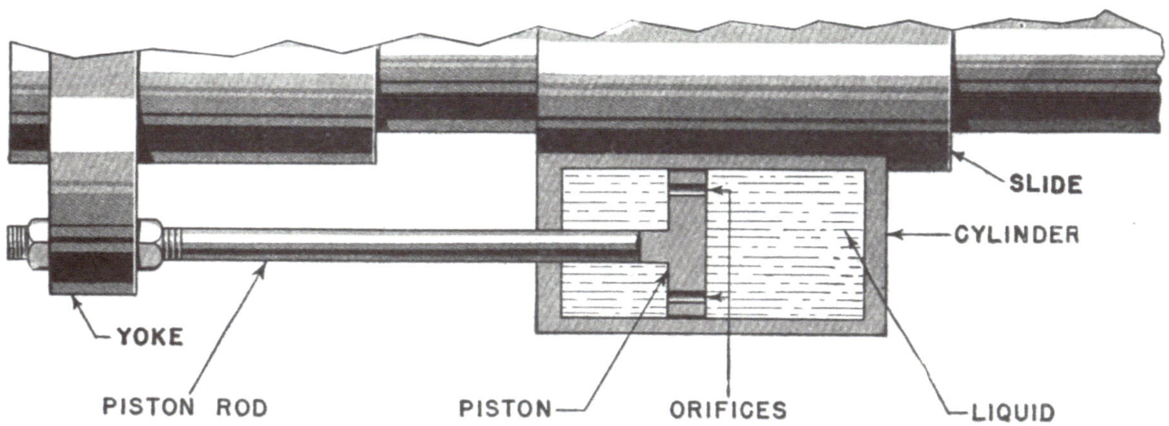

This diagram shows how the movement recoil piston arrests recoil by forcing fluid to move from one side of the piston to the other through small holes. (U.S. Navy)

When the gun is fired, it automatically moved to the 5-degree loading position. The gun captain opened the breech, checked for any burning debris, and wiped the mushroom to ensure the tube from the primer lock was clear. The gun captain here has a cloth wipe wrapped around his arm and is standing on the platform attached to the slide that rotates with the gun. Next the cradle operator rotated the cradle. The arm attached to the cradle and spanning tray caused the spanning tray to extend into the open breech. This photograph shows the cradle while in the process of rotation. This is the right gun of turret no. 2 on USS *Iowa*. (Nat'l Archives)

The rammer operator used the chain rammer to push the projectile all the way into the breech. The rammer chain can only bend one direction (downward). Because the spanning tray prevents it from bending in that direction, the chain can push a 2,700 lb. projectile. (U.S. Navy)

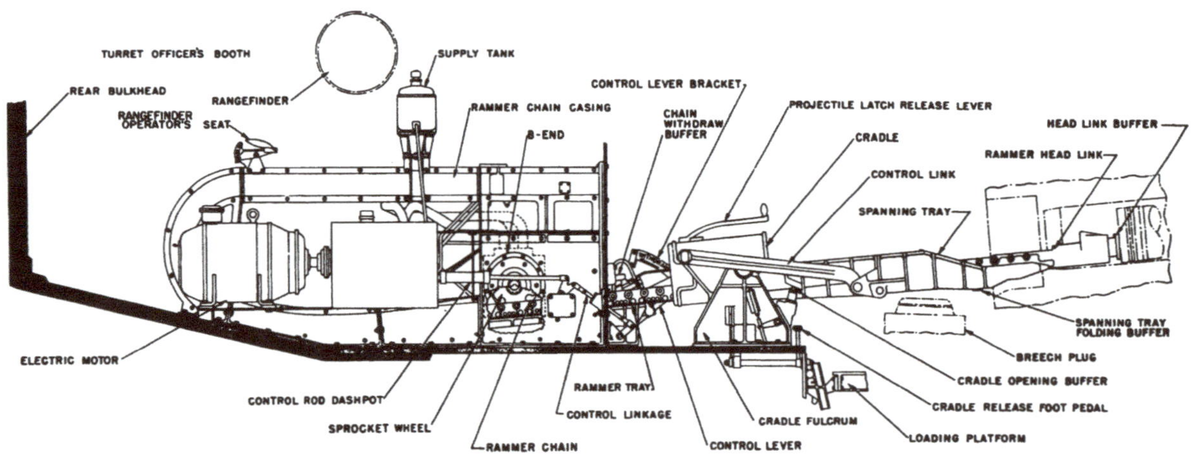

This diagram shows the arrangement of the rammer, cradle, spanning tray, and gun during loading. (U.S. Navy)

The rammer operator opened the door to the powder host. The powder hoist door created a ramp to the spanning tray. The powder hoist has two trays that held three powder bags each. The powder hoist trays rotated to roll the powder bags in to the spanning tray. In this photograph, the first three bags of powder are being positioned in the spanning tray. The gun captain and cradle operator had to push the first three powder bags forward so they were out of the way for the hoist operator to unload the final three powder bags. The rammer operator is assisting in this photograph.

When all six bags were in place, the rammer operator had to slowly push them into the breech. While all this was going on, the primer man below inserted a primer into the primer lock in the breech. The gun captain could then close breech and signal the gun was ready.

The pads containing black powder are visible at the ends of the powder bags. These had to be aligned towards the breech end of the gun. The quilting on the pads ensured there was powder spread over the entire pad. (Nat'l Archives)

The right gun of turret no. 2 in the process of firing on USS *Iowa*. The large tubes at the top of the gun are the counter-recoil cylinders mounted on the slide. They pushed the gun back into battery after recoil was complete. The smaller rods alongside link the yoke to the pistons in the cylinders. Poking up from the center is the buffer that shut off the elevation motor when the gun reached maximum depression. (Nat'l Archives)

Fire Rooms

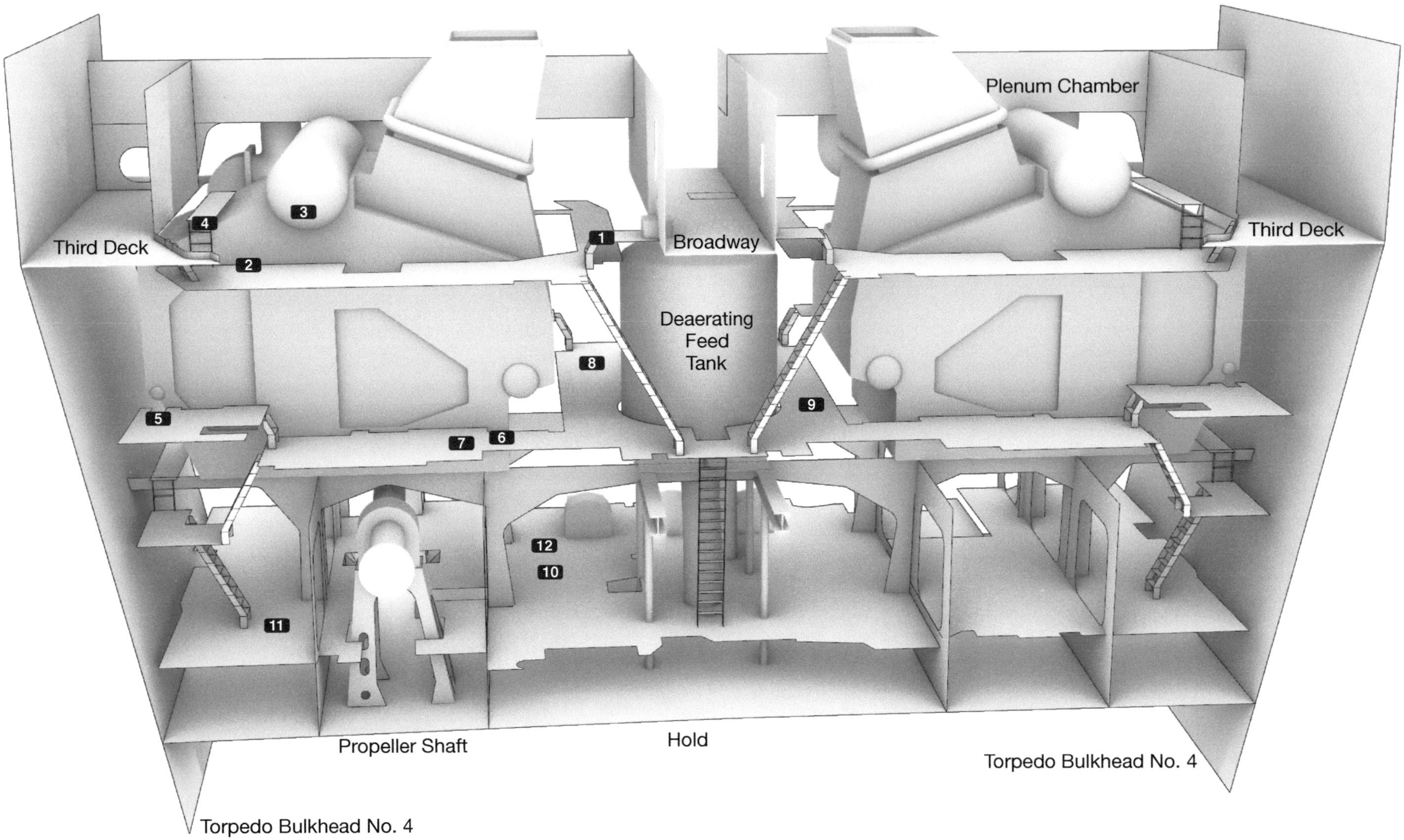

Fire Room No. 2
Looking Aft

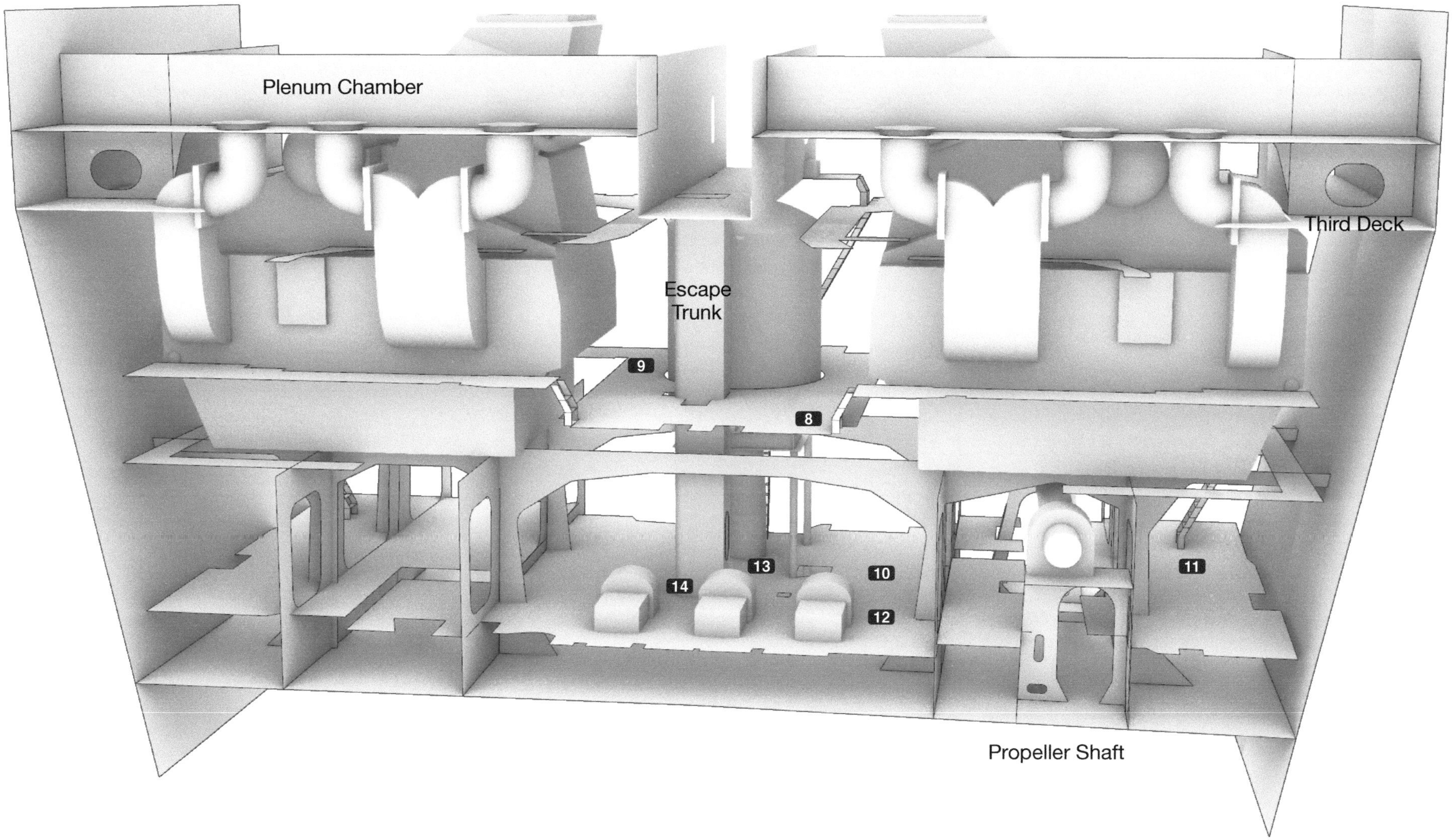

Fire Room No. 2
Looking Forward

(1) This a panorama view of starboard side of fire room no. 2 taken from the Broadway entrance. The right side of the image is looking aft and the left side is to starboard. The fire room is an open space that extends from the hold to splinter deck. In the fire and engine rooms there are platforms that give access to the machinery instead of decks. In the fire rooms there are about a dozen levels of platforms. The fire room platforms are in three groups (upper, middle, and lower). The step up at the far left is where third deck resumes at the starboard side.

One of the two M-Type Babcock & Wilcox boilers in the fire room fills most of the photograph. The uptake segment at the center passes through splinter deck and ends at a second deck armored grate. Loops of pipe are visible below the uptake. This is a heat exchanger called an *economizer* that cools the uptake gases while heating the feed water. The two vertical tubes are periscopes (see photograph (7)).

The bulkhead to Broadway runs along the far right. Part of the plenum chamber that supplies the boilers with air is visible at the right along the overhead.

Each fire room normally supplied steam to the corresponding engine room immediately aft. However, in an emergency fire rooms 1 and 2 could be cross connected to engine rooms 1 or 2 and fire rooms 3 and 4 could be cross connected to engine rooms 3 or 4.

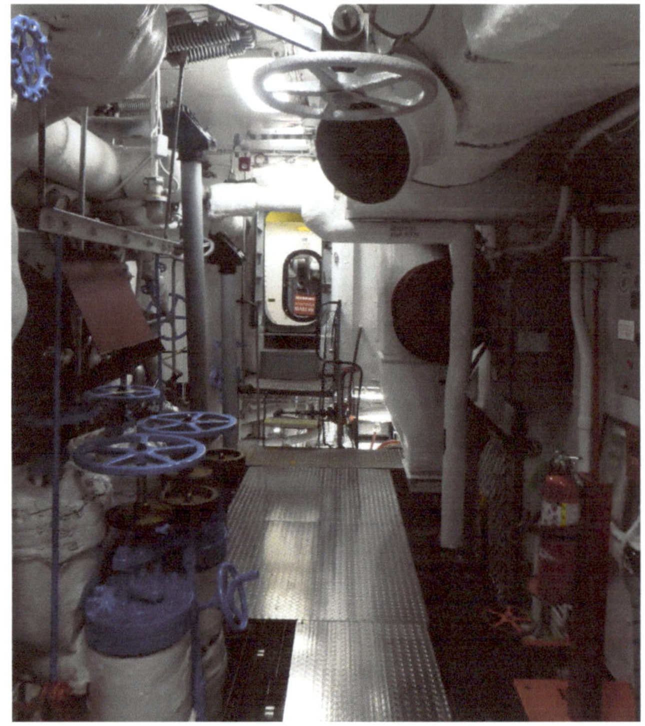

(2) This view is in the opposition direct of that in photograph (1). The door at the center leads to Broadway. One can see that Broadway forms a tunnel that passes through the fire room. A boiler is to the right. The ventilation air intake is at the upper right.

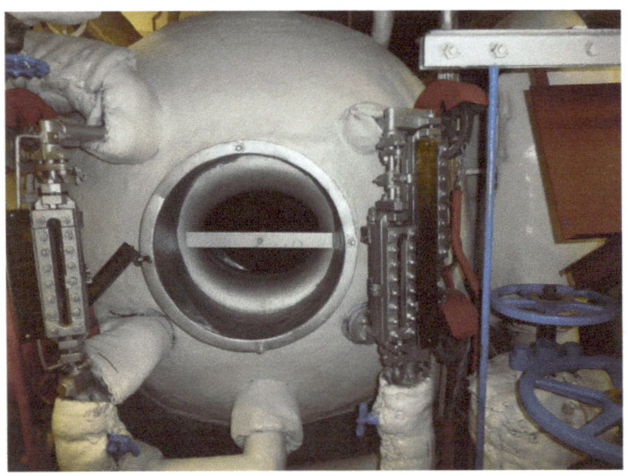

(3) The steam drum sits at the top of the boiler. The lower half of the steam drum is half filled with pure water that feeds by gravity into the saturated steam side of the boiler. Saturated steam collected at the top of the drum where it was routed to other systems.

(4) Platforms at various levels give access to the boilers for maintenance. The highest platform runs along the top of the boiler at the outer side.

(5) This is a panorama view of the forward end of fire room no. 2 looking to port showing the middle level. At this level one can see the full width of the fire room. The starboard boiler is to the left. The port boiler is at the far end of the platform. The arrangement of the port boiler is mirrored from that of the starboard boiler. Normally, only one boiler was operated at a time for best fuel efficiency. This would also allow maintenance to be done on the idle boiler. Both boilers would be fired up when high speed was needed.

The circular structures on the face of the boiler are burners. The burners are in two groups. One group of four burners is clearly visible to the left. The other group of five burners is partially visible at the center. The boiler has two separate boiler units internally. The inboard unit of the boiler produces saturated steam. Steam is saturated when it is close to its boiling point where water molecules can shift between liquid and gas. The boiling point increases with pressure. The temperature of saturated steam in the boilers was around 400–500 degrees F. The outboard unit of the boiler produces superheated (also called *dry*) steam. Saturated steam flowed from the steam drum into the superheater where the temperature could be increased up to 850° F. The effect of this arrangement was that the saturated steam unit controlled how much steam was produced and the superheating unit controlled the temperature (and the pressure) of the steam. *Iowa*-class battleships originally burned bunker oil. This fuel was so viscous that steam heaters in the fuel system were required to allow it to flow to the boilers. During the 1980's reactivations, *Iowa*-class battleships were converted to use distillated marine (similar to diesel) fuel and the heaters were removed.

(6) This is a closer view of the burners on the saturated side of the boiler.

A. Fuel supply line
B. Fuel cut off valves
C. Fuel lines to burners
D. Burner Barrel (each burner)
E. Fuel flow adjustment (each burner)
F. Air Door adjustment (each burner)
G. Observation window (each burner)
H. Ignition port. The boiler was lit by inserting a rod with a flame at the end through this port.

The burner barrel is a tube that extends into the boiler. The barrel has an atomizer at the inside end (at right) to direct fuel into the boiler. A selection of atomizers was available to give different fuel burn rates. The burner barrels had to be replaced and cleaned every few hours or when the burn rate needed to be changed.

(7) If the amount of oxygen relative to fuel flowing into the boiler (the stoichiometric ratio) were correct, its exhaust should be close to clear. If there were more fuel than air, the exhaust becomes black and when there was more air than fuel, it became white. Periscopes allowed the boiler room crew to see the smoke coming out of the uptake so that they could ensure the mixture is correct.

The structure in the background center is the deaerating tank. Next to the boilers, the deaerating feed tank is the largest piece of equipment in the fire rooms. The purpose of the deaerating tank is to remove oxygen and carbon dioxide dissolved in the boiler water. The distilling plants can remove the solids—but not the gasses—dissolved in the water. If these are not removed, under the high heat of the boiler oxygen causes rust and carbon dioxide combines with water to create carbonic acid that erodes piping. The deaerating tank used a combination of mechanical action and heat provided by steam from the boilers to force gases out the feed water. The deaerating tank also served as the water reservoir for the boiler system.

(9) Sailor art is located throughout the ship. Such artwork tends to become more *risqué* in the more inaccessible areas of the ship. Here the final crew of fire room no.2 created a tribute on the deaerating tank. "BT" indicates a *boiler technician* and "FN" means *fireman*.

(8) This panorama view was taken from the aft end of the starboard boiler. The view to the left is forward and the view to the right is to starboard. Part of the deaerating tank is visible at the left side. Access panels to the boiler were removed during the mothballing process. The boiler tubes are visible through the opening. The class of boiler used in *Iowa*-class battleships was called an *express type boiler* because it heated water flowing through small, two-inch tubes. Smaller tubes create a greater surface area for move efficient heating. The drawback of small tubes is that they are more prone to clogging from solids dissolved in the water. Small tube boilers were impracticable in service until ships were equipped with systems that could remove the solids from the water. The two curved air blowers feeding the saturated and superheating units are at the rear of the boiler. The blowers contain electrically-powered turbines that force air from the plenum chamber into the boiler.

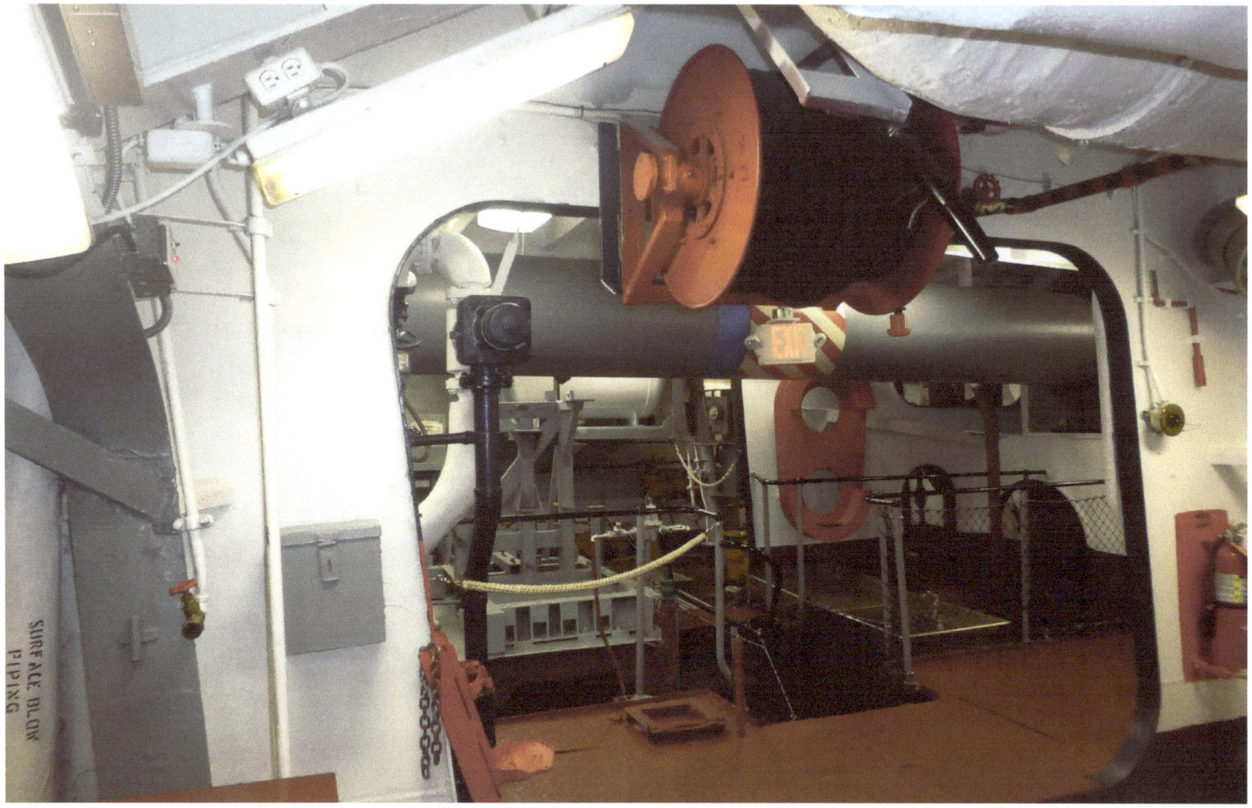

(10) The upper levels of the different fire rooms are very similar in layout. The lowest levels differ because propeller shafts. The propeller shaft from engine room no. 1 passes through fire room no. 2. Fire room no. 3 has two shafts and fire room no. 4 has three shafts passing through. This is the lower level of fire room no. 2 showing the outboard port shaft looking starboard and aft. The platforms here are elevated above the hold. The diagonal stripes indicate to the crew when the shaft is turning and there is a recess in the platform to give extra clearance. What appears to be a bulkhead in the foreground is actually the foundation for the boiler. The transverse foundation at the left is not vertical and the opening in the longitudinal foundation is a trapezoid that follows its slope.

(11) The forward shaft bearing in fire room no. 2. Shaft bearings are spaced at regular intervals along each shaft to provide support. The shaft bearings are mounted on top of tower-like foundations. Because the hull curves upward at the stern, the height of these foundations decreases towards the aft end of the shafts.

(12) A panorama view of central area of the lower level of fire room no. 2 looking to port. The forward fire rooms have a fair amount of open space. In the aft fire rooms propeller shafts take up this free space. At the left are the three feed pumps that supply water for the two boilers. There is one extra pump so to allow maintenance on one while operating both boilers. The open door to the escape trunk is at the center. The large blue vertical pipe with an ess-curve is the feedwater output from the deaerating tank. This pipe supplied a feed water storage tank within the inner bottom between the hold and hull shell.

(13) The only way out of the fire room is up. Battle damage to a boiler could result in steam blocking the way out. The escape trunk provided a way to get from the bottom of the fire room to Broadway where the boilers could be shut off in an emergency.

(14) The view up the fire room no. 2 escape trunk showing the open hatch at Broadway . The fire room escape trunks all follow a vertical path.

Engine Rooms

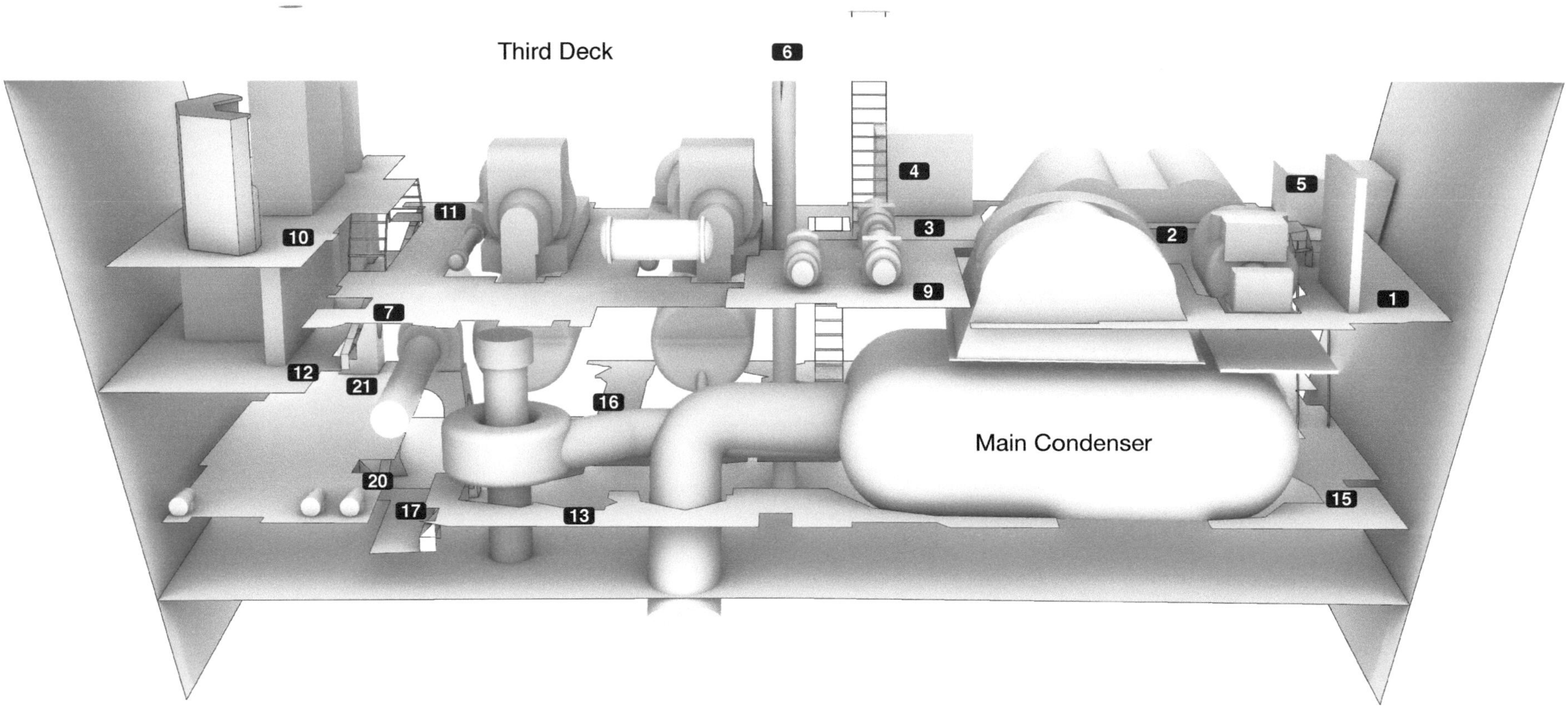

Engine Room No. 2
Looking Aft

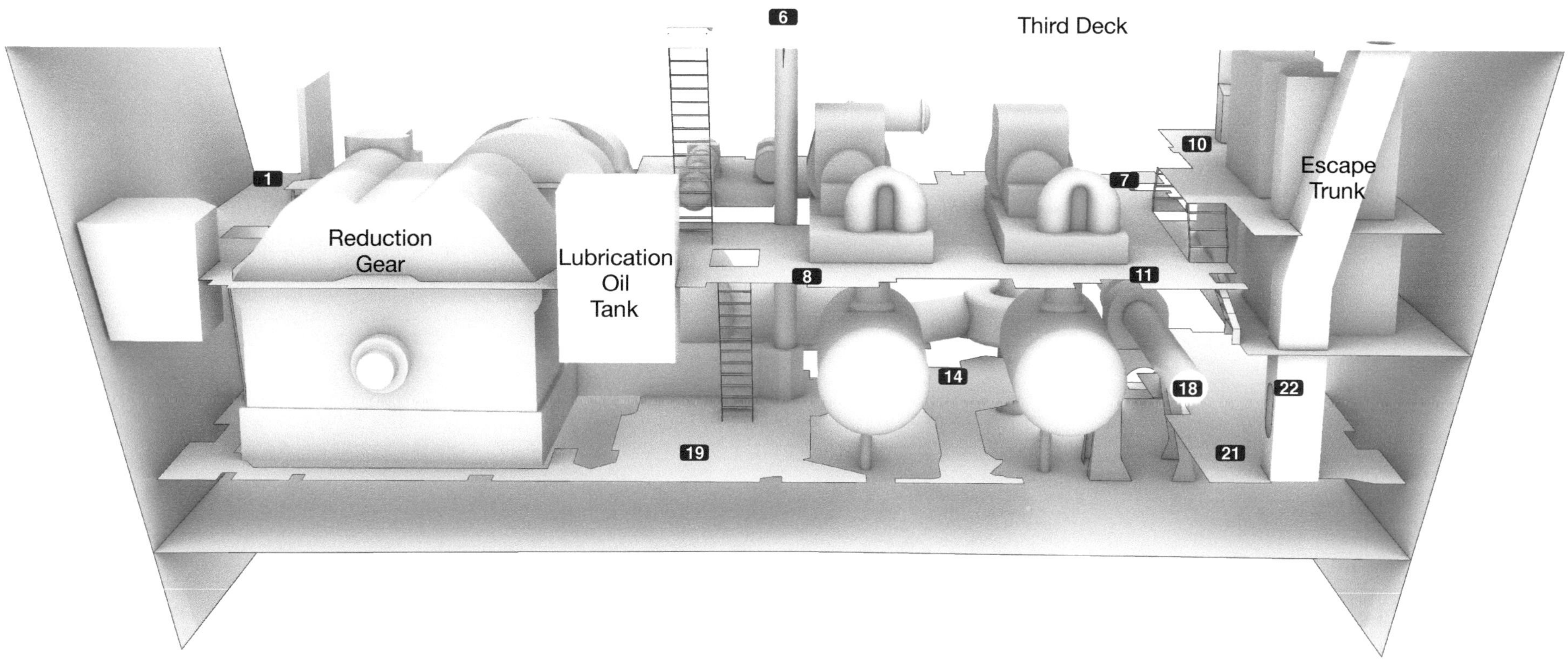

Engine Room No. 2
Looking Forward

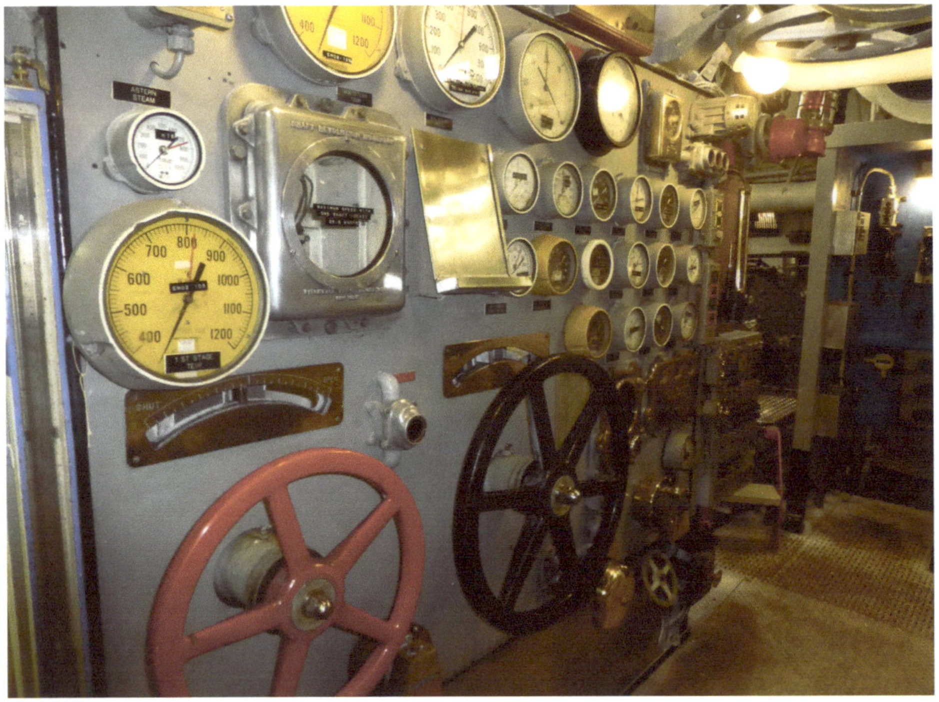

(1) The main turbine gauge board in engine room no. 2. The wheel at the left is the astern throttle and the wheel at right is the forward throttle. Orders were transmitted to the engine room using the telegraph at the right of the panel. For voice communication, there is a telephone booth at the far right. The gauge board is located starboard in engine room nos. 1 & 3 and port in nos. 2 & 4.

The turbine gauge board for engine room no. 3. The reverse throttle is at the far right and the forward throttle is to the left of that. The gauge board is much larger than that of the other engine rooms (compare with photograph (1)). This is because the station for the engineering watch officer was in engine room no. 3. In order for the watch officer to have an overview of the propulsion system, this gauge board duplicates the gauges of the other engine rooms. From this location, the watch officer could see if the other engine rooms were following their orders from the engine order telegraphs or to see if the gauges indicated any problems. The gauge board does not duplicate the controls for the other engine rooms. If an issue were detected here, it had to be resolved by telephone.

(2) This is the high-pressure turbine looking to port. Steam enters through the two pipes at the forward end (to the right), flows aft through the turbine blades to turn a drive shaft, and then exits through the large crossover pipe at the aft end (to the left). The steam exits the high-pressure turbine with much energy remaining so it then flows into a low-pressure turbine to gain additional power. The high-pressure turbine can only drive the propeller shaft in the forward direction. The propulsion machinery for USS *New Jersey* and USS *Wisconsin* were supplied by Westinghouse and those for USS *Iowa* and USS Missouri were made by General Electric.

(4) This large tank supplied the lubrication oil for the propulsion system.

(3) The need to align the propulsion machinery with its propeller shaft creates a much greater variation in layout in the engine rooms than there is in the fire rooms. This is a wide view of the upper level in engine room no. 2 looking to port. The reduction gearbox is at the left and the low-pressure turbine at the right. The high-pressure turbine is partially visible behind the low-pressure turbine and the telephone booth shown in photograph (1) is at the end of the narrow platform running between the gearbox and the turbines.

The crossover pipe from the high-pressure turbine enters the low-pressure turbine at its center. The lower pressure turbine has two sets of turbine blades with the crossover pipe entering between them. When operated forward, the steam divides and flows outward from the center in both directions so that each stream flows through a set of blades that turn a drive shaft. When operated in reverse, steam enters the turbine through the smaller pipes at the ends and drives separate sets of blades located at the ends of the turbine that are connected to the same drive shaft. Spent steam exits the bottom of the low-pressure turbine where it goes into the main condenser.

The turbines operate around 4,000 r.p.m. If the propellers turned at that speed they would boil water rather than move the ship. The reduction gearbox reduces the output speed from the high-pressure turbine by about 24:1 and the low-pressure turbine by about 19:1. Note the chains secured to the overhead that were used to move machinery for repair.

(5) The lubrication oil settling tank was used to remove water and other impurities from so that the oil could be reused.

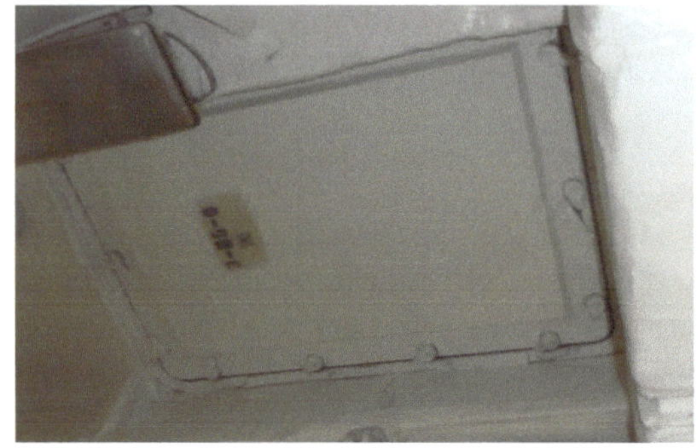

(6) This large hatch in the overhead leads to Broadway. Large machinery could be moved from the second deck workshops using the overhead monorails to this hatch where it could be lowered into the engine room.

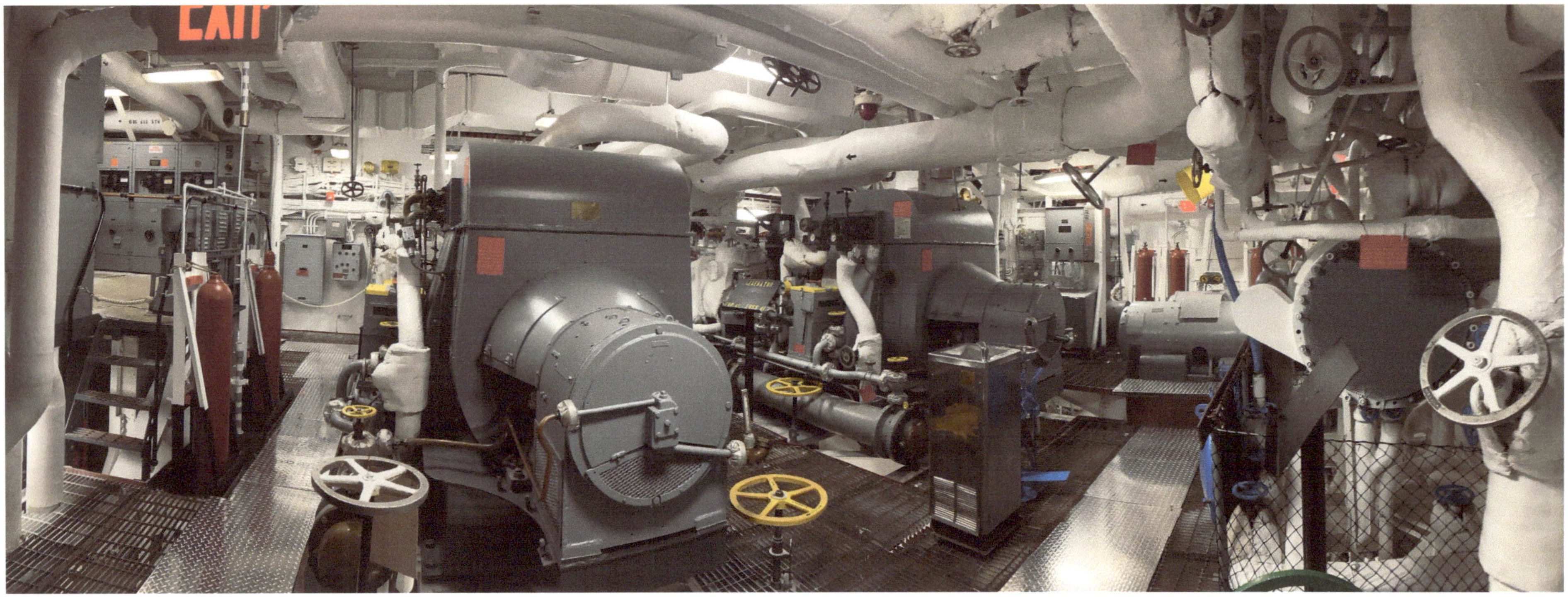

(7) The port side of engine room no. 2 is an electrical power plant. This panorama view looking aft shows the two steam powered ship's service turbo-generators at the center. The electrical switchboards are located at split-levels shown at the left. The air ejector is at the extreme right. The air ejectors are used to removed air and gases from the condensers located on the lower engine room platform.

(10) This is the engine room no. 2 electrical switchboard control panel with part of the switchboard in the background. This is where the electrical output from the turbo-generators was managed. The electrical switchboards in engine rooms nos. 1 & 3 are located on the port side and those in engine rooms nos. 2 & 4 are on the starboard side.

(8) This is the aft end of the turbo-generator units where the turbines are located. Some of the boiler steam was directed through pipes running along to the overhead into turbines at the aft end of the unit. The turbines turn the shaft for the generators at the forward end of the unit. The spent steam exits at the aft end underneath and flows downward into auxiliary condensers directly below. Part of the distribution switchboard is visible in the background.

(9) These are degaussing units that are part of the ship's protection against magnetic mines. The large mass of the steel in a battleship naturally creates a giant magnet. The purpose of these units it to create a D.C. current flowing through cables looped around the ship that reduces the ship's magnetic field. The air ejector is at the far right and the turbo-generators are in the background center.

(11) This photograph shows how the electric switchboard components are located on split-levels from the other engine room platforms. The engine room escape trunk is at the far right.

(12) The lower electrical switchboard platform is located between the upper and lower engine room platforms. The lower electrical platform is 3'-4" below the upper engine room platform.

(13) The lower platform in the engine rooms is about three feet above the hold. Within the machinery spaces, the hold slopes outwards for drainage. To keep machinery level, it has to be mounted on foundations erected over the hold. The main condenser is at the lower engine room platform, directly below the drive turbines. The view here is to port. The lower platform is elevated about three feet above the hold. The condenser is a heat exchanger that uses sea-water to cool the spent steam exhaust leaving the low-pressure turbine. The main circulating pump draws water from an induction sea chest near the centerline, through the condenser, and out through a discharge sea chest at the port side. The largest pipe shown leads to the induction sea chest. The smaller pipe branching off at the left leads to the main circulating pump.

The rectangular panel resting against the condenser was part of the bulkhead that the museum cut away so visitors could move directly between the fire and engine rooms.

(14) The main circulating pump looking to starboard.

(15) The port side of the lower platform in engine room no. 2 looking aft. The main condenser is to the left. The discharge sea chest is immediately below the end of the condenser. The group of valves in the center are for fuel oil and those at the sides of the platform are for seawater.

(16) These are the auxiliary condensers that serve the turbo-generators located directly above them. These operate the same way as the main condenser except that they are much smaller.

(17) The starboard outboard shaft continues aft through engine room no. 2. The platform below the shaft is lowered to give greater clearance. A workbench is in the background to the left with a shaft bearing opposite.

(18) The watertight seal for the shaft at the aft bulkhead of engine room no. 2.

(19) A panorama view of the lower platform in engine room no. 2 looking to port. The propeller shaft emerges from the gear box at the left. The two main lubrication oil pumps are in the foreground. Part of the lubrication oil tank is visible above. The lubrication oil strainer is partially visible behind the pipes. To the right, the foundation for the low-pressure turbine wraps around the main condenser that pokes out at the right.

(20) These pumps supply seawater for cooling in the air conditioning system. The view is to starboard.

(21) The entrance to the escape trunk in engine room no. 2 (and no. 4) is at the starboard side aft. The escape trunks for engine rooms nos. 1 & 3 are at the port side.

The main distilling plans are located in the forward emergency diesel generator room. Engine room no. 3 has a distilling plant that can also create boiler feed water from sea water.

(22) The escape trunk slopes outboard to starboard and emerges in the radio transmitter room at third deck.

Appendix

Abbreviations used in this book and others commonly found in U.S. Navy plans.

3M	Maintenance and Material Management
AA	Anti-Aircraft
A/C	Air Conditioning
ABL	Armored Box Launcher
ABV	Above
ABV BL	Above Baseline
AERO	Aerological
AG	Armored Grating
ALA	Aluminum Alloy
AL AL	Aluminum Alloy
AL ALY	Aluminum Alloy
AL C	Aluminum Corrugated
AL CORR	Aluminum Corrugated
AL P	Aluminum Pilaster
AMM	Ammunition
AMMN	Ammunition
AP	Aft Perpendicular
APPD	Approved
ARMD	Armored
AT	Air Tight
ATQA	Air Tight Quick Acting
ATMOS	Atmospheric
ATMOSPH	Atmospheric
AVN	Aviation
BAL	Balanced
BATT	Battery
BATTY	Battery
BEL	Below
BET	Between
BKTS	Buckets
BH	Bulkhead
BHD	Bulkhead
BKT	Bracket
B_L	Baseline
BOSN	Boatswain
BP	Bolted Plate
BR	Bridge
BU ORD	Bureau of Ordnance
BU C&R	Bureau of Construction and Repair
C TO C	Center to Center
CAL	Caliber
CANV	Canvas
CAPT	Captain
CEC	Combat Engagement Center
CG	Cleaning Gear
CHKD	Checked
CHT	Collection and Holding Tank (sewage)
CIC	Combat Information Center
CIWS	Close-In Weapons System
C_L	Centerline
CL	Class
CLASS	Classified
CLEAN	Cleaning
CLEAN GR	Cleaning Gear
CMAA	Command Master-at-Arms
COAM	Coaming
COMP	Companion
COMPT	Compartment
CON	Conning
COND	Conditioning
CONT	Control
CONTWR	Conning Tower
CORR	Corrugated
CORR AL	Corrugated Aluminum
COV	Covered or Cover
CPO	Chief Petty Officer
C&R	Construction and Repair
CRS	Corrosion Resistant Steel
CT	Conning Tower
CVK	Centerline Vertical Keel
D	Diameter
DAPA	Drug and Alcohol Prevention Advisor
DASH	Drone Anti-submarine Helicopter
DC	Damage Control
DECAN	Decanned or Decanning
DECON	Decontaminating
DECONTAMIN	Decontaminating
DET	Detachment
DIA	Diameter
DIR	Director
DIST	Distributing
DK	Deck
DN	Down
DOCUM	Document
DESTRUCT	Destruction
DWL	Designed Waterline
ECM	Electronic Countermeasures
EEBD	Emergency Escape Breathing Device

ELEC	Electrical
EM	Expanded Metal, Electronic Materials, or Electromagnetic
EMERG	Emergency
ENG	Engineering
ENG	Engine
EQUIP	Equipment
EXCL	Excluding
EXEC	Executive
EXH	Exhaust
EXP	Expansion
FACCON	Facilities Control
FAS	Fueling at Sea
FAT	Flush Air-Tight
FATQA	Flush Airtight Quick Acting
FATQAS	Flush Airtight Quick Acting Scuttle
FDN	Foundation
FFT	Flush Fume-Tight
FL	Floor
FLT	Flight
FNTMH	Flush Non-Tight Manhole
FNTQA	Flush Non-Tight Quick Acting
FMTBP	Flush Flame-Tight Bolted Plate
FMTBPHM	Flush Flame-Tight Bolted Plate Manhole
FMTMH	Flush Flame-Tight Manhole
FNTMH	Flush Non-Tight Manhole
FNTQA	Flush Non-Tight Quick Acting
FO	Fuel Oil
FORD	Forward
FOTBPMH	Flush Oil-Tight Bolted Plate Manhole
FOTHBMH	Flush Oil-Tight Hinged Bolted Manhole
FOUND	Foundation
FP	Forward Perpendicular
FW	Fresh Water
FWT	Flush Watertight
FWTBPMH	Flush Watertight Bolted Plate Manhole
FWTHBP	Flush Watertight Hinged Bolted Plate
FWTQA	Flush Watertight Quick Acting
FP	Forward Perpendicular
FR	Frame
FRM	Fire Room
FT	Fume-Tight
FTMJ	Fume Tight Metal Joiner
FZ	Fire Zone
GALV	Galvanized
GEN	Generator
GR	Gear
HBP	Hinged Bolted Plate
HBPMH	Hinged Bolted Plate Manhole
HGT	Height
HIFR	Helicopter Inflight Refueling
HTR	Heater
HTS	High Tensile Steel
HWTBP	Hinged Watertight Bolted Plate
HYD	Hydraulic
IC	Interior Communications
IFF	Identification Friend or Foe
INBD	Inboard
INCL	Inclined or Inclusive
INCL LAD	Inclined Ladder
INCLN LAD	Inclined Ladder
INCLN	Included or Inclusive
ID	Inside Diameter
INSUL	Insulation or Insulated
INTER	Interchangeable
LAB	Laboratory
LAD	Ladder
LAV	Lavatory
LBRY	Library
LH	Lightening Hole
LIEUT	Lieutenant
LIGHT	Lightening
LGL	Longitudinal
LKR	Locker
LONGL	Longitudinal
LT	Light
LR	Ladder Rungs
LUB	Lubrication
MAA	Master at Arms
MACH	Machine or Machinery
MCHRY	Machinery
MAG	Magazine
MARDET	Marine Detachment
MAX	Maximum
MEAS	Measured
MECH	Mechanical
METCAL	Metrology and Calibration
MG	Motor Generator
MH	Manhole
MHO	Manhole Over
MIN	Minimum
MJ	Metal Joiner
MJFT	Metal Joiner Fume-Tight
MJL	Metal Joiner Louvre
MJSP	Metal Joiner Solid Panel
MLD	Molded
MLD B	Molded Base line

MLD BL	Molded Base Line
MS	Medium Steel
MT	Flame-Tight
NAV	Navigating
NAVR	Navigator
NO	Number
NT	Non-Tight
NT AL AL	Non-Tight Aluminum Alloy
NTS	Non-Tight Scuttle
NO	Number
OBA	Oxygen Breathing Apparatus
OD	Outside Diameter
OFCE	Office
OFF	Officer
OFFS	Officers
OHT	Oral Health Therapy
OPEN	Opening
OPER	Operated
ORD	Ordnance
OT	Oil Tight
OTHBPMH	Oil-Tight Hinged Bolted Plate Manhole
OTHMH	Oil-Tight Hinged Manhole
OUTBD	Outboard
(P)	Port
PAO	Public Affairs Officer
PBX	Private Branch Exchange
PCG	Portable Cleaning Gear Locker
PERF	Perforated
PH	Pilot House
PL	Plate
PLT	Plate
PLTG	Plating
PLATF	Platform
PORT	Portable
PREP	Preparation
PROJ	Projection
PROJ	Projectile
PROV	Provisions
P/S	Port and Starboard
PTN	Partition
PUB	Publications
PWDR	Powder
PYRO	Pyrotechnic
QA	Quick Acting
QAD	Quick Acting Door
QAS	Quick Acting Scuttle
QAWT	Quick Acting Watertight Door
QTRS	Quarters
R	Radius
RAD	Radius
RAS	Replenishment at Sea
RATO	Rocket-assisted Takeoff
RCVG	Receiving
RDR	Radar
REFR	Refrigeration
RMTMH	Raised Flame-Tight Manhole
ROTBPMH	Raised Oil-Tight Bolted Plate Manhole
RPR	Repair
RWT	Raised Watertight
RWTBPHM	Raised Watertight Bolted Plate Manhole
RWTHMH	Raised Watertight Hinged Manhole
RWTMH	Raised Watertight Manhole
RM	Room
RPV	Remotely Piloted Vehicle
RS	Ready Service
(S)	Starboard
SATCOM	Satellite Communications
SBROC	Super Blooming Rapid Offboard Countermeasures
SD	Supply Department
SD PROOF	Sound Proof
SEC	Secondary
SEW	Sewage
SHR	Shower
SHWRS	Showers
SITE	Shipboard Information Training and Entertainment
SPECS	Specifications
SPG	Spring
SR	State Room or Senior
STA	Station
STBD	Starboard
STAN	Stanchion
STL	Steel
STRM	Storeroom
STS	Special Treatment Steel
STWG	Storage
STRUCT	Structural
SUP	Supply
SUPER	Superstructure
SUPERSTR	Superstructure
SVCE	Services
SWBD	Switchboard
TAC	Tactical
TACAN	Tactical Air Navigation
TAO	Tactical Action Officer
TK	Trunk
TLI	Tank Level Indicator

TNG	Training
TRAIN	Training
TRANS	Transverse
TRANS	Transmitter
TRK	Trunk
TURR	Turret
UPS	Uninterruptible Power Supply
UR	Urinal
VEG	Vegetable
VENT	Ventilation
VEST	Vestibule
VL	Vertical Ladder
WC	Water Closet
WDO	Window
WH	Water Heater
WKS	Workshop
WL	Water Line
WR	Washroom
WRSR	Wardroom Stateroom
WRWC	Wardroom Water Closet
WT	Watertight
WTBP	Watertight Bolted Plate
WTHBP	Watertight Hinged Bolted Plate
WTHMH	Watertight Hinged Manholes
WTR	Water
WTS	Watertight Scuttle
XFR	Transfer
XMTR	Transmitter
XO	Executive Officer

Bibliography

- Bureau of Naval Personnel, *Naval Ordnance and Gunnery*, USGPO, 1944
- Bureau of Naval Personnel, *Naval Ordnance and Gunnery*, USGPO, 1955
- Norman Friedman, *U.S. Battleships: An Illustrated Design History, Naval Inst. Press*, 1986
- Norman Friedman, *Naval Firepower*, Naval Institute Press, Naval Institute Press, 2008
- William H. Garzke, Jr. & Robert O. Dulin, Jr., *Battleships: United States Battleships, 1935-1992*, Naval Inst, Press, 1995
- Richard A Landgraff, *A History of the Long Beach Naval Shipyard*, CreateSpace, 2017
- Malcolm Muir, *The Iowa Class Battleships: Iowa, New Jersey, Missouri & Wisconsin*, Blandford, 1987
- *Aircraft Procedures for Air-Capable Ships NATOPS Manual*, Naval Air Systems Command, 2012
- Buford Rowland & William B. Boyd, U.S. Navy Bureau of Ordnance in World War II, U.S. Gov. Printing Office
- Robert F. Sumrall, *Iowa Class Battleships: Their Design, Weapons and Equipment*, Naval Inst. Press, 1989
- *16-Inch Three Gun Turrets*, Ordnance Systems Command, OP-769, 1967
- *5-Inch Twin Gun Mounts Mark 28, 32 and 38 All Mods. Description and Instructions*, Bureau of Ordnance, OP 805, 1957
- W.E. Swanson, *Modern Shipfitter's Handbook*, Cornell Maritime Press, 1941
- *War Log:* USS *New Jersey World War II Cruise Book*, 1945
- USS *New Jersey Korean War Cruise Book* 1950–51
- USS *New Jersey Cruise Book*, 1988
- Anthony G. Williams, *Rapid Fire*, Airlife, 2000

www.ingramcontent.com/pod-product-compliance
Lightning Source LLC
LaVergne TN
LVHW071514220926
848486LV00009BA/374